Physical Processes in Earth and Environmental Sciences

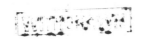

Dedicated to our parents
Cruz Arlucea
Norman Leeder
Evelyn Patterson
Albino Pérez

Physical Processes in Earth and Environmental Sciences

Mike Leeder

Marta Pérez-Arlucea

Blackwell Publishing

350 Main Street, Malden, MA 02148-5020, USA
9600 Garsington Road, Oxford OX4 2DQ, UK
550 Swanston Street, Carlton, Victoria 3053, Australia

First published 2006 by Blackwell Publishing Ltd

1 2006

Library of Congress Cataloging-in-Publication Data
Leeder, M. R. (Mike R.)
Physical processes in Earth and environmental sciences/Mike Leeder,
Marta Pérez-Arlucea.
p. cm.
Includes bibliographical references and index.
ISBN-13: 978-1-4051-0173-8 (pbk. : acid-free paper)
ISBN-10: 1-4051-0173-3 (pbk. : acid-free paper)
1. Geodynamics. 2. Earth sciences–Mathematics.   I. Pérez-Arlucea,
Marta. II. Title.
QE501.L345 2006
550–dc22
2005018434

A catalogue record for this title is available from the British Library.
Set in 9.5/12 Galliard
by NewGen Imaging Systems (P) Ltd, Chennai, India
Printed and bound in the United Kingdom
by T J International Ltd, Padstow, Cornwall

The publisher's policy is to use permanent paper from mills that operate a
sustainable forestry policy, and which has been manufactured from pulp
processed using acid-free and elementary chlorine-free practices.
Furthermore, the publisher ensures that the text paper and cover board
used have met acceptable environmental accreditation standards.

For further information on
Blackwell Publishing, visit our website:
www.blackwellpublishing.com

Contents

Preface

As we began to write this book in the wet year of 2001, Marta's apartment overlooking the Galician coast of northwest Spain was beset by winter storms as frontal depressions ran in from the Central Atlantic Ocean over the lush, vegetation-covered granitic outcrops surrounding the Rias Baixas. It was here in Baiona Bay on March 1, 1492 that such winds blew "La Pinta" in with the first news of Cristabel Colon's "discovery" of the Americas. Now, as then, the incoming moist, warm winds of mid-latitude weather systems are forced upward to over 1,000 m altitude within 10 km of the coastline causing well over a meter of rain to fall per year (2 m in 2001). Analyses of stream waters from far inland reveal telltale chlorine ions transported in as aerosols from sea spray. Warm temperatures and plentiful rains enable growth of the abundant vegetation that characterizes this *España Verde*. High rates of chemical reaction between soil, water, and granite bedrock cause weathering to penetrate deep below surface, now revealed as never before in deep unstable cuttings along the new *Autopista* to Portugal. The plentiful runoff ensures high rates of stream discharge and transport of water, dissolved ions, and sediment back down to the sea. Storms are accompanied at the sea surface by trains of waves generated far out into the Atlantic whose periodic forms are dissipated as kinetic energy of breaking water upon coastal outcrops. The winter winds gusting over the foreshore mould beach sand into dunes, where untouched by urbanization. Even so, winter storms at high spring tides wash over everything and beat the car parks, tennis courts, *paseos*, and *lidos* back into some state of submissiveness prior to the *concello* workmen tidying them all up again in time for summer visitors. Now and again a coastal defense wall falls under the strain and is undercut to helplessness on the beach below. Neither are the rocky outcrops themselves stable, despite their age (300–400 My) and general solidity, for we are not far distant from plate boundaries and faults; the plaster in one of our walls has cracks from a small earthquake whose epicenter was 30 km away at Lugo in 1997. And previous Galician generations would have felt the 1700s Lisbon earthquake much more strongly!

Environment is *medio ambiente* in the Spanish language, somehow a more apposite and elegant term than the English. You, our reader, will have your own *medio ambiente* around your daily life and in your own interactions with landscape, atmosphere, and hydrosphere. Some environments will be dramatic and potentially dangerous, perhaps under the threat of active volcanic eruption, close to an active plate boundary or close to a floodplain with rising river levels. In order to understand the outer Earth and to manipulate or modify natural environments in a sensitive and safe way it is necessary to have a basic physical understanding of how Earth physical processes work and how the various parts of the Earth system interact physically – hence our book. It is written with the aim of explaining the basic physical processes affecting the outer Earth, its hydrosphere and atmosphere. It starts from basic physical principles and aims to prepare the reader for exposure to more advanced specialized texts that seldom explain the basic science involved. The book is cumulative and unashamedly linear in the sense that it gradually builds upon what has gone previously. Topics in simple physics and mathematics are introduced from the point of view of particular examples drawn from Earth and environmental science. The book is distinctive as an introductory University/College text for several reasons. It

1 begins from basic physical principles and assumes little prior advanced physical or mathematical background, though the reader/student will be expected to have proceeded further as the book goes on;

2 deals with *all* aspects of the outer part of Earth, bringing together the physical principles that govern behavior of solids (rock, ice), liquids (water, magma), and gases (atmosphere);

3 gives certain derivations from first principles for important physical principles;

4 gives specially drawn and collated figures containing most physical explanation by graphs, formulae, and physical law.

In our general introduction, "Planet Earth and Earth Systems," we try to set out the delights and challenges faced by environmental and Earth scientists as they grapple with a diverse and complex planet. We point forward in this chapter and try to engage the nonspecialist in the wonders of the natural physical world. In Chapter 2 we introduce the fundamental principles of "States and Motion," giving examples from the environmental and Earth sciences wherever possible. Chapter 3, "Forces and Dynamics," gets more serious on dynamics and we make frequent reference to material in the maths Appendix and in the Cookies sections at the end of the book. In both this chapter and the succeeding Chapter 4, "Flow, Deformation, and Transport," we discuss the general principles of fluid flow, solid deformation, and thermal energy transfers before discussing specific processes of melting, magma production, volcanic activity, and plate tectonics in Chapter 5, "Earth Interior Processes and Systems." The physical processes at work in the atmosphere, ocean, and land form the basis for the final Chapter 6, "Earth Exterior Processes and Systems." In both these last chapters we lay emphasis on the processes that act across the different layers and states that make up the outer Earth, a theme we emphasized early on.

We are rather humble about what we offer. It is not a "Bible" and certainly not the answer to understanding the universe! We offer a unified view of the very basics of the subject perhaps. We offer signposts and guidelines for further reading and database searches. We give a maths refresher. We put more involved or challenging derivations in our Cookie boxes at the end of the book. We try to combine some physical processes with interesting data about the Earth. We use rates of change a lot but the book doesn't "do" calculus, so it is mostly a pre-calculus excursion into the physical world. We stop at the 660 km mantle discontinuity below and at the 12 km troposphere boundary above. Why? Because we can't do everything! Finally, we don't do chemistry. Not because we don't like it or think its not important, but because, again, you can't do everything. *Cybertectonic Earth* surely does combine physics and chemistry, but that is another project.

Finally, we have spent so much time drawing and redrawing our figures, selecting images, and carefully considering the content of their headings; they are meant to be read with just as much attention and enthusiasm as the regular text. We often put key items and explanations into them. It is so much easier to follow complicated topics with them, rather than lots of boring words. Well, we hope you enjoy reading and looking at the diagrams and considering the simple equations as much as we did writing, drawing, and assembling them . . . it's time to walk the dog . . . adios!

Mike and Marta
Brooke and Nigran

Acknowledgments

Sources, credits, and inspiration for illustrations

Many illustrations in this text are the creation of the authors or of their colleagues and friends. Many have also been assembled, simplified, annotated, and redrawn by the authors (using Adobe Illustrator™), often from disparate original sources, including papers from the scientific literature, previously published texts, and websites of noncopyright and governmental organizations. The remainder are often directly reproduced, chiefly NASA, SPL, USGS, NOAA, USDA, AGU. We acknowledge the following sources or inspiration for our figures:

Main Text

Fig. 1.1 NASA/JPL images; 1.2 *Nature* **350**, 55; 1.6 USDA at Kansas State University; 1.7, 1.9 USGS; 1.10 K. West/Montserrat Volcanic Observatory; 1.11 NOAA/M. Perfit; 1.15, 1.16 www.waterhistory.org; 1.17 EOS **83**, 382; 1.19 I. Stewart *Does God Play Dice?* (Penguin, 1989); 1.20 A. Berger; 2.1, 2.2, 2.3 B. Flowers & F. Mendoza *Properties of Matter* (Wiley, 1970); 2.4 NASA image; 2.5 E. Linacre & B. Geerts *Climates and Weather Explained* (Routledge, 1997); 2.6 *Ocean Circulation* (Open University, 2001); 2.7 D. Turcotte & G. Schubert *Geodynamics* (Cambridge, 2002); 2.9 A.Vardy *Fluid Principles* (McGraw-Hill, 1990); 2.11 Pond and Pickard *Introductory Dynamical Oceanography (Pergamon,* 1983); 2.15 M. Pritchard & M. Simons *Nature* 418, 167; 2.16 R. McCluskey *Journal of Geophysical Research* **105**, 2000, 5695; 2.19 M. Van Dyke *An Album of Fluid Motion* (Parabolic Press, 1982); 3.13-3.18 R. Fishbane et al. *Physics for Scientists and Engin*eers (Prentice-Hall, 1993); 3.19 USAF/Bulletin of Volcanology, **30**, 337; 3.21 British Meteorological Ofice; 3.28 iceberg 3.31 USDA image; 3.35, 3.36 *Ocean Circulation* (Open University, 2001); 3.38, 3.39 Pond and Pickard, *Introductory Dynamical Oceanography (Pergamon,* 1983); 3.43 R. Roscoe *British Journal of Applied Physics*, **3**, 267 and M. Leeder *Sedimentology and Sedimentary Basins* (Blackwell, 1999); 3.48 R. Falco *Physics of Fluids*, **20**, 124; 3.50-3.52; J.R.D. Francis *A Textbook of Fluid Mechanics* (Arnold, 1969); 3.53, 3.54, 3.56A M. Van Dyke *An Album of Fluid Motion* (Parabolic Press, 1982); 3.58 G. Davis & S. Reynolds *Structural Geology* (Wiley, 1996); 3. 74 R. Twiss & E. Moores *Structural Geology* (Freeman, 1992), 3.79 P. Molnar & P. Tapponier (Science, **189**, 419), 394 N. Price & J. Cosgrove *Analysis of Geological Structures* (Cambridge, 1990), 3.101, D. Griggs et al., Geological Society of America, Memoir 79, 3.102 A F. Donath American Scientist, **58**, 54, 4.6, 4.7, 4.9 O. Reynolds *Proceedings of the Royal Society* 1883; 4.8 M. Van Dyke *An Album of Fluid Motion* (Parabolic Press, 1982); 4.11A ibid; 4.12 R.A.Bagnold Physics of Wind-blown Sand and Desert Dunes (Chapman Hall, 1954); 4.13 A. Grass, *Journal of Fluid Mechanics*, **50**, 233; 4.16 D. Tritton *Physical Fluid Dynamics* (Oxford 1988); 4.17 M. Coward *Journal of the Geological*

Society London **137**, 605; 4.18B-D S. Kline, *Journal of Fluid Mechanics,* **30**, 741; 4.18E M. Head *Journal of Fluid Mechanics,* **107**, 297; 4.19 J. Best *Turbulence; Perspectives on Flow and Sediment Transport* (Wiley, 1993); 4.20-4.22 A. Grass, *Journal of Fluid Mechanics,* **50**, 233; 4.23 D. Tritton *Physical Fluid Dynamics* (Oxford 1988); 4.24 A. Grass, *Journal of Fluid Mechanics,* **50**, 233; 4.28 image courtesy of A. Cherkaoui; 4.33 M. Samimy et al. *A Gallery of Fluid Motion* (Cambridge, 2003); 4.31 M. Van Dyke *An Album of Fluid Motion* (Parabolic Press, 1982), R. Gibbs *Journal of Sedimentary Petrology,* **41**, 7; 4.35 W. Chepil *Proceedings Soil Science Society of America* **25**, 343; USDA/Kansa State University; M. Miller *Sedimentology,* **24**, 507; 4.44 M. Van Dyke *An Album of Fluid Motion* (Parabolic Press, 1982); 4.45 J.S. Russell British Association for the Advancement of Science, 1845, 311; 4.48 M. Van Dyke *An Album of Fluid Motion* (Parabolic Press, 1982), D. Tritton *Physical Fluid Dynamics* (Oxford 1988); 4.49-4.52 R. Tricker, *Bores, Breakers, Waves and Wakes* (Mills and Boon, 1964); 4.53 D. Tritton *Physical Fluid Dynamics* (Oxford 1988); 4.56 K. Tietze; 4.60 H. Makse; 4.62 A.C. Twomey/SPL; 4.65, 4.66 T. Gray et al. *Sedimentology,* **52**, 467; 4.67 Edwards *Sedimentology,* **41**, 437; 4.69A USGS; 4.69B Nichols *Sedimentology,* **41**, 233 4.73 R. Gimenez; 4.80 J. Suppe *Principles of Structural Geology* (Prentice- Hall, 1985);4.84-4.87 R. Twiss & E. Moores *Structural Geology* (Freeman, 1992); 4.91 R. Twiss & E. Moores *Structural Geology* (Freeman, 1992); 4.106 W. Hafner, *Bulletin Geological Society of America,* **62**, 373; 4.109B, 4.110B, 4.112B NASA;4.120, 121 J. Ramsay *Folding and Fracturing of Rocks* (McGraw-Hill, 1967); 4.124-4.126 R.Twiss & E. Moores *Structural Geology* (Freeman, 1992); 4.127 USGS; 4.131, 4.133 - 4.137 B. Bolt *Inside the Earth,* (Freeman, 1982); 4.138, 4.140 USGS; 4.141 B. Bolt *Inside the Earth,* (Freeman, 1982) 4.142C USGS; 4.143-4.144 R. Fishbane et al. *Physics for Scientists and Engineers* (Prentice-Hall, 1993); 4.150 J.G. Lockwood, *Causes of Climate* (Arnold, 1979); 4.151-4.152 D. Tritton *Physical Fluid Dynamics* (Oxford 1988); 4.153-4.159 M. Van Dyke *An Album of Fluid Motion* (Parabolic Press, 1982); 5.2 EOS Transactions AGU;5.3 J.G. Moore/USGS; 5.4 www.stromboli.net; 5.5 Hatch et al. *Petrology of the Igneous Rocks* (George Allen and Unwin, 1961); 5.10 D. Latin in *Tectonic Evoution of the North Sea Rifts* (Oxford, 1990); 5.12, 5.13, 5.14 I. Kushiro, in *Physics of Magmatic Processes,* (Princeton, 1980); 5.15 USGS; 5.16, 5.17 H.S. Yoder, *Generation of Basaltic Magma* (National Academy of Sciences, 1976); 5.19 J. Elder, *The Bowels of the Earth* (Oxford, 1976); 5.20 USGS; 5.21A New Mexico School Mines; 51.21B C. Tegner & http:77www.geo.au.dk/English/ research/minpetr/mcp/pge/; 5.23 I. Kushiro, in *Physics of Magmatic Processes,* (Princeton, 1980); 5.24 H. Shaw, in *Physics of Magmatic Processes,* (Princeton, 1980); 5.25A J. Elder, *The Bowels of the Earth* (Oxford, 1976); 5.27 USGS; 5.29, 5.30, 5.31, 5.31 R. Cas & J. Wright *Volcanic Successions* (Allen & Unwin, 1988); 5.32 USGS; 5.33 A. Matthews & J. Barclay *Geophysical Research Letters,* **31** (5), LO 5614; 5.34 USGS/JPL/NASA; 5.35 A. Cox & R.B. Hart *Plate Tectonics; How it Works* (Blackwell, 1986); 5.36 N.Pavoni EOS **86**/10; 5.37, 5.38; 5.39 C. Fowler *The Solid Earth* (Cambridge, 1990); 5.40, 5.41 D. Turcotte & G. Schubert Geodynamics (Cambridge, 2002); 5.42 D. Forsyth and S. Uyeda *Geophysical Journal of the RoyalAstronomical Society* **43**, 163; A. Cox & R.B. Hart *Plate Tectonics; How it Works* (Blackwell, 1986); 5.43 M. Leeder Journal of the Geological Society, London **162**, 549; 5.44A R.Twiss & E. Moores *Structural Geology* (Freeman, 1992); 5.45 J. Dewey & R. Shackleton *Transactions of the Royal Society* **327**, 729; 5.46 P.Silver & R. Carlson Annual Reviews Earth and Planetary Sciences, 16, 477; 6.1 J.G. Lockwood *Causes of Climate* (Arnold, 1979); 6.2, 6.3 J.T. Kiehl, *Physics Today,* Nov issue, **36**, 1994; 6.4 Inspired by A. Matthews; 6.5, 6.6 E. Linacre & B. Geerts *Climates and*

Weather Explained (Routledge, 1997); 6.7 F.H. Ludlam *Clouds and Storms* (Penn State University, 1980); 6.9 J.G. Lockwood *Causes of Climate* (Arnold, 1979); 6.10 P. Brimblecombe and T. Davies, in *Encyclopaedia of Earth Sciences* (Cambridge, 1982); 6.12 J. Imbrie and K.P. Imbrie *Ice Ages: Solving the Mystery* (MacMillan, 1979); 6.13 Carl Friehe; 6.15 M.D. Powell et al. *Nature*, **422**, 279, 2003; 6.16 *Ocean Circulation* (Open University, 2001); 6.17 *SEAWIFS Project*, Nasa/Goddard SFC; 6.18 H.E. Willoughby *Nature*, **401**, 649; 6.19 I. Wilson, *Geographical Journal*, **137**, 180; 6.20A *SEAWIFS Project*, Nasa/Goddard SFC; 6.20C N.J. Middleton et al. in *Aeolian Geomorphology* (Allen and Unwin, 1986); 6.21, 6.22, 6.23 *Ocean Circulation* (Open University, 2001); 6.24 *TOPEX-Poseidon* satellite image; 6.25 *Ocean Circulation* (Open University, 2001); 6.26 NOAA; 6.27 L. Fu et al. *EOS* **84**, 241; 6.28 Pond and Pickard, *Introductory Dynamical Oceanography (Pergamon*, 1983); 6.29, 6.30 Schmitz & McCartney, *Reviews Geophysics*, **31**, 29; 6.31 Mulder et al. *EOS*, 22 Oct 2002; 6.32 P.E. Biscaye & S.L. Eittreim, *Marine Geology*, **23**, 155; 6.33 D. Swift *Shelf Sediment Transport* (Dowden, Hutchinson & Ross, 1972); 634, 6.35 C. Nittrouer & L.D. Wright, *Reviews Geophysics*, **32**, 85; 6.36 R. Haworth, in *Offshore Tidal Sands* (Chapman Hall, 1982); 6.37 *Waves, Tides & Shallow-Water Processes* (Open University, 1999); 6.38, 6.39 N. Wells, *The Atmosphere and Ocean* (Taylor and Francis, 1986); 6.40 *Waves, Tides & Shallow-Water Processes* (Open University, 1999); 6.41 *Offshore Tidal Sands* (Chapman Hall, 1982); 6.42 W. Duke Journal of Sedimentary Petrology, **60**, 870; 6.44 C. Galvin Journal of Geophysical Research, **73**, 3651; 6.45 *Waves, Tides & Shallow-Water Processes* (Open University, 1999); 6.47 M. Longuet-Higgins & R. Stewart *Deep-Sea Research* **11**, 529; 6.47, 6.48 Bowen et al. *Journal of Geophysical Research*, **73**, 256; 6.49 Pritchard and Carter, in *The Estuarine Environment* (American Geological Institute, 1971); 6.50 R. Kostaschuk et al., *Sedimentology*, **39**, 205; 6.51 I. Grabemann & G. Krause, *Journal of Geophysical Research*, **94C**, 14373; 6.52, 6.53 A. Mehta *Journal of Geophysical Research*, **94**, 14303; 6.54 Landsat Image; 6.57 M. Leeder et al. Basin Research **10**, 7; 6.61 Chikita et al. *Sedimentology*, **43**, 865; 6.62 Wetzel, *Limnology* (Saunders, 1983); 6.65 J. Bridge; 6.67 J Best; 6.68B N.D. Smith; 6.69 I. Wilson, *Geographical Journal*, **137**, 180; 6.73 J. Dixon; 6.75, 6.76 I. Wilson, *Geographical Journal*, **137**, 180; 6.77, 6.78 NASA; 6.79 Alley et al. *Nature*, **322**, 57; 6.81 Harbor et al. *Geology*, **25**, 739; 6.82 EOS, exact source unknown; 6.83 NASA/Scott Polar Research; 6.84 http://glaciers.pdx.edu/gdb/maps/home.php.

Cookie figures

2 Pond and Pickard, *Introductory Dynamical Oceanography (Pergamon*, 1983); 3,4 J.R.D. Francis *A Textbook of Fluid Mechanics* (Arnold, 1969); 5, 7 M.W. Denny, *Air and Water* (Princeton, 1993); 8 P. Rowe, *Proceedings Royal Society London* **269**, 500; 9 R. Bagnold, *Proceedings Royal Society* **225**, 49.

We also thank referees Jenni Barclay, Adrian Mathews, Chris Paola, and Dave Waltham for their different perspectives on a wide-ranging subject, for help in making us focus our approach in this difficult endeavor and for rescuing us from some errors. Also thanks to Ian Francis, Delia Sandford, and Rosie Hayden of the Blackwell team, for their faith in the project and for their great help in its evolution from plan to execution.

1 Planet Earth and Earth systems

1.1 Comparative planetology

1.1.1 Lateral thinking from general principles

Physical processes on Earth and other planets must obey the same basic physical laws, depending in detail on the nature of the particular planetary environment, for example physical composition and gravity. While this book is obviously concerned with Earth processes, it would be narrow-minded of us not to pause for a moment right at the start and make some comparisons between Earth and our three nearest neighbor rocky planets. This turns out to be the beginning of a stimulating intellectual and practical exercise. Why so? An anecdote will help explain our point.

In the early 1970s, the desert explorer, soldier, and hydraulics engineer R. A. Bagnold, who helped create the scientific discipline of *loose-boundary hydraulics*, was contacted by NASA to undertake consultancy regarding ongoing orbital and future lander missions to Mars. The background to this strange request from the world's most prestigious space outfit to a retired brigadier of engineers was that NASA scientists had been appalled and intrigued by the enormous planetary dust storm that covered the planet for the first 2 months of the Mariner 9 mission. Although the storms died down in late January 1972, revealing a fabulous dune-covered landscape, like the Sahara in places, NASA wondered if the planetary winds were so severe that ground conditions would be inimical to survival of the planned lander mission. This was an especial worry in the face of the failure of contemporary Russian Mars 3 orbiter and lander missions: the latter had arrived in the middle of sandstorms and never transmitted more than a few seconds of data back to Earth.

Bagnold's work was the key here. Working from first physical principles and making use of breakthroughs in fluid dynamics achieved in the 1920s and 1930s, he had discovered, by judicious use of experiment, field observation, and theory, the immutable physical laws that governed the transport of sand and silt particles in the Earth's atmosphere, especially in the concentrated layers close to the ground surface during sandstorms. NASA asked Bagnold advice on how to modify his earthbound physical laws for application to the Red Planet. Bagnold and his collaborator C. Sagan had to take due account of the Martian atmosphere, surface, and rock properties, such as were then known: they had to find accurate values for gravitational acceleration, air density, rock density, and surface wind velocity. Then they had to calculate the likely extent and severity of sand blasting, dust transport, and possible effects on the landers. The results are of continued interest in view of plans to land humans on Mars early this millennium.

1.1.2 Earth in context

How to characterize a planet (Fig. 1.1)? There are intrinsic properties of solid size (diameter d) and mass (m) from which we can compute mean planetary density (ρ_p) and gravitational acceleration (g). Then the nature of any atmospheric envelope, its surface pressure (p_s), and temperature (t_s). Also its mass, composition, and thickness. Astronomical information includes distance from the Sun, rate of planetary spin (length of day L_d), rate of revolution about the Sun (length of year L_y), and inclination of the equator with respect to orbit (I_e). The regularity and eccentricity of the orbit are of additional interest. We wish to know the mean chemical compositions of the solid and gaseous components and whether the planet has internal layering that might separate distinctive functioning

From Sun to (·10⁶ km)

↓

57.9 Mercury

Atmospheric gases %:
H_2, He, Ne.
Mariner 10 1974–75.
Highly eccentric orbit.
Extreme surface
temperature variations.
500–600 km outer silicate
"crust." Partially molten
Fe core: weak magnetic field

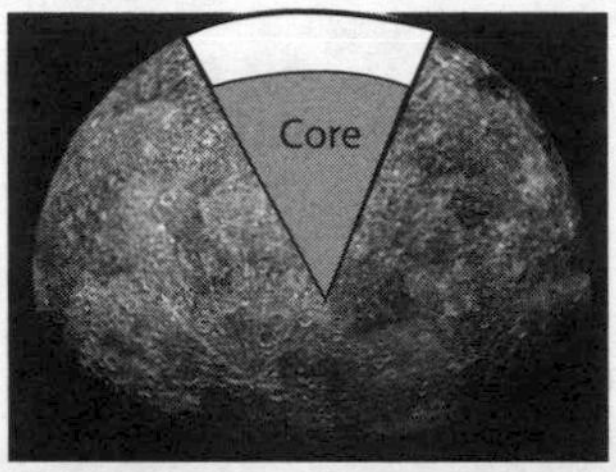

d	4,880 km
m	$3.30 \cdot 10^{23}$ kg
ρ_p	5,427 kg m^{-3}
g	3.701 m^2 s^{-1}
L_d	58.6 days
L_y	88 days
l_e	0
t_s	167°C (−170 to 430)
p_s	$2 \cdot 10^{-6}$ atm

↓

108.2 Venus

Atmospheric gases %:
CO_2 96%, N_2 3.4,
H_2O 0.14.
H_2SO_4 clouds
Radar mapped by
Magellan.
Low orbital eccentricity.
"Runaway" greenhouse
atmosphere.
Spectacular extinct
volcanoes and rift valleys.
Some active hot spots.
Convecting molten
mantle. No magnetic
field

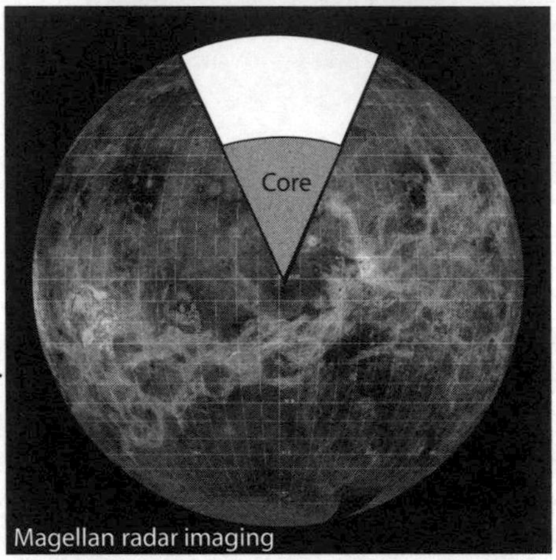

Magellan radar imaging

d	12,100 km
m	$4.87 \cdot 10^{24}$ kg
ρ_p	5,204 kg m^{-3}
g	8.87 m^2 s^{-1}
L_d	243 days (retrograde)
L_y	225 days
l_e	2°
t_s	420–485°C ("runaway" greenhouse)
p_s	c.90 atm

Our moon

↓

149.6 Earth

Atmospheric gases %:
N_2 78.1%, O_2 21.0,
Ar 0.9, CO_2 0.03.
Densest planet in solar
system. Strong magnetic
field. Molten Fe outer
core. Convecting silicate
mantle. Drives plate
tectonics in lithosphere.
71% water covered, but
oceans less known than
Venus. Surface life.
Oxygen-rich
atmosphere. Soil

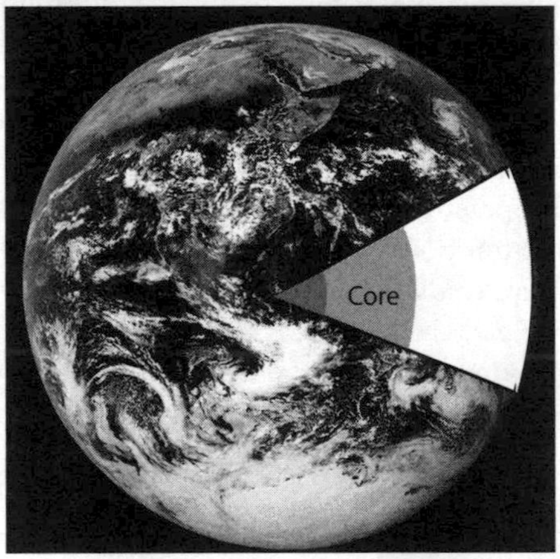

d	12,756 km
m	$5.98 \cdot 10^{24}$ kg
ρ_p	5,515 kg m^{-3}
g	9.81 m^2 s^{-1}
L_d	1 earth day
L_y	365 days
l_e	23° 26′
t_s	15°C (greenhouse)
p_s	1 atm

↓

228.0 Mars

Atmospheric gases %:
CO_2 95.3, N_2 2.7, Ar 1.6,
O_2 0.13. High orbital ellipticity.
Long rifts, extinct volcanic forms.
Southern half cratered. Molten
mantle, no active plate tectonics.
No surface water today but
runoff earlier in history.
Oxidative aqueous rock
weathering. Perennial polar
water-ice caps. Wind blown dust

d	6,794 km
m	$6.42 \cdot 10^{23}$ kg
ρ_p	3,933 kg m^{-3}
g	3.69 m^2 s^{-1} (Fe + S core)
L_d	1.03 earth day
L_y	687 days
l_e	24°
t_s	−55°C (no greenhouse)
p_s	0.006 atm (strong wind)

Fig. 1.1 Comparative planetology.

subparts of the whole. Layering might indicate active plate tectonics. Signs of tectonics and earthquake activity would come from surface zones of active faulting or folding, and any volcanic activity from eruptive clouds or surface traces of recent volcanic flows. Naturally, as Earthlings, we are curious to know whether there is water around and we would thus be looking for signs of erosion by flowing water, ice, or snow formation and movement. Also, whether the atmosphere contains any oxygen from photosynthetic processes. In addition to all these properties and current processes, conscious of the vast span of time the planetary systems have existed, we wish to know something of the history of the planet. We would try to "read the rocks" as geologists do on Earth in order to see whether the current planetary state has evolved over time.

The summaries given in Fig. 1.1 come from the results of >4,000 years of study: from astronomy, astrophysical and astrochemical analysis, satellite remote sensing, remote sampling from sondes and probes, physical sampling from remote landers. Recent spectacular discoveries (2003–5) concerning the undoubted evidence for previous Martian surface water flow, permafrost, and perennial polar ice caps simply serve to make us humble in the face of ignorance concerning the nature of our own Solar System. With this in mind, we now turn to the physical nature of planet Earth, again bearing in mind the oft-quoted fact that we know more about the nature of the planetary surface of Venus (from systematic satellite-based radar data) than we do about the motions, equilibrium, and interactions of our own oceans.

1.2　Unique Earth

We generalize here pointing out the major features of Earth that combine to make it unique within the Solar System:

1 Solid Earth is *multilayered*, the various layers (Section 1.5) fractionated according to chemical composition and physical properties. The fundamental subdivisions into crust, mantle, and core probably date from quite early in planetary history. Of interest here is how the layers have preserved their identity in the face of deep mixing during the operation of the plate tectonic cycle over the past 3 Gy (Fig. 1.2).

2 Although the three states (phases) of matter, gaseous, liquid, and solid, dominate in the atmosphere, oceans, and

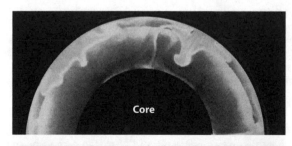
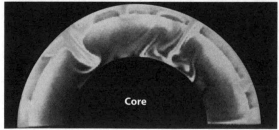

Fig. 1.2 Global cycling of rock, sediment, and water by plate tectonics. Here, lithospheric plates "sink" into the lower mantle during numerical "experiments."

solid Earth, respectively, there are mixtures of phases everywhere. These mixtures are made at planetary layer interfaces (Section 1.5) and their reactions are of fundamental importance in the workings of the Earth system. We note dust particles and raindrop nuclei in the atmosphere, sedimentary particles in water, and gas volatiles in magma and lava. Most Earth layers are thus more or less *multiphase*. For this reason, they present special problems in terms of investigating physical processes.

3 Many adjacent Earth layers move relative to each other. Their interfaces are thus prone to mixing due to processes like diffusion (Section 4.18) and bulk shearing caused by relative motions. An important feature for Earth evolution is how rapidly the different layers communicate and intermix, for example, the reactions of the ocean to changes in mean atmospheric temperature due to warming or cooling (Fig. 1.3), tropical storm reaction to changes in ocean surface temperature, and the physical and chemical effect of descending cool lithospheric plates on the deep mantle.

4 A consequence of the tendency of layers to intermix is that they must have evolved to some steady state with time, or be still evolving.

5 Earth's atmospheric oxygen is a by-product of photosynthesis. Oxygen levels evolved rapidly about 2.5 Ga from previously very low levels. Oxygen nurtures animal life but at the same time is hostile to the many mineral phases of rocky Earth that crystallized under anoxygenic conditions.

6 Earth has abundant surface water, stratified oceans, and a thoroughgoing hydrological cycle that encompasses abundant near-surface and surface life forms.

7 Earth's solid surface is dominated by horizontal and vertical motions associated with the motion of external

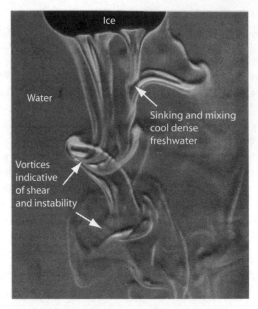

Fig. 1.3 Water circulation in the oceans is aided by density contrasts due to temperature and salinity variations, illustrated here by the downflow of dense water from melting ice.

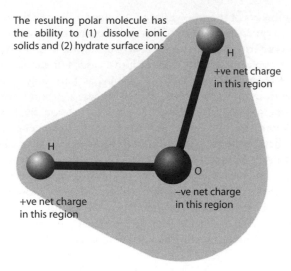

Fig. 1.4 In the water molecule, the oxygen atom is strongly electronegative, attracting a higher electron density than the two bonding hydrogen atoms. Charge field is outlined by grayscale line.

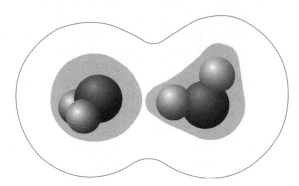

Fig. 1.5 Water molecules aggregate in clusgers of four (tetrahedral form) due to hydrogen bonding: a pair (dimer) is shown bonded opposite. Each net +ve charged hydrogen atom attracts a net −ve charged oxygen from an adjacent molecule. The tetrahedral clusters cause high surface tension and capillary pressure. The increased frequency of hydrogen bonding below 4°C causes the anomalous expansion of water.

lithospheric plates moving above a flowing and convecting upper mantle. Major consequences of plate tectonics are volcanic eruptions and mountain building by the growth of folds and faults caused by deformation: both phenomena are associated with earthquakes.

1.2.1 Water, plant life, and plate tectonics

The thermal history and temperature equilibrium of Earth has allowed copious water to remain at and close to the planetary surface. The other rocky planets closer to the Sun may have lost their water by solar boiling early on. The Martian landscape still bears telltale signs of extensive past surface water transfer (channels, subaqueous ripple structures). Today water is stored as permafrost at the poles and as seasonal water vapor in weathered surface layers (regolith).

Water has a distinctive polar molecular structure (Figs 1.4 and 1.5). It is important for the following reasons:

1 Water vapor is the most important *greenhouse gas*, absorbing infrared radiation in several absorption bands from reflected incoming short wavelength solar radiation. Water vapor in clouds thus plays an important role in the process of climate regulation because of its feedback role in reflecting and absorbing incoming and outgoing radiation.

2 Its very high *thermal heat capacity* causes ocean currents to flow and, in conjunction with the atmosphere, enables heat to be transferred tremendous distances meridionally.

3 The coexistence of solid, liquid, and gaseous phases at the Earth's surface enables rapid heat transfers to be made as the phases are forced to change from one to another in response to motions of the atmosphere and oceans, for example, the *latent heat* released as water vapor forms liquid rain droplets.

4 It provides life's medium via cellular development and photosynthesis.

5 In its liquid state, together with oxygen and carbon dioxide, it helps cause continental rock weathering and ocean crust alteration: it is thus responsible for global elemental redistribution called *geochemical cycling*.

6 Streams and waves of water physically transport weathered material far and wide over the planetary surface.

7 Water present in the rocks below the ocean floor is pushed under the surface during plate destruction. At depths it reduces rock melting points, thus permitting volcanism at island arcs and also manufacture of continental land masses over the last >3 Gy.

1.2.2 The planetary evolutionary consequences of water and plate tectonic cycling: Cybertectonica

As we noted earlier, since about 2.5 Ga, the plant biosphere has produced an oxygenated atmosphere that has allowed the subsequent evolution of animal life and the oxidative release of elements locked up in certain mineral phases, especially iron (from ferrous to ferric and back again), and other elements essential to efficient cellular metabolism. Lithospheric plates, lubricated and melted by water fluxes, recycle all accumulated elemental deposits from ocean water to sediment to atmosphere over timescales of 10^6–10^8 years. The absence of recycling over such time periods would have meant that all atmospheric and oceanic primary production and deposition of elements like carbon would simply have accumulated subsequently. Thus global cycling requires both flux and reservoir; it is plate tectonics that supplies the necessary renewal of the reservoir through the working of what we term *Cybertectonic Earth*.

1.3 Earth systems snapshots

1.3.1 Dust storm from the 1930s "Dustbowl"

Atmospheric winds pick up millions of tons of silt and clay from the land surface annually. Atmospheric turbulence initially suspends this finer sediment, leaving sand particles to travel as denser bedload "carpets" close to ground surface. Transfer to the middle atmosphere along frontal air masses results in long distance transport, then deposition from dry suspension to form sediment accumulations called *loess*. The finest sediment, together with any pollutants picked up en route, remains aloft for years, circumnavigating the globe many times, eventually depositing due to "rain-out." Deposition in the oceans contributes vitally to the input of elements, such as iron, necessary for the efficient metabolism of phytoplankton. An increasing frequency of dust storms in East Asia in recent years has brought back memories of the infamous "Dust Bowl" of the western USA in the 1930s (Fig. 1.6). The relative importance of human environmental degradation versus regional climate change in both cases is not known for sure, though a combination of causes is likely. Export of dust across the Pacific to the western USA may in future lead to intergovernmental cooperation to alleviate environmental hazard.

1.3.2 River canyon cutting uplifting plateaus

Water collects in the upstream catchment to run down the major river channel tributary, collecting sediment and water from countless other tributaries and hillslopes as it does so. The power of the water flow enables a certain magnitude of sediment to be transported close to the bed where it is able to erode bedrock by abrasion and hence to form a valley. The abrasional process is rapid compared with the lateral mass wasting of the valley walls that are kept up by periodic layers of more resistant rock. The River Colorado (Fig. 1.7) has thus been able to keep pace with regional uplift of the entire Colorado Plateau area caused by tectonic processes in the Earth's interior: the end result is the spectacular Grand Canyon.

1.3.3 Desert flash flood from overland flow

The silt- and mud-laden *flash flood* in Arizona, USA (Fig. 1.8) has developed tumultuous upstream-migrating waves of turbulence. The flood started as thunderstorm precipitation 24 h earlier. The dry, compacted earth and rock outcrops in the upstream drainage catchment intercepted the rainfall but low permeability of the rocky surface soil and the absence of vegetation led to conditions whereby water was unable to infiltrate the soil sufficiently quickly to prevent development of overland

Fig. 1.6 Dust storm front with typical overhanging head composed of lobes and clefts. Prowers Co, Colorado 1937.

Fig. 1.7 Grand Canyon, Arizona: "Mother of all canyons."

Fig. 1.9 San Andreas Fault, California.

Fig. 1.8 Flash flood showing upstream-migrating waves typical of supercritical flow. Arizona, August 1982. Flow top to bottom.

flow. Downstream coalescence of overland flow into newly eroded rill channels and then into larger tributary channels concentrated the runoff as a flash flood. The flood built up far from the source of the rainfall and took a community of campers by surprise, though fortunately there was no lasting damage in this case.

1.3.4   Earthquake fault along plate boundary

The linear surface trace of perhaps the world's most famous active fault, the San Andreas fault of southern California, cuts the landscape (Fig. 1.9). Its eroded and gullied scarp is witness to periodic catastrophic ruptures of the Earth's lithosphere in response to long-term stress build up as the North American plate (right) slides relentlessly past a subportion of the Pacific plate (left). The surface displacement, fault trace length, and orientation of the fault provide information concerning the tectonic stresses responsible, while the energy release as seismic waves during earthquakes gives insight into the Earth's interior structure.

1.3.5   Volcanic eruption in island arc

Molten magma at 600–1,200°C is produced as lithospheric plates separate at mid-ocean ridges or at island arcs inboard of subduction zones where plate is returned to middle Earth. The magma reaches Earth's surface and interacts with the hydrosphere and atmosphere at 20°C. Outgassing, quiescent lava extrusion, local explosive fire fountains, upward-directed explosions of tephra, growth and collapse of rock lava domes, and lateral flow of gas mixed with incandescent scoriae (pyroclastic flows, Fig. 1.10) are all alternative eruption scenarios: the exact outcome is dependent upon the type of magma and its near-surface interaction. Aerosols and fine ash from Plinian explosive eruptions may enter the top atmospheric boundary layer where they reflect shortwave solar energy back into space, causing temporary global cooling over a year or so.

Fig. 1.10 Pyroclastic flow descending Montserrat volcano, West Indies. Flow moving to left: note various scales of mixing eddies on upper surface shear layer with the atmosphere.

Fig. 1.11 Black smokers venting metal sulfide at Monolith vent site on the Juan de Fuca ocean ridge.

1.3.6 Black smoker at mid-ocean ridge

Black smokers were discovered as recently as 8 years after the first lunar landing, emphasizing human ignorance of the very basic cycling of Earth's hydrosphere and lithosphere. Eruption of lava from volcanic vents along the mid-ocean ridges attests to magma melt present at shallow levels and thus to high geothermal gradients. Seawater in the cracks and interstices of the surrounding ocean crust is drawn in to the ridge crest where it reissues as superheated water through vents: it represents about 35 percent of the total heat input from crustal rocks into the oceans. Chemical oxidation reactions between the hot waters and cool ambient seawater cause metal sulfide production (the black smoker particles visualizing the flows in Fig. 1.11). Chemical reactions (sulfur reduction) in the hot (17–40°C) vented waters provide energy for chemoautotrophic bacteria that form the basic member of a food chain reaching to abundant specialized metazoan life, the famous giant worms and clams of the so-called vent community.

1.4 Measuring Earth

Humans have for long measured the features on Earth, beginning with rod and knotted rope and ending with satellite GPS and Total Station Surveying. According to Herodotus (*c*.2,484–*c*.2,420 ka) and later authors, the word geometry, literally meaning "measuring the Earth," originated from the necessity of accurately and rapidly surveying land-holding boundaries destroyed by the annual Nile flood. Nowadays we use a huge array of techniques to remotely measure natural features, like ocean currents, atmospheric phenomena, and lithospheric plates, to name but a few. Yet there are still a lot of things we do not know about Earth. Here we briefly review the progress of whole Earth measurements over the past millennia.

1.4.1 Earth's shape

Earth's surface was deduced to be curved everywhere and the planet essentially spherical (Fig. 1.12) because (1) large ships can be seen to gradually disappear from the hull upward as they travel toward a distant sea horizon; (2) all other celestial bodies are of this shape (Babylonian discovery); (3) during lunar eclipses, the Earth's shadow is curved (Hellenistic discovery); (4) star constellations vary slowly according to latitude, some rising, some falling as position shifts for a given time (Hellenistic discovery).

1.4.2 Earth's diameter/circumference

Knowing Earth to be spherical and with a thorough understanding of Euclidean geometry, Eratosthenes of Cyrene (Fig. 1.13; Egypt, died 2.198 ka) observed that at summer solstice, Syene in Upper Egypt lay directly under the Sun. He then determined that at his workplace in the great library of Alexandria, a shadow of angle 7.5° was cast by a vertical pole at solstice. He reasoned that if the sun's rays were parallel, the Earth's circumference must lie in proportion to the longitudinal distance between Cyrene and Alexandria, as 7.5° lay to 360°. His logic was impeccable and despite longitude being a little off, the circumferential estimate was accurate to *c*.10 percent.

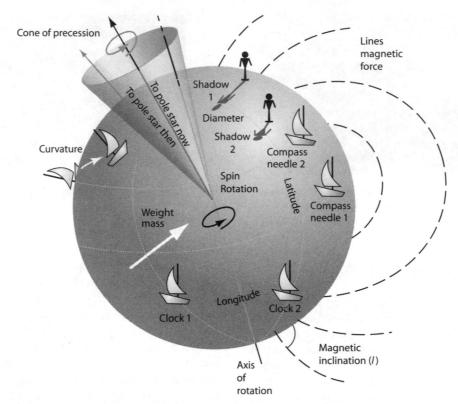

Fig. 1.12 Classic techniques to measure Earth features.

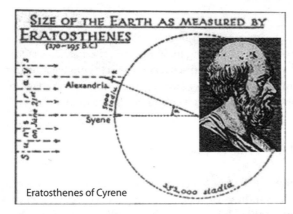

Fig. 1.13 Vintage sketch of Eratosthene's scheme for calculation of Earth's circumference.

1.4.3 Earth's mass

Newton's Law of Gravitation, also known as the inverse-square law, says that gravity is the product of any body's mass, m, times a universal constant, g, divided by the square of the radius, r. In symbols $g = Gm/r^2$. Earth's radius is circumference divided by 2π, from Euclid's formula. Knowing the values of g, G (from a famous experiment by Cavendish), and r, the value of m is computed as about $6 \cdot 10^{21}$ tons.

1.4.4 Earth's density

Knowing mass from Newton and volume via Eratosthenes and Euclid to be approximately $1.08 \cdot 10^{12}$ km^3, we can get Earth's mean density, ρ, as about 5,500 kg m^{-3}. The fact that this is so much more than that of either water (1,000 kg m^{-3}) or typical crustal rock (granite at 2,750 kg m^{-3}) provided the first clue to early geoscientists that the planet must be very dense internally, most probably due to a central core of dense metal.

1.4.5 Latitude

Chinese astronomers and navigators estimated latitude by the systematic variation of shadow lengths, from fixed vertical gnomon, observed at noon and by measuring the height of Polaris above the horizon. This star is vertically above the North Pole and at 0° at the equator. Later they mapped stars close to the South Pole for the same purpose in the southern hemisphere. They were also able to use Polaris to correct for secular magnetic variations in the magnetic compasses that they invented. Portuguese mariners first determined latitude from Euclidean geometry by the angular height of the midday Sun above the horizon adjusted for time of year. Gilbert (c.0.4 ka) discovered geomagnetism and the latitudinal dependence of the magnetic inclination (Fig. 1.14). He measured magnetic latitude, λ, by observing the inclination, I, of compass needles and making an approximation to the relation we now calculate as $\tan \lambda = 0.5 \tan I$.

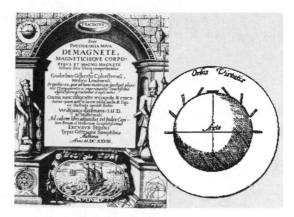

Fig. 1.14 William Gilbert's epoch-making book preface with his sketch of magnetic inclination variation around Earth's surface.

1.4.6 Longitude

Accurately knowing time from shadow lengths, Chinese astronomers and navigators (c.0.58 ka) computed longitude accurately from determining the onset of lunar eclipses at different locations and made corrections for orbital eccentricity and obliquity of the ecliptic. In the Western maritime tradition, longitude on the ocean was computed with the aid of the accurate clock invented by Harrison (c.0.22 ka). This was set for reference to Greenwich time at zero longitude; local time at the latitude in question then being estimated by sighting the Sun's zenith (maximum angular distance above the horizon), corresponding to local noon.

1.4.7 Eccentric rotation of the orbital axis

Hipparchus of Rhodes (c.2.12 ka) compared the position of Polaris with that of Thuban in Draco, used as pole star by Egyptian/Babylonian astronomers. The effect gives the 22 ky precession of equinoxes cycle.

1.4.8 Earth fluxes

As the earliest example, the Egyptians set up "Nilometers" to measure Nile water levels (Fig. 1.15). These were like modern flood gauges and measured height in cubits above low water. The annual record of the Ethiopian-sourced flood peaks were carefully preserved, for comparative purposes doubtless, although the time series were lost. Fortunately, later Nilometers and their records built by Arab and other dynasties (Fig. 1.16) have survived (they were used for tax purposes: the higher the flood, the more

Fig. 1.15 The Umayyad period Nilometer on Roda Island designed and built by the Turkestani astronomer Alfraganus.

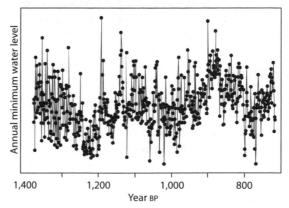

Fig. 1.16 World's oldest time series from the Roda Nilometer. Such records provide important evidence to evaluate paleoclimate proxies over medium-term time scales.

buoyant the economic prospects and the higher the tax!) and give us invaluable evidence of past climatic variations.

1.4.9 Earth's magnetic field

Although we noted geomagnetism and latitude previously, the most astonishing fact is that our ancestors, perhaps the Ethiopian hominids in the great rift valley at about 1 Ma, had they been capable or interested, could have measured the magnetic pole direction with a lump of magnetite on a string of biltong: they would have seen it pointing to the *South* Pole. Before that, throughout Earth's history, the field has periodically reversed and switched back to normal (normal meaning the present situation). This introduces us to the concept of magnetic reversals.

1.5 Whole Earth

Earth's outer interfaces can all be directly observed or indirectly monitored. For example, the highest mountains penetrate about 40 percent of tropospheric thickness and direct sampling has proved possible from manned or remote balloons, aircrafts, spacecrafts, and satellites. Concerning direct evidence for the composition, state, and temperature of the Earth's interior, we are largely ignorant, despite the efforts of Jules Verne, Satan, and Gandalf. Recourse has been made to instrumental signals transmitted to and from interfaces using artificial or natural sources of energy. A medical analogy is appropriate between an external examination and an internal body scan. Sound traveling at known speed from explosions or earthquakes and electromagnetic radiation provide the energy necessary to scan Earth. Signal processing reveals reflection, refraction, and absorption of parts or all of input energy signals from internal interfaces. Use is also made of geological gifts in the form of rocks from deeply eroded mountain belts that originated under colossal pressures and temperatures and of remote-sensed data on subsurface physical properties. Laboratory experiments are also sources of inspiration. There is a rich inventory of indirect evidence for a well-layered, largely solid planet beneath our feet, though "largely solid" entertains a vast range of subtleties.

1.5.1 Layers of composition, temperature, and state

We list the important layers of composition, state, and temperature below, waiting until future chapters for explanations of the phenomena observed. We use composition to mean chemical or mineral make-up. State refers to whether a particular layer is liquid, solid, or gas. Note the following initial subtleties:

1 A given composition or state may also be layered due to temperature or pressure effects. For example, when swimming in a lake or shallow calm sea under summer sunshine, you will often notice a sharp change in temperature, at a body-length depth or so, where the warmer water above is fairly sharply separated from cooler water below. Changes of mineral phase in the Earth's mantle are related to pressure-induced "repacking" of mineral atomic lattices.

2 Layers also form where single phases of contrasting composition come into contact. Commonly observed examples are jets of freshwater spreading out over salty seawater, a phenomenon observed where certain river deltas meet the sea or where springs discharge seaward to form a "floating lid" of freshwater.

3 Not only do materials of identical or similar chemical composition but of different states also exhibit layering. Familiar examples are ice layers forming on water or the solid crust that quickly forms on flowing molten lava.

Earth has numerous layers due to changing composition, state, and temperature. The brief notes, definitions, and descriptions below are augmented and explained later in this book. They are designed to stimulate interdisciplinarity. For example, you might care to ponder over the history of the different layers and why they have persisted over time.

1.5.2 Earth layers defined by composition

Ionosphere ($>c.$60 km altitude) Layer of concentrated charged particles, electrons and positive ions, formed from atoms and molecules chiefly not only by solar radiation but also by galactic cosmic rays (Table 1.1). It is visible at high latitudes as *aurora luminosity*. The degree of ionization is sensitive to outbursts of solar radiation during sunspot cycles.

Table 1.1 Summary of the features found on >30 major interfaces of Earth.

	Atmosphere (air + water)	Ice	River water	Seawater	Continental crust	Oceanic crust
Atmosphere (air + water)	Atmospheric air masses interact	ABL on ice surfaces	Gaseous exchanges	ABL on ocean surfaces	ABL Weathering and soil plumbing	—
Ice		Ice streams	Glacier termini	Ice shelves Icebergs	Glaciers Ice caps	Icebergs
River water			Channel confluences	Plumes Jets River deltas Estuaries Underflows	Landscape evolution Erosion Transport Deposition	Underflows Turbidity currents
Seawater				Oceanic water masses	Shelf (benthic) seabed interfaces	Ocean (benthic) seabed interfaces
Continental crust					Major crustal faults	Major thrust faults
Oceanic crust						Oceanic transform faults
Mantle						
Lithosphere					Major faults	Major faults
Asthenosphere						Major faults

	Mantle	Lithospheric plate	Asthenosphere	Magma melt	Fe–Ni liquid	Fe–Ni solid
Atmosphere (air + water)	—	As continent	—	Lava and pyroclastic flows	—	—
Ice	—	As continent	—	Jokelhaups	—	—
River water	—	As continent	—	Phreato-magmatic explosions	—	—
Seawater	MOR plumbing	As continent	As mantle	Pillow lavas and volcanic gasses	—	—
Continental crust	Major thrust faults	As continent	As mantle	Volcanoes and their plumbing Plume heads	—	—
Oceanic crust	MOHO Usually no differential motion	As continent	MOR and oceanic volcanoes Plume heads	MOR and oceanic volcanoes Plume heads	—	—

(cont.)

Table 1.1 (*cont.*)

Mantle	Oceanic transform faults	As continent	As above	MOR and oceanic volcanic plumbing Plumes	CMB	—
Lithospheric plate		Major faults	Low velocity zone	As above	Plate graveyard at CMB	
Asthenosphere			Mantle convection currents	Plume heads	CMB	
Magma melt				Magma chamber mixing	CMB plume sources	—
Fe–Ni liquid					Core fluid masses	Inner core boundary

Note the scope for interactions between subdisciplines concerned with studying Earth processes. Even deep planetary interfaces, normally the playgrounds for solid state physicists, have relevance to surface-orientated Earth scientists. ABL: atmospheric boundary layer; MOR: mid-oceanic ridge; CMB: core–mantle boundary; MOHO: Mohorovicic discontinuity.

Ozonosphere (45 km thick) Ozone layer where photons of ultraviolet light cause continuous reversible oxygen disassociation to ozone via the *Chapman cycle*. This is perhaps the first Earth layer to receive focused "environmental" attention due to the discovery in the 1970s of the role of human-produced chlorofluorohydrocarbons in ozone destruction at polar latitudes.

Troposphere (18 km thick) Gas, chiefly N_2 and O_2 with aerosols of solid dust and liquid water.

Oceans (average 3.8 km deep) Liquid water with dissolved ions, chiefly Cl^-, Na^+, SO_4^{--}, Mg^{++}, and Ca^{++}.

Continental crust (average 30 km thick) Crystalline silicate solid, granite-rich. Chief elements are Si, Al, K, Na. Lower boundary with mantle defined at MOHO, marked by increase in density.

Oceanic crust (average 10 km thick) Crystalline silicate solid, basalt-rich. Chief elements are Si, Al, Ca, Fe, and Mg. Lower boundary with mantle defined at MOHO, marked by increase in density.

Mantle (2,858 km thick) Crystalline silicate solid/ plastic rich in Mg and Fe elements. Upper boundary with crust at MOHO, defined by marked decrease in density.

Core (3,480 km thick) Metallic center of Earth, rich in Fe and Ni.

1.5.3 Earth layers defined by state

Atmosphere Gaseous envelope with clouds and weather, containing traces of particles and aerosols.

Cryosphere Solid water in ice caps up to 5 km thick, also seasonal as permafrost.

Hydrosphere Liquid water in oceans, lakes, and rivers. Also as liquid phase in crustal rocks and soils.

Lithosphere Relatively rigid outer 100+ km of mantle and crust, defining moving lithospheric plates.

Asthenosphere (*Low Velocity Zone, LVZ*) 100–200 km thick, partially molten (<1 percent by volume) mantle below lithosphere.

Mesosphere Most of the mantle below the *LVZ*, as in Section 1.5.2.

Outer core Metallic (Fe–Ni) molten liquid.

Inner core Metallic (Fe–Ni) solid.

1.5.4 Outer Earth layers defined by temperature

Thermosphere (85+ km) −80°C upward. Rare gas molecules warmed by extreme ultraviolet radiation.

Mesosphere (*c.*50–85 km) 0°C to −80°C. The ozone effect decays upward.

Stratosphere (*c.*18–50 km) −60°C to 0°C. Warming trend as heat is released by ozone formation.

Troposphere (0–*c.*18 km) 15°C to −60°C. Well mixed, warmed by greenhouse effect. Cooled as greenhouse effect decays upward.

Ocean water layers (North Atlantic Deep Water, etc.) (100s–1,000s m) Vary in temperature by up to a few degree centigrades.

1.5.5 Further notes on some key interfaces

Ocean seawater : atmosphere This is a complex interface, circulating on a rotating spherical Earth, strongly modified by the polar cryosphere and global sea level dependent upon polar exchanges. Tides and waves raised on the ocean surface generate kinetic energy that is dissipated on shallow continental shelves and coastlines, causing sediment transport, deposition, and erosion. Atmospheric winds act upon the ocean's surface and gases mix into the ocean surface. Water vapor from the ocean surface is transported, the resulting rainfall washing out continent-derived silicate dusts. Heat energy is exchanged, but ocean water seldom reaches temperatures >30°C. Solar radiation penetrates 10s of meters into the ocean surface waters causing massive photosynthetic reactions in the presence of seawater nutrients – biological gases are added and organic carbon produced.

Continental crust : atmosphere/hydrosphere Humans in their billions live on, modify, and pollute this delicate interface. Silicate and other minerals are weathered by aqueous chemical and biochemical reactions involving atmosphere and biosphere. The weathered *regolith* is mixed with organic breakdown products to form soil that acts as both valve and filter, regulating the two-way flow of materials to surface and near-surface ecosystems. Surface water runs off, over and through rock and soil, sometimes eroding it. Lakes collect on the interface and rivers run through it, sometimes cutting down deeply as tectonics locally uplifts the surface. The rivers flood their channels in response to extreme rainfall. Atmosphere flows over the interface, sometimes eroding it, transporting soil particles as sand and silt, the latter lifted into the atmospheric boundary layer as a dusty aerosol. Effects of climate change and war wreak havoc at some interface margins.

Mid-ocean ridge crust : ocean seawater A new plate forms from the cooling of molten magma rising from the mantle. Cool seawater penetrates the cracked hot rocks of the ridge flanks and is redistributed below the surface, some reemerging as jets of superheated water, the famous "black smokers" noted in Section 1.3.6, containing dissolved ions scavenged from the hot rock. Far from the penetrating influence of solar radiation, entire specialized ecosystems live around the smokers and in the hot rocks as chemoautotrophs, obtaining their energy from bacterially induced reactions taking place between the hot fluid and cool ambient seawater.

Magma/atmosphere/hydrosphere This most spectacular interface occurs during volcanic eruptions when molten magma from Earth's interior crust or mantle approaches the surface. The magma is generated at lithospheric plate boundaries or by hot plumes rising from the core–mantle boundary. It may vary widely in the concentration of dissolved gases and also chemically in ways that ultimately control the style of volcanic eruption. The sudden or gradual juxtaposition of silicate melt, temperature 600–1,200°C, containing variable amounts of dissolved gases under very great pressure, with atmosphere or surface water at *c*.20°C and 1 bar pressure results in various possible combinations of slow and fast lava eruption, explosive fragmentation, and vertical and lateral volcanic blasts. Volcanic hazard prediction involves second-guessing the resulting eruption style from basic principles.

Mantle : crust This is the famous MOHO interface between the mantle and the crust, detectable by either refracted or reflected seismic waves, the wave speeds increasing by 25 percent across it. This variation of transmissibility matches that predicted between silicate crustal rocks rich in feldspar and quartz, and silicate upper mantle rich in the mineral olivine. To their great delight, geologists also directly recognize the MOHO (a very sharp interface) within mountain ranges where gigantic faults have thrust it up toward the surface during past plate collisions. Fragments of mantle may also turn up in volcanic vents.

Lithospheric plate : asthenospheric mantle This is the fundamental dynamic interface that enables plate tectonics to operate. Rigid lithospheric plates slide around the surface of the Earth (velocities of a few centimeters per year) on a lubricating layer containing a tiny proportion of partially molten rock. The partially molten zone is termed the low velocity zone (LVZ), since the melt slightly slows down by (>1 percent) the passage of certain kinds of seismic waves across it.

Outer core : mantle This interface is between mantle silicate rocks transmitting all kinds of seismic waves at predictable velocities and a metallic liquid core, the latter demonstrated by widespread wave refraction and disappearance of certain key wave types. The interface is the ultimate site of submerged lithosphere plate, the so-called "slab graveyard." According to some, the interface layer periodically melts the slabs and erupts molten silicate material upward like a lava lamp. These plumes rise to the surface of the Earth to form voluminous volcanic eruptions (oceanic islands like Hawaii and Iceland are thought to overlie such plumes).

1.5.6 Summary of Earth's interfaces

We can distinguish interfaces between compositionally distinct layers (e.g. the crust–mantle interface at the MOHO), layers of distinct states of different materials (e.g. the ocean–atmosphere boundary), and layers of distinct states of the same material (e.g. cryosphere–hydrosphere). The majority also features differential movement across them: these are *boundary layers* (discussed later in this book). Compare and contrast this dynamism with the static interfaces of the Moon and those thought to be present on other rocky planets.

1.6   Subtle, interactive Earth

Prior to the now widespread acceptance of diversity, complexity, and interactions among the natural and environmental sciences, it was common for physicists and engineers to look upon Earth as a gigantic machine. They only had to understand how all the various parts worked and by a process of integration put all the components together so that Machine-Earth would then be understood; it would do work at rates set by the outflow of heat energy from the interior and the input of external solar energy. While individual parts of the Earth system undoubtedly benefit from this approach, and ultimately we will fully understand Machine-Earth, interactions between subparts are so various that they create a richer and more complex Earth than we often realize. It is this complexity that often attracts students of the Earth and environmental sciences today. Crucially, present Earth does not, and indeed cannot, sample all possible scenarios of complexity that might arise, for this is time dependent and stochastic, that is, it depends upon a degree of chance. Time is certainly something Earth has in plenty; the planet has operated as a coherent body for >4.5 Gy, and during this time it has experienced many of the combinations of interacting variables that matter, energy, space, and time can throw at it (though not all – witness the "runaway greenhouse" climate of Venus). It is the task of the geoscientist to try to decipher the past history of Earth, informing the Earth/environment "process-engineers" who study the present system of the exact nature of time dependency. Our intention here is not to confuse the mechanistic issues that form the content of the following sections of this book, but instead to draw the reader's attention to some interesting consequences of interaction.

1.6.1   Inter- and intra-system feedbacks

In everyday life, feedback is a message considered returned from audience to speaker or operator. In nature, feedback is also a return of usually energy within or between functioning systems. Thus a growing ice cap in a cooling climate further cools the surrounding atmosphere by positive feedback due to enhanced shortwave solar energy reflectivity (albedo) from the shiny white ice and snow-covered surface (Fig. 1.17). Can you think of an example of negative feedback?

1.6.2   Fluid turbulence

Turbulence is a characteristic of fluid flow (Fig. 1.18) that dominates many Earth systems, chiefly atmosphere, rivers, and oceans. It is a difficult subject; a famous physicist is said to have remarked that when he eventually arrived in Heaven he would want to ask the Omnipotence about the origin and nature of two phenomena: quantum electrodynamics and turbulence. But he only expected to be able to understand the former! This is because we cannot predict the *exact* magnitude of local turbulent flow velocity at any point in time or space because the velocity depends on that which existed previously; it is both the cause and result of the turbulent motion. Instead, recourse is made to statistically determined quantities, like mean velocity and deviations from this mean expressed as fluctuations.

1.6.3   Chaos theory – randomness in deterministic systems (butterflies and cyclones)

Two solutions to certain differential equations using parameters differing by only 0.001 are initially identical,

Fig. 1.17 Expanding ice sheets give positive feedback to cooling because of their high albedo/reflectivity of shortwave incoming radiation.

but as the iteration continues, the solutions diverge far apart (Fig. 1.19). Hence the initially ridiculous-seeming idea that the flap of butterfly wings where weather is generated may influence subsequent events far away as the weather system develops. The metaphor simply says that random elements exist and evolve in mechanistic systems.

1.6.4 Earth's orbital wobbles

In our heliocentric Solar System, Earth orbits elliptically in the plane of ecliptic, spinning on an axis inclined obliquely to this plane. Ellipticity, orientation, and tilt of the spin axis fluctuate over time spans of 10^4–10^5 years (Fig. 1.20), reinforcing each other and causing predictable changes in Sun–Earth distance. These cause tiny variations in solar radiation intensity, significantly affecting delicate controls on climate.

Fig. 1.18 Turbulent eddies (at scale $c.10$ m here) give unpredictable consequences for velocity or stress distributions.

1.6.5 Motion on a sphere: Curved paths and angular momentum

The geometry of curved surfaces is termed non-Euclidean. On rotating, curved Earth's surface, observed motions of the high wind or ocean water over the rotating surface also curve (Fig. 1.21); The rotation of the Earth induces gradients in the magnitude of spin and angular momentum that act to influence the motion. This does not apply to fast-moving objects held to the surface by frictional contact, like automobiles, low-level wind, or river flow.

1.6.6 Cycling

The Law of Mass Conservation says that cyclic trading is material (element, compound) transferred between different states and/or locations at characteristic rates, the whole maintaining constant mass. Cycling includes mass reservoirs and the fluxes between these as pathways of mass per time. Plate tectonics controls the rock cycle and the long-term rates of carbon and water cycling.

1.6.7 Biosphere versus tectonics as planetary agent

Energy for life processes comes from sunlight or abiotic reactions such as those associated with ocean-ridge convection. The only known source of the abundant oxygen in our atmosphere is plant photosynthesis. Cycling over outer Earth is mediated, but not controlled, by living organisms. Rather than mother *Gaia*, we have a long-term Earth system controlled by tectonics called *Cybertectonica*.

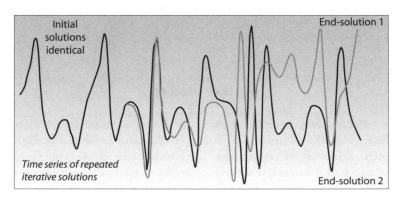

Fig. 1.19 Solutions to certain differential equations using initial values that differ by only 0.001 may end up utterly different.

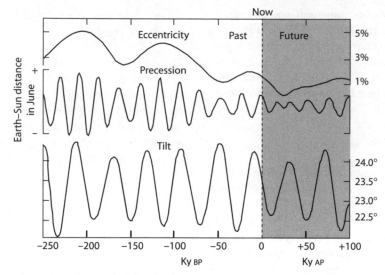

Fig. 1.20 Orbital wobbles due to variations in planetary eccentricity, precession, and axial tilt as hindcasted and forecasted by astrophysicists. Note that our immediate future looks a little less "wobbly."

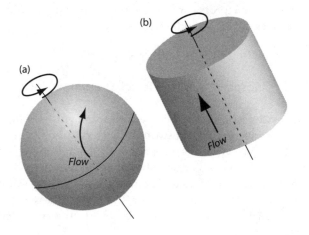

Fig. 1.21 Only flows over rotating spherical curved surfaces are deviated due to gradients in spin. (a) Sphere: spin variable, flow deviated and (b) cylinder: spin constant, no deviation.

1.6.8   Dangers of correlating x with y

In mathematics, two variables can exist such that the value of y depends *exactly and only* on the value of x (in shorthand, $y = f(x)$, as in $y = x^2$; a cue to consult our excellent Math appendix). In the natural world, correlation is less than perfect and certainly does not identify cause and effect. Thus highly correlated plots of increased atmospheric CO_2 against increasing mean global temperature do not *a priori* imply that the CO_2 is *causing* the positive correlation: some other factor may be involved, for example, increase in CO_2 may be the *result* of temperature increase caused by something else (e.g. variations in solar output). Keep your mind open.

Further reading

R. A. Bagnold's somewhat varied career is described in his autobiography, *Sand, Wind and War* (University of Arizona Press, 1991). Much of interest in comparative planetology is in R. Greeley and J. D. Iverson's *Wind as a Geological Process* (Cambridge, 1985). Aspects of Earth measurement may be found in K. Ferguson's *Measuring the Universe* (Headline, 1999), S. Pumfrey's *Latitude and the Magnetic Earth* (Icon, 2002), D. Sobel's *Longitude* (Fourth Estate, 1996), and G. Menzies' superb account of medieval Chinese navigation, *1421* (Bantam, 2002). The story of mid-ocean ridge vents and their associated life forms is beautifully told in C. L. Van Dover's *The Octopus's*

Garden (Addison Wesley, 1996). Aspects of Earth's history and Earth's position in the Universe is told by C. Allègre's *From Stone to Star* (Harvard, 1992), and, engagingly for the very beginner, in B. Bryson's *A Short History of nearly Everything* (Doubleday/Black Swan, 2003). Pioneering efforts to date the Earth are beautifully set out in C. Lewis' biography of Arthur Holmes, *The Dating Game* (Cambridge, 2000). Chaos theory is stimulatingly told by I. Stewart in *Does God Play Dice?* (Penguin, 1989). The tectonic-free parable of Gaia is told by its inventor, J. Lovelock, in *Gaia* (Oxford, 1979) while P. Westbroek adds a geological perspective in *Life as a Geological Force* (Norton, 1992). The history and origins of oxygen are told in N. Lane's *Oxygen* (Oxford, 2002). Information on Nilometers and other aspects of ancient measurement are at www.waterhistory.org.

2 Matters of state and motion

2.1 Matters of state

Earth contains each of the states of matter – *solid, liquid* and *gas* (Fig. 2.1); we have noted already that Earth's surface is unique among the planets of the Solar System in its abundance of solid, liquid, and gaseous water. The mass and energy transfers that accompany changes of state of water dominate physical conditions in the troposphere, hydrosphere, and cryosphere and have great importance in modulating more "rocky" processes such as flow and partial melting in the upper lithosphere. Yet matters of state are more subtle and interesting than just a simple threefold division; for example, consider the initially rather strange idea of *granular fluids* noted below.

populated by relatively few individual molecules in constant random high-speed motion. The spaciousness of gases explains their low density and high compressibility. Gas pressure is a measure of the intensity of the random collisions of gas molecules with some rigid wall or container. Thermal properties include low conductivity and low specific heat capacity. These result from the inefficiency of kinetic energy exchange due to rare intermolecular collisions and the lack of transmittable molecular oscillations and rotations to cause conduction of heat energy as temperature is increased. Gases have low potential energies due to their lack of interaction with neighboring molecules. They are thus good insulators and poor heat carriers.

2.1.1 Gases

The physical properties of gases are low density, absence of rigidity, and very high compressibility. These reflect the "openness" of space within a gas volume (Figs 2.2 and 2.3),

2.1.2 Liquids

Liquids are typically of the order of 10^3 times denser than gases, with low compressibility and no rigidity. Liquid molecules within a given volume are relatively close

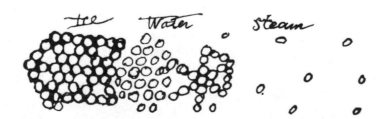

Fig. 2.1 States of water. Joule's famous "doodle," done in an 1847 notebook, of the three states of matter for H_2O. Sketched as he realized that gas molecules must be the most widely spaced and have independent motions. Around this time, Avogadro's number was becoming reliably known. Its huge magnitude (there are $6 \cdot 10^{23}$ molecules in a single mole of every substance) and the tiny size of atoms gives the appearance of continuous matter at the scale of human eyesight. This *continuum* approach is still the most useful for analysis of fluid and solid properties in the bulk.

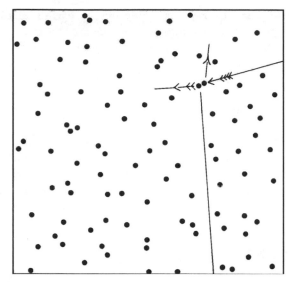

Fig. 2.2 Gas molecules collide. In 1883 Kelvin drew this conceptual diagram of colliding gas molecules to scale. The molecules are represented in an area 1 μ^2 and one molecule thick. By this time, Avagadro's number was reliably known and the diameter of atoms was calculated to be of order 10^{-7} mm or 10^{-4} μ. Atoms were suspected from the evidence of the Brownian motion of tiny clay flakes (not pollen grains as widely stated) floating in water, the moving molecules colliding with the clays and imparting momentum. But it was not until 1906, when Einstein developed a theory of molecular collisions to explain the motions and the nature of *diffusion*, that the atomic theory was universally accepted. Kinetic theory (Section 4.18) tells us that molecules have speeds of order $10^2 - 10^3$ ms^{-1}. An air molecule at standard temperature and pressure has a mean speed of 470 ms^{-1}.

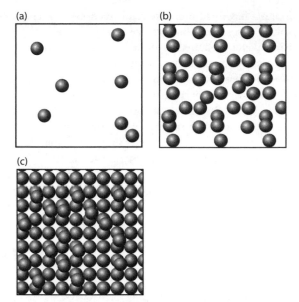

Fig. 2.3 The increasing packing and regularity of molecules in solids, liquids, and gases. (a) molecules in Gaseous 20 Å cubic volume; (b) molecules in a Liquid 20 Å $\times$ 20 Å $\times$ 3 Å volume; (c) molecules in a Solid 20 Å $\times$ 20 Å $\times$ 3 Å volume.

together, though the pattern is generally disordered (Fig. 2.3) and unsteady, with the individual molecules still moving randomly at the same high speeds as in gases but with motion also in the form of oscillations and transmitted vibrations. The closeness of neighboring molecules explains the increased density and the difficulty of compression. The vibrational and oscillatory molecular motions explain the greater thermal conductivity and specific heat compared to gases. Liquid molecules are close enough to have large potential energies.

2.1.3   Solids

In solids the molecules or ions are very closely packed and thus generally rigid (Fig. 2.3). Solid and liquid phases of the same substance (e.g. ice and water) have similar densities and compressibilities, emphasizing the close packing of molecules in each state. X-rays reveal that many solids with clear melting points form crystalline aggregates with almost regular internal frameworks or lattices if formed slowly from the liquid state. These crystals have planar

boundaries that interlock at certain angles to form the whole aggregate. Under normal conditions solid molecules undergo no net translational motions, but move rapidly *in situ* by oscillation about some mean position. Solid molecules thus have high potential energies and the sensitive temperature control of oscillation frequency means they can be good heat conductors. Metals may be regarded as solid lattices of ions permeated by an electron "gas." The higher thermal conductivity of metals compared to other crystalline solids like silicates and chlorides is explained by the extreme mobility of these electrons that carry heat energy, along with their intrinsic electrical energy, with them.

2.1.4   Earth's complex substances

Many of Earth substances are mixtures:
1 Earth's atmosphere is a mixture of gases. Despite gases being good insulators and poor heat carriers, efficient heat transport is due to ease of bulk movement and heat release during changes of state.
2 Rock is a mixture of different crystalline, rarely glassy, solids, each component having certain distinctive material properties, melting points, and chemical composition.
3 Sediment avalanches are aggregates of solids but their flow behavior resembles both fluids and gases, with frequent granular collisions. Such aggregates are termed *granular fluids*.

4 Magma has small but important fractions of pressurized dissolved gases, including water vapor.

5 River water contains suspended solids, while the atmosphere carries dust particles and liquid aerosols.

6 Seawater has *c.*3 percent by weight of dissolved salts and also suspensions of particulate organic matter.

Solid Earth substances may break or flow:

1 Ice fragments when struck, yet deformation of boreholes drilled to the base of glaciers also shows that the ice there flows, while cracking along crevasses at the surface.

2 Earth's mantle imaged by rapidly transmitted seismic waves behaves as a solid mass of crystalline silicate minerals. Yet there is ample evidence that in the longer term ($>10^3$ years) it flows, convecting most of Earth's internal heat production as it does so. Even the rigid lower crust is thought to flow at depth, given the right temperature and water content.

2.1.5 Timescales of *in situ* reaction

The lesson from the last of the above examples is that we must appreciate characteristic *timescales* of reaction of Earth materials to imposed forces and be careful to relate state behavior to the precise conditions of temperature and pressure where the materials are found *in situ*.

2.2 Thermal matters

2.2.1 Heat and temperature

Heat is a more abstract and less commonsense notion than temperature, the use of the two terms in everyday speech being almost synonymous. We measure temperature with some form of heat sensor or thermometer. It is a measure of the energy resulting from random molecular motions in any substance. It is directly proportional to the *mean* kinetic energy, that is, mean product of mass times velocity squared (Section 3.3), of molecules. Heat on the other hand is a measure of the *total* thermal energy, depending again on the kinetic energy of molecules, and also on the number of molecules present in any substance.

It is through *specific heat, c,* that we can relate temperature and heat of any substance. Specific heat is a finite capacity, sometimes referred to as *specific heat capacity*, in that it is a measure of how much heat is required to raise the temperature of a unit mass (1 kg) of any substance by unit Kelvin ($K = °C + 273$). It is thus also a *storage* indicator – since only a certain amount of heat is required to raise temperature between given limits, it follows that only this amount of heat can be stored. In Box 2.1, notice the extremely high storage capacity of water, compared to the gaseous atmosphere or rock.

Temperature change induces internal changes to any substance and also external changes to surrounding environments, for example,

1 Molten magma cools on eruption at Earth's surface, turning into lava; this in turn slowly crystallizes into rock.

2 Glacier ice in icebergs takes in heat from contact with the ocean, expands, and melts. The liquid sinks or floats depending upon the density of surrounding seawater.

3 Water vapor in a descending air mass condenses and heat is given out to the surrounding atmospheric flow.

In each case temperature change signifies internal *energy change*. Changes of state between solid, liquid, and gas require major energy transfers, expressed as *latent heats* (Box 2.1). We shall further investigate the world of *thermodynamics* and its relation to mechanics later in this book (Section 3.4).

Substances subject to changed temperature also change volume, and therefore density; they exhibit the phenomenon of *thermal expansion* or *contraction* (Box 2.1). This arises as constituent atoms and molecules vibrate or travel around more or less rapidly, and any free electrons flow around more or less easily. If changes in volume affect only discrete parts of a body, then thermal stresses are set up that must be resisted by other stresses failing which a net force results. Temperature change can thus induce motion or change in the rate of motion. Stationary air or water when heated or cooled may move. Molten rock may move through solid rock. A substance already moving steadily may accelerate or decelerate if its temperature is forced to change. But we need to consider the complicating fact that substances (particularly the flow of fluids) also change in their resistance to motion, through the properties of viscosity and turbulence, as their temperatures change. We investigate the forces set up by contrasting densities later in this book (e.g. Sections 2.17, 4.6, 4.12, and 4.20).

2.2.2 Where does heat energy come from?

There are two sources for the heat energy supplied to Earth (Fig. 2.4). Both are ultimately due to nuclear reactions. The external source is thermonuclear reactions in the Sun. These produce an almost steady radiance of shortwave energy (sunlight is the visible portion), the

Box 2.1 Some thermal definitions and properties of earth materials

Specific heat capacities, c_p, units
of $J\,kg^{-1}\,K^{-1}$, at standard T and P.

Air	1,006
Water vapor (100°C)	2,020
Water	4,182
Seawater	3,900
Olive oil	1,970
Iron	106
Copper	385
Aluminum	913
Silica fiber	788
Carbon (graphite)	710
Mantle rock (olivine)	840
Limestone	880

Specific Heat Capacity , c_p, c_v, is the amount of heat
required to raise the temperature of 1 kg of substance
by 1 K . Subscripts refer to constant volume or pressure

Coefficients (multiply by 10^{-6}) of
linear thermal expansion, α_l, units
of K^{-1} at standard T and P.

Iron	12
Copper	17
Aluminum	23
Silica fiber	0.4
Carbon (graphite)	7.9
Crustal rock (to 373 K)	7–10

Coefficients (multiply by 10^{-4}) of
cubical thermal expansion, α_v, units
of K^{-1} at standard T and P.

Water	2.1
Olive Oil	7.0
Crustal rock	0.2–0.3

Thermal conductivity, λ, units of
$W\,m^{-1}\,K^{-1}$ at standard T and P.

Air	0.0241
Water	0.591
Olive oil	0.170
Iron	80
Copper	385
Aluminum	201
Silica fiber	9.2
Carbon (graphite)	5
Mantle rock (olivine)	3–4.5
Limestone	2–3.4

Thermal Conductivity is the rate of flow of heat
through unit area in unit time

Thermal Diffusivity, κ, units $m^2\,s^{-1} \times 10^{-6}$ at
standard T and P.

Air	21.5
Water	0.143
Mantle rock	1.1

Thermal diffusivity indicates the rate of dissemination
of heat with time. It is the ratio of rate of passage of
heat energy (conductivity) to heat energy storage
capacity (specific heat per unit volume) of any material

Heat flow required for vaporization, L_v,
units of $kJ\,kg^{-1}$. Sometimes termed *latent
heat of vaporization*, more correctly it is the
specific enthalpy change on vaporization
(see Section 3.4).

water to water vapor (and vice versa)	2,260

Heat flow required for fusion, L_f,
units of $kJ\,kg^{-1}$. Sometimes termed *latent
heat of fusion*, more correctly it is the specific
enthalpy change on fusion (see Section 3.4).

Ice	335
Mg Olivine	871
Na Feldspar	216
Basalt	308

Heat flow produced by crystallization,
(multiply by 10^4) units of $J\,kg^{-1}$.

Basalt magma to basalt	40
Water to ice	32

GEOTHERMAL HEAT
65 mW m⁻²

HEAT ENERGY is required
for *life, plate motion, water
cycling, weather, and
convectional circulations*

SOLAR HEAT
1,367 W m⁻²

Fig. 2.4 Heat energy available to drive plates is minuscule when compared with that provided by solar sources for life, the hydrological cycle, weather, etc.

average magnitude of which on an imaginary unit surface placed at the uppermost surface of Earth's atmosphere facing the sun is now approximately 1,367 W m^{-2}. This *solar constant* is the result of a luminosity which varies by >0.3 percent during sunspot cycles, possibly more during mysterious periods of negligible sunspots like the Maunder Minimum (300–370 years BP) coincident with the Little Ice Age. At any point on Earth's surface, seasonal variations in received radiation occur due to planetary tilt and elliptical orbit, with longer term variations up to 1 percent due to the Croll–Milankovitch effect (Section 6.1).

Internal heat energy comes from two sources. A minority, about 20 percent, comes from the "fossil" heat of the molten outer core. The remainder comes from the radioactive decay of elemental isotopes like ^{238}U and ^{40}K locked up in rock minerals, especially low density granite-type rocks of the Earth's crust where such elements have been concentrated over geological time. However, the total mass of such isotopes has continued to decrease since the origin of the Earth's mantle and crust, so that the mean internal outward heat flux has also decreased with time. Today, the mean flux of heat issuing from interior Earth is around 65 mW m^{-2} (Fig. 2.4), though there are areas of active volcanoes and geothermally active areas where the flux is very much greater. The mean flux outward is thus only some $4.8 \cdot 10^{-5}$ of the solar constant. To make this contrast readily apparent, the total output of internal heat from the area enclosed by a 400 m circumference racetrack would be about 1 kW, of the same order as that received

by only 1 m^2 area of the outer atmosphere and equivalent to the output of a small domestic electric bar heater. The heat energy available to drive plates is thus minuscule (though quite adequate for the purpose) by comparison with that provided to drive external Earth processes like life's metabolism, hydrological cycling, oceanographic circulations, and weather.

2.2.3 How does heat travel?

Radiative heat energy is felt from a hot object at a distance, for example, when we sunbathe or bask in the glow of a fire, in the latter case feeling less as we move further away. The heat energy is being transported through space and atmosphere at the speed of light as electromagnetic waves.

Conductive heat energy is also felt as a transfer process by directly touching a hot mass, like rock or water, because the energy transmits or travels through the substance to be detected by our nervous system. In liquids we feel the effects of movement of free molecules possessing kinetic energy, in metals the transfer of free electrons, and in the solid or liquid state as the atoms transmit heat energy by vibrations.

Convection is when heat energy is transferred in bulk motion or flow of a fluid mass (gas or liquid) that has been externally or internally heated in the first place by radiation or conduction.

2.2.4 Temperature through Earth's atmosphere

The mean air temperature close to the land surface at sea level is about 15°C. Commonsense might suggest that the mean temperature increases the further we ascend in the atmosphere: like Icarus, "flying too close to the sun," more radiant energy would be received. In the lower atmosphere, this commonsense notion, like many, is soon proved wrong (Fig. 2.5) either by direct experience of temperatures at altitude or from airborne temperature measurements. The "greenhouse" effect of the lower atmosphere (Sections 3.4, 4.19, and 6.1) keeps the surface warmer than the mean – 20°C or so, which would result in the absence of atmosphere. Although a little difficult to compare exactly, since the Moon always faces the same way toward the Sun, mean Moon surface temperature is of about this order (varying from +130°C on the sunlit side to −158°C on the dark side). Due to the declining greenhouse effect, as Earth's atmosphere thins, temperature declines upward to a minimum of about −55°C above

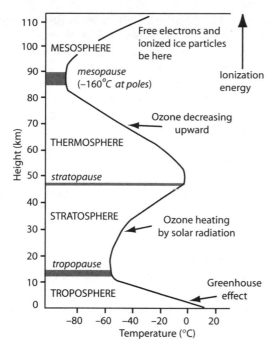

Fig. 2.5 Mean temperature gradients for atmosphere.

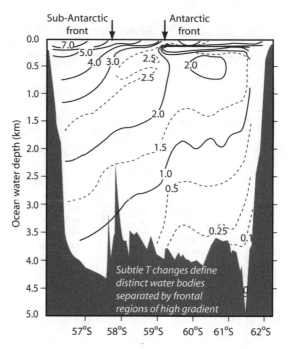

Fig. 2.6 Section across Drake Passage between South America and Antarctic to show oceanic temperature (°C): depth field.

the equator at 12–18 km altitude. The mean *lapse rate* is thus some 4°C km^{-1}. The temperature minimum is the *tropopause*. Above this, temperature steadily rises through the *stratosphere* at about half the tropospheric lapse rate, to a maximum of about 5°C at 50 km above the equator. This is because stratospheric temperatures depend on the radiative heating of ozone molecules by direct solar shortwave radiation. Another rapid dip in temperature through the *mesosphere* to the *mesopause* at about 85 km altitude reflects the decrease in ozone concentration. Above this the positive 1.6°C km^{-1} lapse rate in the *thermosphere (ionosphere)* to 400 km altitude is due to the ionization of outer atmosphere gases by incoming ultra-shortwave radiation in the form of γ-rays and x-rays. Beyond that, in space at 32,000 km, the temperature is around 750°C.

2.2.5 Temperature in the oceans

Earth's oceans have an important role in governing climate, since the specific heat capacity of water is very much greater than that of an equivalent mass of air. So, ocean water has a very high thermal inertia, or low diffusivity, enabling heat energy produced by high radiation levels in low-latitude surface waters to be transferred

widely by ocean currents. Thermal energy is lost as water is evaporated (see *latent heat of evaporation* explained in Section 3.4) by the overlying tropospheric winds but this is eventually returned as *latent heat of condensation* (Section 3.4) to heat the atmospheres of more frigid climes. But it is a mistake to assume that the oceans are of homogenous temperature. Distinct ocean water masses are present that have small but significant variations in ambient temperature (Fig. 2.6), which control the density, and hence buoyancy of one ocean water mass over another. Those illustrated for the Southern Ocean show the subtle changes that define fronts of high temperature gradient.

2.2.6 Temperature in the solid Earth

The gradient of temperature against depth in the Earth is called a *geotherm*. The simplest estimate would be a linear one and it is a matter of experience that the downward gradient is positive. We could either take the geotherm to be the observed gradient in rock temperature or that measured in deep boreholes (below *c.*100 m) and extrapolate downward, or take the indirect evidence for molten iron core as the basis for an extrapolation upward. The mean near-surface temperature gradient on the continents

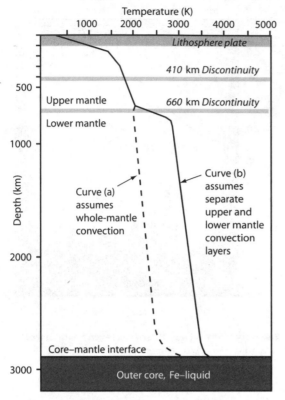

Fig. 2.7 Mean temperature gradient (geotherm) for solid Earth.

is about 25°C km^{-1} and although linear for the very upper part of the crust directly penetrated by humans, such a gradient cannot be extrapolated further downward since widespread lower crustal and mantle melting would result (or even vaporization in the mantle!) for which there is no evidence. We therefore deduce that (Fig. 2.7)

1 The geothermal gradient decreases with depth in the crust; that is, it becomes nonlinear.

2 The high near-surface heat flow must be due to a concentration of heat-producing radioactive elements there.

Concerning the temperature at the 3,000 km radius core–mantle boundary (CMB), metallurgy tells us that iron melts at the surface of the Earth at about 1,550°C. Allowing for the increase of this melting temperature with pressure, the appropriate temperature at the CMB may be approximately 3,000°C, yielding a conveniently easy to remember (though quite possibly wrong) mantle gradient of c.1°C km^{-1}.

2.3 Quantity of matter

2.3.1 Mass

We measure all manner of things in everyday life and express the measured portions in kilograms; we usually say that the portions are of a certain "weight." On old-fashioned beam balances, for example, kilogram or pound "weights" are used. These are of standard quantity for a given material so that comparisons may be universally valid. In science, however, we speak of all such estimates of bulk measured in kilograms as *mass* (symbol m). The bigger the portion of a given material or substance, the larger the mass. We can even "measure" the mass of the Earth and the planets (see Section 1.4). We must never speak of "weight" in such contexts because, as we shall see later in this book, weight is strictly the effect of acceleration due to gravity upon mass. Mass is independent of the gravitational system any substance happens to find itself in. So when we stand on the weighing scales we should strictly speak of being "undermass" or "overmass."

Newton defined mass, what he termed "quantity of matter" succinctly enough (Fig. 2.8). Here is a nineteenth-century English translation of the original Latin: "Quantity of matter is the measure of it arising from its density and bulk conjointly," that is, gravity does not come into it.

2.3.2 Density

The amount of mass in a given volume of substance is a fundamental physical property of that substance. We define *density* as that mass present in a unit volume, the unit being one cubic meter. The units of density are thus kg m^{-3} (there is no special name for this unit) and the dimensions ML^{-3}. The unit cubic meter can comprise air, freshwater, seawater, lead, rock, magma, or in fact anything (Fig. 2.8). In this text ρ will usually symbolize fluid density and σ, solid density (though beware, for we also use σ as a symbol for stress, but the context will be obvious and well explained). Sometimes the density of a substance is compared, as a ratio, to that of water, the quantity being known as the *specific gravity*, a rather

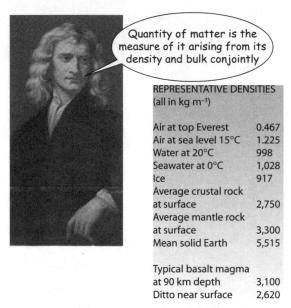

Quantity of matter is the measure of it arising from its density and bulk conjointly

REPRESENTATIVE DENSITIES
(all in kg m⁻³)

Air at top Everest	0.467
Air at sea level 15°C	1.225
Water at 20°C	998
Seawater at 0°C	1,028
Ice	917
Average crustal rock at surface	2,750
Average mantle rock at surface	3,300
Mean solid Earth	5,515
Typical basalt magma at 90 km depth	3,100
Ditto near surface	2,620

Fig. 2.8 Density may vary with state, salinity, temperature, pressure, and content of suspended solids.

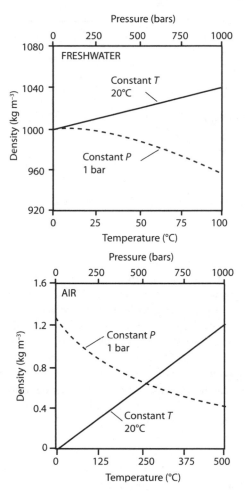

Fig. 2.9 Variation of density of freshwater and air with temperature and pressure.

confusing term. Density is regarded as a *material property* of any *pure* substance. The magnitude of such a property under given conditions of temperature and pressure is invariant and will not change whether the pure substance is on Moon, Mercury, or Pluto, as long as the conditions are identical. Neither does the value change due to any flow or deformation taking place.

2.3.3 Controls on density

Note the emphasis on "given conditions" in Section 2.3.2, for if these change then density will also change. Temperature (T) and pressure (p) can both have major effects on the density of Earth materials. We have already sketched the magnitudes of temperature change with height and depth in the atmosphere, ocean, and within solid Earth (Section 2.2). These variations come about due to variable solar heating by radiation, radioactive heat generation, thermal contact with other bodies, changes of physical state, and so on. Pressure varies according to height or depth in the atmosphere, ocean, or solid Earth (Section 3.5). All of these factors exert their influence on the density of Earth materials. Why is this? Referring to Section 2.1, you can revisit the role of molecular packing upon the behavior of the states of matter. The loose molecular packing of gases means that they are compressible and that small changes in temperature and pressure have major effects upon density (Fig. 2.9). Temperature

also has significant effects on both liquid (Fig. 2.9) and solid density whereas pressure has smaller to negligible effects upon liquid and solid density in most near-surface environments, becoming more important at greater depths. There are also important effects to consider in cold lakes due to the *anomalous expansion of pure water* below approximately 4°C. This means that water is less dense at colder temperatures. As salinity increases to that of seawater the temperature of maximum density falls to about 2°C. In the deep oceans and deep lakes, for example, Lake Baikal, an additional effect must be considered, the *thermobaric effect*. This is the effect of pressure in decreasing the temperature of maximum density.

The case of seawater density is of widespread interest in oceanography since natural density variations create buoyancy and drive ocean currents. Its value depends upon temperature, salinity (Fig. 2.10), and pressure. The covariation with respect to the former two variables is shown in Fig. 2.11. It is convenient to express ocean water density,

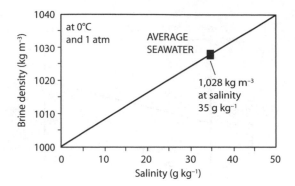

Fig. 2.10 Variation of seawater density with salinity.

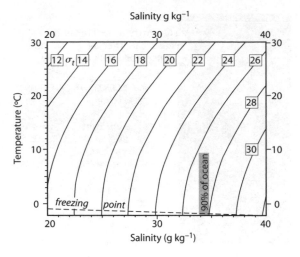

Fig. 2.11 Covariation of seawater density (as σ_t) with salinity and temperature.

ρ, as the excess over that of pure water at standard conditions of temperature and pressure. This is referred to as σ_t and is given by $(\rho - 1{,}000)$ kg m^{-3}. This variation is usually quite small, since over 90 percent of ocean water lies at temperatures between -2 and $10°C$ and salinities of 20–40 parts per thousand (g kg^{-1}) when the density σ_t ranges from 26 to 28 (Fig. 2.11). It is difficult to measure density *in situ* in the ocean, so it is estimated from tables or formulae using standard measurement data on temperature, salinity, and pressure. Detailed measurements reveal that the rate of increase in seawater density with decreasing temperature slows down as temperature approaches freezing: this is important for ocean water stratification at high latitudes when it is more difficult to stratify the very cold, almost surface waters without changes in salinity.

Finally, our definition of density deliberately refers to the "pure" substance. As noted in Section 2.1, many Earth materials are rather "dirty" or impure, due to natural suspended materials or human pollutants. The turbid suspended waters of a river in flood, a turbidity current, or the eruptive plume of an explosive volcanic eruption are cases in point. The changed density of such suspensions (see Fig. 2.12) is a feature of interest and importance in considering the flow dynamics of such systems.

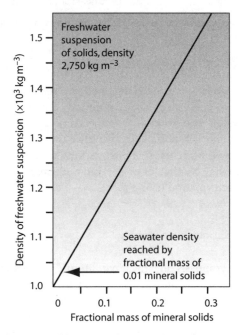

Fig. 2.12 Variation of freshwater density with concentration of suspended mineral solids.

2.4 Motion matters: kinematics

2.4.1 Universality of motion

All parts of the Earth system are in motion, albeit at radically different rates (Box 2.2); the study of motion in general is termed *kinematics*. We may directly observe motion of the atmosphere, oceans, and most of the hydrosphere. Glaciers and ice sheets move, as do the permafrost slopes of the cryosphere during summer thaw. The slow motion of lithospheric plates may be tracked by GPS and by signs of motion over plumes of hot material rising from the deeper mantle. Magma moves through plates to reach the surface, inflating volcanoes as it does so. The

Earth's surface has tiny, but important, vertical motions arising from deeper mantle flow. Spectacular discoveries relating to motions of the interior of the Earth have come from magnetic evidence for convective motion of the outer core and, more recently, for differential rotation of the inner core. Some Earth motions may be regarded as steady, that is to say they are unchanging over specified time periods, for example, the movement of the deforming plates and, presumably, the mantle. Other motion, as we know from experience of weather, is decidedly unsteady, either through gustiness over minutes and seconds or from day to day as weather fronts pass through. How we define unsteadiness at such different timescales is clearly important.

2.4.2 Speed

Faced with the complexity of Earth motions we clearly need a framework and rigorous notation for describing motion. The simplest starting point is rate of motion measured as *speed*; generally we define speed as increment of distance traveled, δs, over increment of time, δt. Speed is thus $\delta s / \delta t$, length traveled per standard time unit (usually per second; units LT^{-1}). In physical terms, speed is a *scalar quantity*, expressing only the magnitude of the motion; it does not tell us anything about where a moving object is going. Thus a speeding ticket does not mention the direction of travel at the time of the offense. Further comments on scalars are given in the appendix.

Box 2.2 Typical order of mean speeds for some Earth flows ($m\,s^{-1}$)

Jet stream	30–70
High latitude front	7–10
Gale force wind	19
Storm force wind	>26
Hurricane	33
Hurricane grade 4	46–63
Gulf stream	1–2
Thermohaline flow	0.5–1
Tidal Kelvin wave at coast	15
Equatorial ocean surface	
Kelvin wave	200
Tsunami	200
Spring tidal flow	2
Mississippi river flood	2
Alpine valley glacier	$3.2 \cdot 10^{-6}$ (10 m a^{-1})
Antarctic ice stream	$3.2 \cdot 10^{-4}$ (1,000 m a^{-1})
Lithospheric plate	$1.6 \cdot 10^{-9}$ (0.05 m a^{-1})
Pyroclastic flow	>100
Magma in volcanic vent	$8.3 \cdot 10^{-3}$ (30 m h^{-1})
Magma in 3 m wide dyke	10^{-3} (3.6 m h^{-1})
Magma in pluton	10^{-8} (0.3 m a^{-1})

2.4.3 Velocity

A practical analysis of motion needs extra information to that provided by speed; for example, (1) it is of little use to determine the speed of a lava flow without specifying its direction of travel; (2) a tidal current may travel at 5 ms^{-1} but the description is incomplete without mentioning that it is toward compass bearing 340°. V*elocity* (symbol *u*, units LT^{-1}) is the physical quantity of motion we use to express both direction and magnitude of any displacement. A quantity such as velocity is known generally as a *vector*. A velocity vector specifies both distance traveled over unit of time *and* the direction of the movement. Vectors will usually be written in bold type, like *u*, in this text, but you may also see them on the lecture board or other texts and papers underlined, $\underline{u}$, with an arrow, $\vec{u}$ or a circumflex, $\hat{u}$. Any vector may be resolved into three orthogonal (i.e. at 90°) components. On maps we represent velocity with *vectorial arrows*, the length of which are proportional to speed, with the arrow pointing in the direction of movement (Fig. 2.13). With vectorial arrows it is easy to show both time and space variations of velocity, and to calculate the relative velocity of moving objects. Further comments on vectors are given in the appendix.

2.4.4 Space frameworks for motion

Both scalars and vectors need space within which they can be placed (Fig. 2.14). Nature provides space but in the lab a simple square graph bounded by orthogonal x and y coordinates is the simplest possibility. The points of the compass are also adequate for certain problems, though many require use of three-dimensional (3D) space, with three orthogonal coordinates, x, y, z. This 3D space (also any two-dimensional (2D) parts of this space) is termed Cartesian, after Descartes who proposed it; legend has it that he came up with the idea while lazily following the path of a fly on his bedroom ceiling. Using the example of the velocity vector, *u*, we will refer to its x, y, z components as u, v, w. The motion on a sphere taken by lithospheric plates and ocean or atmospheric currents is an angular one succinctly summarized using polar coordinates (Fig. 2.13c) or in the framework provided by a latitude and longitude grid.

2.4.5 Steadiness and uniformity of motion

Consider a stationary observer who is continuously measuring the velocity, *u*, of a flow at a point. If the

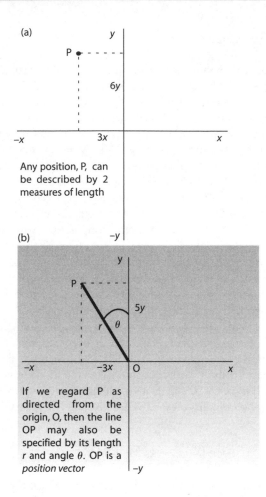

(a)

Any position, P, can be described by 2 measures of length

(b)

If we regard P as directed from the origin, O, then the line OP may also be specified by its length r and angle θ. OP is a *position vector*

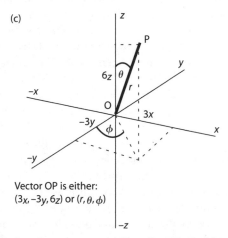

(c)

Vector OP is either: $(3x, -3y, 6z)$ or (r, θ, ϕ)

Fig. 2.13 Coordinate systems: (a) Two dimensions; (b) two dimensions with polar notation, and (c) three dimensions.

velocity is unchanged with time, t, then the flow is said to be *steady* (Fig. 2.14a). Mathematically we can write that the change of u over a time increment is zero, that is, $\delta u / \delta t = 0$.

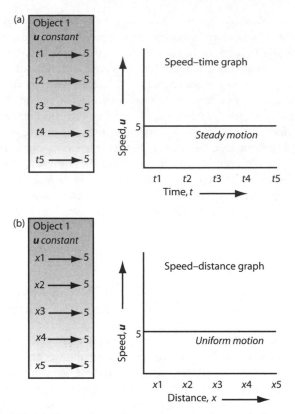

Fig. 2.14 (a) Vectors for steady west to east motion at velocity $u = 5$ ms^{-1} for times $t1 - t5$. (b) Vectors for uniform west to east motion at velocity $u = 5$ ms^{-1} for positions $x1 - x5$.

The description of steadiness depends upon the frame of reference being fixed at a local point. We may take instantaneous velocity measurements down a specific length, s, of the flow. In such a case the flow is said to be *uniform* when there is no velocity change over the length, that is, $\delta u / \delta s = 0$ (Fig. 2.14b).

This division into steady and uniform flow might seem pedantic but in Section 3.2 it will enable us to fully explore the nature of acceleration, a topic of infinite subtlety.

2.4.6 Fields

A *field* is defined as any region of space where a physical scalar or vector quantity has a value at every point. Thus we may have scalar speed or temperature fields, or, a vectorial velocity field. Crustal scale rock velocity (Figs 2.15 and 2.16), atmospheric air velocity, and laboratory turbulent water flow are all defined by fields at various scales. Knowledge of the distribution of velocities within a flow field is essential in order to understand the dynamics of the material comprising the field (e.g. Fig. 2.16).

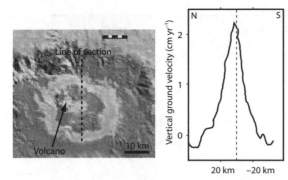

Fig. 2.15 Vertical crustal velocity around Hualca Hualca volcano, southern Peruvian Andes: surface deformation as seen by satellite radar over about four years.

Note the high uplift rates and:
1 Concentric grayscale variations indicate uplift relative to surrounding areas. Maximum uplift is seen due east of the volcanic edifice. Note symmetrical uplift rate and constant uplift gradients.
2 Uplift appears steady over the four years.
3 Surface swelling is due to melting, magma recharge, or hot gas/water activity about 12 km below surface, but significantly offset from volcano axis.
4 Volcano may be actively charging itself for a future eruption.

2.4.7 The observer and the observed: stationary versus moving reference frames

You know the feeling; you are stationary in a bus or train carriage and the adjacent vehicle starts to move away. For a moment you think you are moving yourself. You are confused as to exactly where the fixed reference frame is located – in your space or your neighbors in the adjacent vehicle. Well, both spaces are equally valid, since all space coordinate systems are entirely arbitrary. The important thing is that we think about the differences in the velocity fields witnessed by both stationary and moving observers and understand that one can be exactly transformed into another. Motion of one part of a system with reference to another part is called *relative motion*. Examples are (1) the relative motion of a crystal falling through a magma body that is itself rising to the surface; (2) two lithosphere plates sliding past each other (Fig. 2.16); (3) a mountain or volcano rising (Fig. 2.15) due to tectonic forces but at the same time having its surface lowered by erosion so that a piece of rock fixed within the mountain is being both lifted up and also exhumed (brought nearer to the surface) at the same time.

The flow field seen by a stationary viewer is known as the fixed spatial coordinate, or *Eulerian*, system. Analysis is done with respect to a control volume fixed with respect to the observer and through which fluid or other mass passes. Velocity measurements at different times are thus gained from different fluid "particles" and must therefore be averaged over time to give a time mean velocity.

The flow field seen by a moving viewer is known as the moving spatial coordinate, or *Lagrangian*, system. Analysis is done with respect to Cartesian axes and flow control volumes moving with the same velocity as the flow. Velocity measurements at different times are thus gained from the same fluid "particles" and the time average velocity is that gained over some downstream distance.

Most flow systems benefit by an Eulerian treatment. Certainly for fluids, the mathematics is easier since we consider dynamical results "at a point," rather than the devious fate of a single fluid mass. Adopting a Eulerian stance, any velocity is a function of spatial position coordinates x, y, z, and time; we say in short (appendix), $u = f(x, y, z, t)$.

2.4.8 Harmonic motion

We speak of harmony in everyday life as the experience of mutually compatible levels of being. In music the term applies to the contrasting levels or frequencies of sound that bring about a harmonious combination. Harmonic motion deals with the periodic return of similar levels of some material surface relative to a fixed point; it is best appreciated by reference to the displacement of surface water level during passage of a surface wave, or as illustrated in Fig. 2.17, of the passage of a fixed point on a rotating wheel. The wave itself has various geometrical terms associated with it, period, T, for example, and can be considered mathematically most simply by reference to a sinusoidal curve.

2.4.9 Angular speed and angular velocity

Consider curved (rotating) motion (Fig. 2.18a); in going from a to b in unit time a particle sweeps out an arc of length s, subtending an angle ϕ with the center of curvature, radius r. We can talk about a constant quantity for the traveling particle as $\delta\phi/\delta t$, the *angular speed*, ω, usually measured in radians per second (a radian is defined as $360/2\pi$ degrees). The *linear speed, u,* of the rotating particle is the product of angular speed of the particle and its radial distance from the center of curvature, that is, $u = r\omega$.

Angular velocity (Fig. 2.18b,c) has both magnitude and direction and is thus a vector, denoted Ω. It has units of radians per second. The angular velocity of rotation of

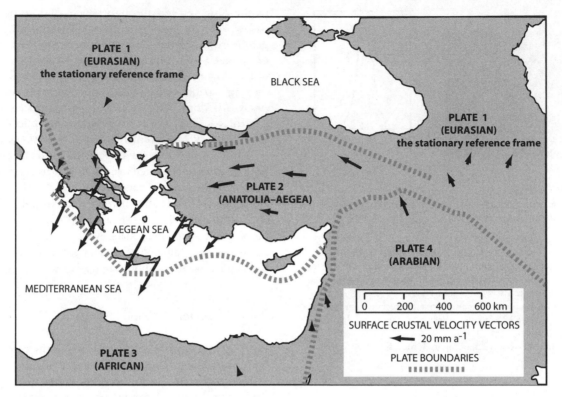

Fig. 2.16 Horizontal surface velocities of the lithospheric plates making up the eastern Mediterranean and Asia Minor. Data derived from satellite geodesy platforms (GPS) averaged over a few years and stated with reference to a stationary Eurasian plate reference frame.

Notes:
1 Contrasts in velocity vectors between different plates and sharp discontinuities present across plate boundaries.
2 Evidence for systematic east to west acceleration (implying crustal strain) and anticlockwise spin (vorticity) of the Anatolia–Aegea plate.

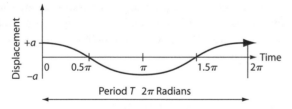

Fig. 2.17 Harmonic motion. A wave has periodic, often sinusoidal, motion. The example is a curve traced out in time, best imagined as the track to a point on a moving wheel.

Earth is $7.29 \cdot 10^{-5}$ rad s^{-1}. In order to give angular velocity its vectorial status, the direction is conventionally taken as a normal axis to the plane of the rotating substance, Ω pointing toward the direction in which a right-handed screw would travel if screwed in by rotating in the same direction as the rotating substance (Fig. 2.18b). For example, in the case of clockwise flow in the xy plane, the axis is in the vertical sense, Ω pointing downward and thus of negative sign. Vice versa for anti-clockwise flow. We can denote the position of any rotating

particle by means of the position vector, r. This leads to the important result that the angular velocity vector, Ω, and the linear velocity vector, u, of the water at position vector, r, are at right angles to each other (Fig. 2.18c). Vector geometry relates the linear velocity vector, u, to the *vector product* of the angular velocity vector and the position vector (i.e. $u = \Omega \times r$).

2.4.10 Vorticity

Vorticity is related to angular motion and is best envisaged as "spin," or rotation; it is the tendency for a parcel of fluid or a solid object to rotate. It is sometimes given the symbol, ω, but in oceanographic contexts more usually, ζ, a convention we follow subsequently. Vortical motions occur all around us: the whole solid planet possesses vorticity (appropriately termed planetary vorticity), on account of spin about its own axis; lithospheric plates and crustal blocks may also slowly spin (Fig. 2.16); the whole atmosphere and atmospheric cyclones and anticyclones

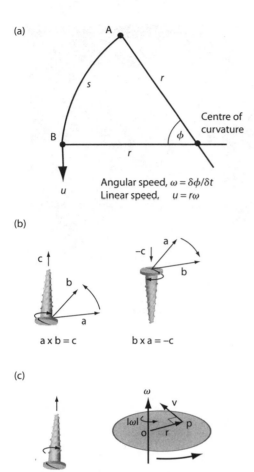

Fig. 2.18 To illustrate curved motion angular speed and velocity. (a) Angular speed, (b) angular velocity conventions, and (c) angular velocity.

rotate; spinning eddies of fluid turbulence are readily observed in rivers and from satellite images in ocean currents. Fluid vorticity is termed *relative* or *shear vorticity* and is due to velocity differences, termed *velocity gradients*, across a fluid element (Section 1.19). It can be shown (Section 3.8) that rigid body vorticity is twice the angular velocity, that is, $\zeta = 2\Omega$. Finally, vorticity must be conserved according to the principle of the Conservation of Absolute Vorticity (see Section 3.8).

2.4.11 Visualization of flow

No dynamical analysis may be confidently begun without some idea of actual flow pattern. In everyday life the gusting eddies of a wind are picked out by the motion of autumn leaves or by the swirling pattern of snow or sleet across a road or field. In the same way in the lab, flow visualization introduces some marker into a flow which is then photographed (Fig. 2.19). Considering the Eulerian case, a

photograph of a continuously introduced dye will yield a *streakline*, the locus of all fluid elements that pass through. A photograph of an instantaneously introduced dye or of reflective particles will yield a *pathline*. For a steady flow it is possible to construct an overall flow map by drawing *streamlines*. These are lines drawn such that the velocity of every particle on the line is in the direction of the line at that point. Numerous examples of flow visualization are given in the text that follows (see in particular Figs 3.53–3.55).

2.4.12 Flow without dynamics: "Ideal" flow along streamlines

From the definition of a streamline quoted above it is obvious that streamlines cannot cross and that it is possible to define a volume of fluid bounded by streamlines along its length. Such an imaginary volume is termed a *stream-tube* (Fig. 2.20). If the discharge into and out of a stream-tube of any shape is constant, areas of streamline convergence indicate flow acceleration and areas of divergence indicate deceleration. Thus areas of close spacing have higher velocity than areas with wide spacing. Some progress may be made concerning the prediction of streamline positions rather than the experimental visualization considered previously by using concepts of *ideal* (*potential*) flow as applied to fluids in which the molecular viscosity (see Section 3.9) is considered zero. Although such frictionless fluids are far from physical reality, ideal flow theory may be of great help in analyzing motions distant from solid boundaries (i.e. away from boundary layers; see Section 4.3) and in flows where viscous effects are negligible (at very high Reynolds' numbers; see Section 4.5). As subsequent discussions will show, in the absence of shearing stresses in an ideal fluid there can be no rotational motion (vorticity), that is, all ideal flows are considered *irrotational*.

Considering any ideal flow past a bounding (solid) surface, it is apparent that discharge between the boundary and a given streamline must be constant. Thus it is possible to label streamlines according to the magnitude of the discharge that is carried past themselves and a distant boundary. This discharge is known as the *stream function*, ψ, of a streamline (Fig. 2.20). The magnitude of ψ is obviously unique to any particular streamline and must be constant along the streamline. Velocity is higher when streamline spacing is closer and vice versa (Cookie 2.1).

Another useful method of analyzing ideal flow arises from the concept of *velocity potential lines*, symbol ϕ. These imaginary lines are drawn normal to streamlines (Fig. 2.20). They define a flow field, as defined in Section 2.4 and are best

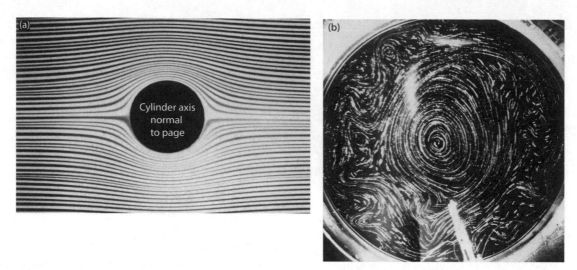

Fig. 2.19 Flow visualization photos. (a) Dye introduced continuously into flow through jets at left define streamlines of laminar flow around a stationary solid cylinder. (b) Streak photograph of aluminum flakes on the surface defines a pattern of convection in a counterclockwise rotating cylinder pan that is being heated at the outside rim and cooled in the center. Flow pattern is analogous to the circulation of the upper atmosphere.

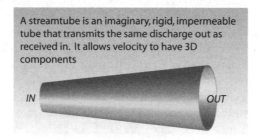

A streamtube is an imaginary, rigid, impermeable tube that transmits the same discharge out as received in. It allows velocity to have 3D components

Streamlines, Ψ_{1-2} define a 2D section through the streamtube. They allow velocity, **u**, to have two components: u and v in this case.

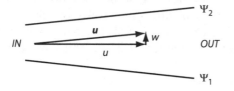

The discharge in and the discharge out are identical. As the streamlines diverge the flow velocity must lessen down tream, vice versa for convergence. So velocity is proportional to streamline spacing

Fig. 2.20 Streamtubes, streamlines, and potentials.

Equipotential lines, ϕ, are drawn normal to *streamlines*, Ψ, with their spacing proportional to velocity. The closer the lines the faster the flow. The combination of streamlines and equipotential lines defines a *flow net*

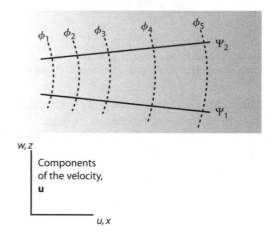

compared to contour lines on a map where the direction of greatest rate of change of height with distance is along any local normal to the contours (gradient of the scalar height). The velocity is the gradient of ϕ (Cookie 1).

If the distance between equipotential lines and streamlines is made close and equal, then the resultant pattern of small squares is known as a *flow net* (Fig. 2.20). Construction of flow nets for flow through various 2D shapes may considerably aid physical analysis. The grid is built up by trial and error from an initial sketch of streamlines between the given boundaries. Then the equipotential lines are drawn so that their spacing is the same as the streamline spacing. Continuous adjustments are made until the grid is composed (as nearly as possible) of squares, and the actual streamlines are then obtained. This is useful because, for example, from the streamline construction one may deduce velocity and, with a knowledge of Bernoulli's equation (Section 3.12), pressure variations. However, it will be obvious to the reader that flow nets are only a rather simple imitation of natural flow patterns. Experimental studies will reveal patterns of flow that cannot be guessed at by potential approaches (e.g. Fig. 2.19b).

2.5　Continuity: mass conservation of fluids

A fundamental principle in fluid flow is that of *conservation*, the interaction between the physical parameters that determine mass between adjacent fluid streamlines. The transport of mass, m, along a streamline involves the parameters velocity, u, density, ρ, and volume, V. These determine the conservation of mass discharge, termed *continuity*.

2.5.1　Continuity of volume with constant density

River, sea, and ocean environments essentially comprise incompressible fluid. They contain layers, conduits, channels, or straits that vary in cross-sectional area, a, while a discharge, Q (units $L^3 T^{-1}$) of the constant density fluid through them remains steady, being supplied from elsewhere due to a balance of applied forces at a constant rate (Fig. 2.21). Generally, if there is cross-sectional area a_1 and mean velocity u_1 upstream, and area a_2 and mean velocity u_2 downstream, the product $Q = ua$ must remain constant (you can check that the product Q has dimensions of discharge, or flux, $L^3 T^{-1}$). We then have the equality $u_1 a_1 = u_2 a_2$ so that any change in cross-sectional area is accompanied by an increase or decrease of mean velocity and there is no change in Q that is, $\Delta Q = 0$. Any changes in u naturally result in acceleration or deceleration. This simplest possible statement of the continuity equation may be used in very many natural environments to calculate the effects of decelerating or accelerating flow (Section 3.2).

To be applicable, continuity of volume has important conditions attached:

1 The fluid is incompressible, so no changes in density due to this cause are allowed.
2 Fluid temperature is constant, so there is no thermally induced change in density.
3 Fluid density due to salinity or suspended sediment content also remains unchanged.
4 No fluid is added, that is, there is no *source*, like a submarine spring or oceanic upwelling.
5 No fluid is subtracted, that is, there is no *sink*, like a permeable bounding layer or thirsty fish.

One natural environment where most of these conditions are satisfied is a length of river channel, where cross-sectional area changes downstream (e.g. Section 3.2).

2.5.2　Continuity of mass with variable density

Consider now a steady discharge of fluid with a variable density that flows into, through, and out of any fixed volume containing mass, m (Fig. 2.22). If that mass changes then the difference, δm, may be due to a change of fluid density, $\delta \rho$, of the fluid within the volume over time and/or space. The fact that density is now free to vary, as

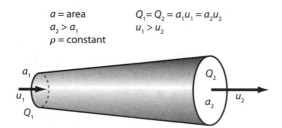

a = area
$a_2 > a_1$
ρ = constant

$Q_1 = Q_2 = a_1 u_1 = a_2 u_2$
$u_1 > u_2$

Fig. 2.21 Continuity of volume: constant density case in 1D.

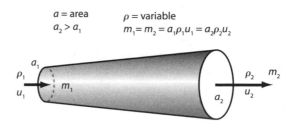

a = area
$a_2 > a_1$

ρ = variable
$m_1 = m_2 = a_1 \rho_1 u_1 = a_2 \rho_2 u_2$

Fig. 2.22 Continuity of mass: variable density case in 1D.

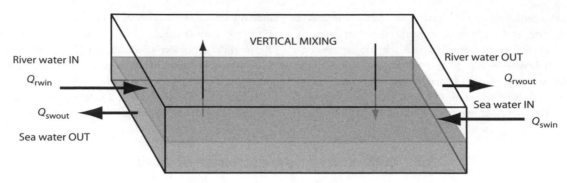

VERTICAL MIXING

River water IN

Q_{rwin}

Q_{swout}

Sea water OUT

River water OUT

Q_{rwout}

Sea water IN

Q_{swin}

Fig. 2.23 Estuarine circulation: example of mass conservation in action.

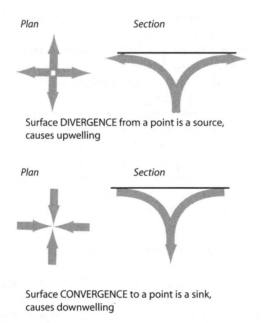

Plan   *Section*

Surface DIVERGENCE from a point is a source, causes upwelling

Plan   *Section*

Surface CONVERGENCE to a point is a sink, causes downwelling

Fig. 2.24 Sources and sinks.

2.5.3 Examples of volume and mass continuity

1 Delta or estuary channels are informative environments within which to consider the workings of continuity (Fig. 2.23). For any control volume the upstream discharge of seawater decreases while the downstream input of fresh river water decreases. A mass balance is brought about by vertical mixing of seawater upward and freshwater downward.

2 It is instructive to apply the 3D volume continuity expression for an incompressible fluid such as that found in an idealized portion of fast-moving ocean, river, or tidal shelf. It is usually fairly straightforward to measure the two mean surface components of the local velocity but more difficult to measure the time mean vertical velocity. We compute this useful parameter from the basic conservation expression in Cookie 3.

3 We finally touch upon *divergence* and *convergence* with respect to sources and sinks. We stated that the continuity expression depends upon the lack of sources or sinks linked to the system in question. Two important cases arise in hydrological, oceanographical, and meteorological flows (Fig. 2.24; see also Cookie 3). Surface divergence of streamlines, most obviously seen when flow is diverging from a point implies that a source is present below the surface, leading to a mass influx. Surface convergence of streamlines to a point implies a sink is present and that downwelling is occurring. An added complication for meteorological flows is that vertical motions of fluid in downwelling or upwelling situations also cause changes of temperature and density, which cause feedback relevant to the stability of a moving air mass.

in the case of compressible gas flow or a thermally varying flow, means there is one more degree of freedom than in the case considered previously; we have: $u_1A_1\rho_1 = u_2A_2\rho_2$, so that any change in net mass outflow per unit time (check the expression gives units MT^{-1}) is now caused by a change in density and/or velocity.

The full algebraic expression for 3D continuity is given in Cookie 2 (the algebra looks hideous but is quite logical).

Further reading

Everyone has their favorite college physics text that explains things to their satisfaction. Our "bible" is P.M. Fishbane et al.'s *Physics for Scientists and Engineers: Extended Version* (Prentice-Hall, 1993). Flowers and Mendoza's *Properties of Matter* (Wiley, 1970) is erudite. Massey's *Mechanics of Fluids* (Van Nostrand Reinhold, 1979) is exceptionally clear.

The math and physics appendices in S. Pond and G. L. Pickard's *Introduction to Dynamical Oceanography* (Pergamon, 1983) and R. McIlveen's *Fundamentals of Weather and Climate* (Stanley Thornes, 1998) are exceptionally clear. More advanced physical derivations are set out in D. J. Furbish's *Fluid Physics in Geology* (Oxford, 1997).

3 Forces and dynamics

3.1 Quantity of motion: momentum

3.1.1 Linear momentum

Momentum, symbol p, is the product of the mass, m, of any substance (gas, liquid, or solid) and its velocity, u. Hence dimensions are MLT^{-1}; there are no special units for momentum. We get the importance of the concept most directly from Newton's Definition 2, translated from the original Latin into the elegant English of the mid-nineteenth century:

> The quantity of motion of a body is the measure of it arising from its velocity and the quantity of matter conjointly.

You may agree with us that the phrase "quantity of motion" (Fig. 3.1) is a good deal more expressive and unequivocal than the term in modern English language usage, "momentum"; the obvious semantic confusion for the beginner is with *moment*, as in moments of forces.

In Spanish, however, *cantidad de movimiento* or "quantity of motion" is a commonly expressed synonym for momentum. We see immediately the significance of the word *conjointly* in Newton's definition, for similar values of $p = mu$ may be achieved as the consequence of either large mass and small velocity or vice versa. It is thus instructive to calculate the momentum of various components of the Earth system; the dual roles of mass and velocity playing off each other can produce some unexpected results (Figs 3.2 and 3.3). For this reason it is also often instructive to express momentum per unit volume, given by $p = \rho u$. Momentum can also be easily related to kinetic energy, E_k (Section 3.3).

Linear momentum is a vector and is orientated through a mass in the same direction as its velocity vector, u. Each of the three Cartesian components of the velocity vector will have its component part of momentum attached to it, that is, ρu, ρv, and ρw.

Fig. 3.1 Newton and his definition of momentum.

I called momentum *"quantity of motion"* – a much more suitable name, don´t you think?

0.5 kg "Bramley" cooking apple falling at velocity 10 m s^{-1}

… is $p = 5$ kg m s^{-1}

1 mm diameter spinning sand grain impacts onto rocky desert floor…

Velocity 2 m s^{-1}
Mass $1.15 \cdot 10^{-5}$ kg

… is $p = 2.3 \cdot 10^{-5}$ kg m s^{-1}

Rebound
(Elastic collision)

Fig. 3.2 On the momentum of apples and sand grains.

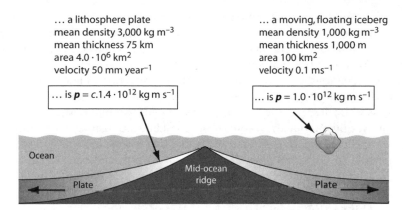

... a lithosphere plate
mean density 3,000 kg m^{-3}
mean thickness 75 km
area $4.0 \cdot 10^6$ km^2
velocity 50 mm year^{-1}

... is $\boldsymbol{p} = c.1.4 \cdot 10^{12}$ kg m s^{-1}

... a moving, floating iceberg
mean density 1,000 kg m^{-3}
mean thickness 1,000 m
area 100 km^2
velocity 0.1 ms^{-1}

... is $\boldsymbol{p} = 1.0 \cdot 10^{12}$ kg m s^{-1}

Ocean

Mid-ocean ridge

Plate   Plate

Fig. 3.3 The momentum of the iceberg (not drawn to scale) is of the same order of magnitude as that of the plate.

3.1.2 Angular momentum

In considering the momentum of rotating solid objects, such as planets, sand grains, or figure skaters, it is necessary to determine the angular momentum (Fig. 3.4) arising from the rotational motion, rather than the linear momentum of the mass. The momentum is thus considered as that arising about a rotational axis. The angular momentum, L, is given by the product of rotational inertia (often called *moment of inertia*) about its rotation axis, I, and its vectorial angular velocity, $\boldsymbol{\omega}$. Thus $\boldsymbol{L} = \boldsymbol{I\omega}$. Notice that the mass term relevant to the determination of linear momentum is here replaced by a rather unfamiliar quantity, rotational inertia. This is a subtle concept arising from the notion of rotational kinetic energy (Section 3.3) and the fact that in rigid body rotation each small element of mass, m, of a solid can be considered to have its own angular velocity of rotation, $\boldsymbol{\omega}$, and therefore kinetic energy, about any rotation axis. All motion is considered about this rotation axis: every small element has its own defined measurable perpendicular distance, R, from this axis and a characteristic speed of $R\boldsymbol{\omega}$. The rotational kinetic energy of the small element is also related to its angular momentum. The rotational inertia and angular momentum of the whole body must be taken as the sum or integral (appendix) of each small element. Particular regular shapes have specific integral solutions, for example, the rotational inertia of a uniform density solid sphere of radius, r, is given by $2/5\,mr^2$. Making use of this expression we can easily calculate the approximate rotational inertia and angular momentum of Earth (ignoring its internal density layering) or of a spinning sand grain (Fig. 3.4).

... Earth's angular momentum, $L = I\omega$, is

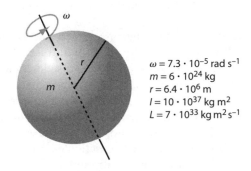

$\omega = 7.3 \cdot 10^{-5}$ rad s^{-1}
$m = 6 \cdot 10^{24}$ kg
$r = 6.4 \cdot 10^6$ m
$I = 10 \cdot 10^{37}$ kg m^2
$L = 7 \cdot 10^{33}$ kg m^2 s^{-1}

... spinning sand grain's angular momentum, $L = I\omega$, is

$\omega = 50$ rad s^{-1}
$m = 1.15 \cdot 10^{-5}$ kg
$r = 1 \cdot 10^{-3}$ m
$I = 4.6 \cdot 10^{-12}$ kg m^2
$L = 2.3 \cdot 10^{-10}$ kg m^2 s^{-1}

Kinetic energy of rotation
$E_k = 0.5\, mR^2\,\omega^2 = 0.5\, I\omega^2$

Fig. 3.4 Angular momentum. The rotational inertia, I, of a solid sphere is $2/5\,mr^2$.

3.1.3 Dynamic significance of momentum and inertia

The big clue concerning the significance of momentum can be approached simply from first principles. We have seen that mass gives us a measure of the quantity of matter present in a solid or fluid; this helps determine an object's *inertia*, its tendency to carry on in the same line of motion or to resist changes in motion. This can be the inertia of a stationary object or that of a steadily moving object to any acceleration. In many relevant physical situations the mass of a given volume element is constant with time and therefore it is the velocity term that determines the *conservation of momentum*. When velocity changes in magnitude or direction, momentum changes. As we shall see a little later, any such change in momentum over time is due to an equivalent force, $F = \mathrm{d}p/\mathrm{d}t$. We expect momentum changes to arise in Nature very frequently: in fluids when an air or water mass changes direction and/or speed due to changes in external conditions; in fluid flow over solid boundaries where a velocity gradient is set up; in the ascent of molten magma that is losing pressure and exsolving gas in bubbles, and so on. The example of colliding solid bodies such as sand grains violently impacting on a desert floor or colliding in a granular fluid brings us to one definition of the conservation of linear momentum for such solid–solid interactions: "... the sum of momenta of an isolated system of two bodies that exert forces on one another is a constant, no matter what form the forces take ...." In other words, the collision of bodies or their interaction leads to no change in overall energy (the production of collisional heat energy is included in the balance). This principle of the conservation of momentum forms the basis of Newton's Third Law (see Section 3.3).

3.2 Acceleration

3.2.1 A simple introduction

Acceleration is a very obvious physical phenomenon; we feel it driving or being driven. In both cases the effect is due to a change of velocity; the more sudden the change, the more reaction we feel. The very fact that a body can "feel" acceleration and is forced to move in response to it means that the phenomenon is somehow connected to that of force. Another kind of universal acceleration concerns a falling body through a frictionless medium, such as a solid through a vacuum, or through air whose resistance to motion is low (and may sometimes be neglected). Here the falling solid is attracted by Earth's gravity field. Some examples of the use of uniform acceleration appropriate to such falling bodies are given in Cookie 4. In other more "resistant" liquids, a steady rate of fall is achieved after an interval such that the downward acceleration due to gravity is quickly balanced by the resistance of the liquid medium.

Acceleration (from now on we use the term without regard for the sign, positive or negative) of the kinds mentioned is most simply imagined as change of velocity over time (Fig. 3.5). Thus in differential form (appendix), $a = \mathrm{d}u/\mathrm{d}t$, with dimensions of LT^{-2}. The standard acceleration due to gravity, g, at sea level is $9.81\ \mathrm{ms}^{-2}$. We stress that natural accelerations may be *extreme* compared with this; for example, turbulent eddies are subject to accelerations of many times gravity (order $\times 10^3 g$) and the Earth's surface is subject to several g acceleration during earthquake motions. By way of contrast, a slow-moving lithospheric plate may change velocity so slowly over such a long time period (of the order of 10^6 years) that the acceleration can be practically neglected.

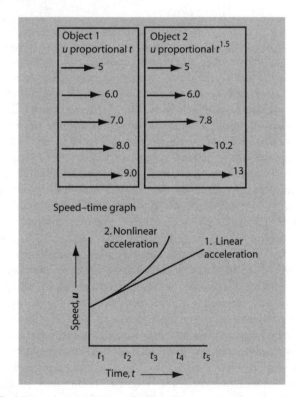

Fig. 3.5 Acceleration. Vectors for W to E motion at velocity, u, for times t_1–t_5.

3.2.2 Complications in moving fluids

Now we consider constant flow or discharge of fluid through conduits, channels, cols, or gates when the passageway has varying cross-sectional area along its length (Fig. 3.6). There is no change of velocity over time at any

A downstream-narrowing channel
with constant discharge (Q)

The field situation depicted involves absolutely steady discharge, so that
at each section AB and CD there is no variation of discharge or velocity
with time ... but there is a change in space

In symbols
$du/dt = 0$

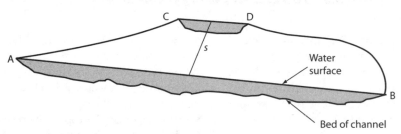

Discharge in, Q_{in}, equals discharge out, Q_{out}. Area of cross-section AB >> area of
cross-section CD. By continuity: mean flow velocity u_{AB} << mean flow velocity u_{CD}.
Therefore a spatial acceleration takes place over distance, s, between cross-sections AB and CD.
Magnitude of the spatial acceleration is $u_{AB}(u_{CD} - u_{AB})/s$

In differential form, udu/ds

The full expression for acceleration is written below. It looks rather complicated, but is not. The numbered terms
are discussed in the text.

$$\frac{Du}{Dt} \equiv \frac{\partial u}{\partial t} + u \frac{\partial u}{\partial x} + w \frac{\partial u}{\partial z} + v \frac{\partial u}{\partial y} \equiv \frac{\partial u}{\partial t} + u \cdot \nabla u$$

term   1   2   3   4

Fig. 3.6 Spatial or advective acceleration and the full expression for total acceleration that includes the unsteady term. The Scale dog is Alaska,
1m long.

place so that $du/dt = 0$ everywhere and the velocity is steady (Section 2.4). From the definition given above, you would therefore expect no acceleration. But there is acceleration along these passageways. Why? The velocity potential lines and streamline constructs in Fig. 2.20 show that velocity must increase or decrease as cross-sectional area changes. Therefore, something is wrong with our simple definition of acceleration. We must allow for accelerations due to *spatial* changes in velocity that affect a fluid cell as it goes from *a* to *b*, where the velocity is different. In this case, we are letting a fluid cell see a change in the velocity field as it travels, in addition to any *local* velocity change. This raises some complications in fluid analysis, since we need to know something about the upstream flow history of any fluid in order to understand its state as it arrives in front of the local observer.

Spatial acceleration, sometimes called *advective acceleration*, the change in velocity, du/ds, along the flow (i.e. as discussed in Section 2.4 for a Lagrangian observer traveling with the flow), is given generally by udu/ds where *u* is

the upstream velocity. You can check the dimensions to see that it really is acceleration.

3.2.3 Total acceleration in moving fluids

It is common to find that both time and space acceleration occur at the same time. To allow for this we make use of term 1 in the equation of Fig. 3.6, designated as *total acceleration*, written Du/Dt, the *substantive* or *total derivative* as we follow the fluid (substantive is used in the same sense as in the "substantive motion" in political debate). It comprises the sum of both time (term 2) and spatial (term 3) accelerations: flows may show either acceleration, or both, or none. It is sometimes also termed the Lagrangian derivative. Term 4 is shorthand for terms 2 and 3 and is explained in the appendix. An important analysis of turbulent flows, done originally by Reynolds (Section 4.5), makes much use of this expression and, although it looks long and cumbersome, it contains a wealth of information about a fluid flow.

3.3 Force, work, energy, and power

We have previously hinted (Section 3.2) that any acceleration or change of momentum implies that an equivalent force must be acting to cause the change. Physical Earth and environmental processes cannot be understood without an appreciation of what forces are, how they arise, and how they operate upon Earth materials.

3.3.1 Weight as a gravity force

We may generalize our definition of force, *F*, as causing an acceleration, *a*, to act upon a mass, *m*. In symbols, $F = ma$. It is clear from this definition that despite a mass being in motion, if there is no acceleration there can be no net force acting, though every moving substance, whether accelerating or not, has momentum. We made a fuss about the appropriate use of the term mass in an earlier section. Spring balances are calibrated by standard masses: their *action* of measurement is not relative, as in a beam balance, but due to the balance of forces between the effect of gravity on the mass suspended by the torsion in the elastic spring.

Weight, the action of gravity on mass (Fig. 3.7), is perhaps the easiest concept of force to begin with, provided we carefully avoid discussion of the true origin of gravity! It is given by the product mg. You can check the dimensions of force from this expression: MLT^{-2}, designated unit, N, for Newtons. This definition means

that a substance does not need to be moving for gravity to exert a force: gravity acts upon everything: moving or stationary. A very accurate spring balance in a constant temperature room at sea level at the equator will thus record a different "weight" for a standard kilogram at the North Pole, or on the top of Everest: in each case because the distance from the center of the Earth to the balance, and hence gravity, is different. The big moonboots of 1960s astronauts had the same mass on the Moon as they had when they were manufactured on Earth or when they were tried out in the desert landscape of New Mexico. It was the vastly reduced gravity on Moon that gave them less weight. Similarly, an average sand grain (mean diameter 1 mm) made of silicate mineral dropping onto the surface of the Martian desert at its terminal velocity has a weight of ratio $g_{mars}/g_{earth} = 3.69/9.78$, about 0.4, to an identical grain in a Sahara desert sandstorm.

3.3.2 Gravitational forces

Forces are due to gravity acting from a distance on the *partial* or *total* mass of any moving or stationary substance, the total force being the sum of those acting on all the partial masses. Thus, we could speak of the force exerted by gravity on an individual apple, volcano, or ice sheet. The concept is of great use when surveying the precise values of

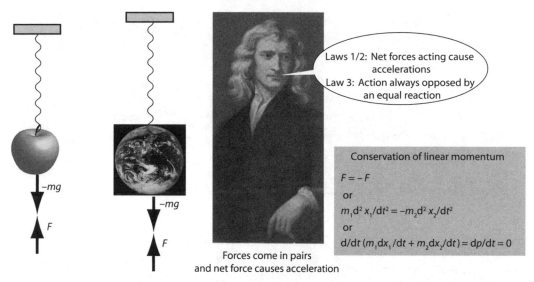

Laws 1/2: Net forces acting cause accelerations

Law 3: Action always opposed by an equal reaction

Conservation of linear momentum

$F = -F$

or

$m_1 d^2 x_1/dt^2 = -m_2 d^2 x_2/dt^2$

or

$d/dt \, (m_1 dx_1/dt + m_2 dx_2/dt) = dp/dt = 0$

Forces come in pairs and net force causes acceleration

Fig. 3.7 The force due to the spring balances the equal and opposite forces due to the product of mass and gravitational acceleration.

gravity at the Earth's surface due to the tiny differences in gravitational accelerations brought about by changes in water depth of the oceans or to rock density under the Earth's surface: these are termed *gravity anomalies*. The gravity forces give rise to a force *field* which shows the spatial variation or *gradient* of the magnitude and direction of the forces.

We are often much more interested in forces that act within or upon masses of material. These give rise to stresses and pressures (such as the hydrostatic pressure examined in Section 3.5) on imaginary planes within a rock or fluid mass. In this way they are defined as forces which one gravitational mass applies to another gravitational mass: we simply need to specify the defined surface plane (with respect to coordinates) on or within a substance over which the gravity force acts. The concept of a surface weight force due to Everest acting on a plane at greater than 7 km elevation is illustrated in Fig. 3.8.

One important force affecting bodies in contact is the *friction force*. This is a contact force that acts to resist sliding (see Sections 4.11 and 4.14 for granular flows and rock masses respectively) and depends upon the nature (roughness, physical state), but not the contact area, of opposing surfaces.

3.3.3 Inertial force and rate of change of momentum

Accelerations act upon mass due to *inertial forces* that cause a change of momentum. For example, in rapidly moving turbulent fluids, though body forces are always present, inertial forces act causing rapidly changing vectors of motion within the fluid. Such forces can also act on any

discontinuity, for example, between adjacent rock layers on a mountainside outcrop or when rapid displacement along faults occurs during earthquakes.

The inertial concept of force comes courtesy of Newton's Second Law, the most essential and general of his three laws and the basis of all mechanics. We give it via the same nineteenth-century translation from the original Latin as previously:

Law 2. Change of motion is proportional to the moving force impressed, and takes place in the straight line in which that force is impressed.

By "change of motion" Newton is referring to the changing "quantity of motion," or momentum (see Section 3.1). This is quite easy to imagine as we are all familiar with the forces that result when moving objects collide with stationary surfaces, the former sometimes losing momentum entirely. Natural examples of total loss would be volcanic bombs hitting soft ground of ash (Fig. 3.9), the base of an earthquake-displaced crustal block (Fig. 3.10) or of momentum transfer as desert rock surfaces are bombarded by bouncing sand grains. Loss, exchange, or change of momentum over time is $d(mu)/dt$. Change of momentum or acceleration requires a force to produce it. We can say all this in simple terms using the Second Law, $F = ma$.

3.3.4 Conservation of linear momentum

When two masses interact with each other, for example, along surface contact due to friction, equal and opposite

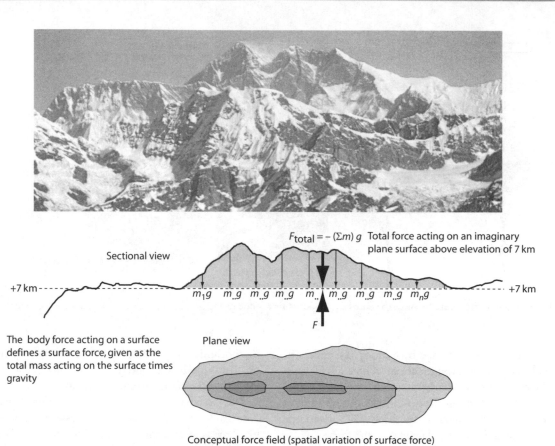

Fig. 3.8 Body force of Everest acting on a horizontal plane at 7 km elevation.

Fig. 3.9 These 20 kg volcanic bombs (b) were thrown upwards 100 m from a volcanic vent to attain an impact velocity of some 44 m s^{-1} in moist soft volcanic ash, whereupon their momentum p:

$$p = mu = 880 \text{ kg m s}^{-1}$$

was totally destroyed upon impact. The resultant force, about 196 N, was sufficient for the bombs to penetrate and deform the layered ash to a depth of some 0.4 m. Tenerife, Islas Canarias.

Fig. 3.10 This normal fault scarp formed as the ground to the right dropped quickly downward, with a smaller motion upward of the ground to the left (person standing). The 1.5 m displacement affected a $c.10$ km^3 crustal block, giving a force at the base of the block of some $2.75 \cdot 10^{14}$ N, equivalent to work done in a second or so of some $4.13 \cdot 10^{14}$ J. 1983 Borah Peake earthquake, Idaho.

forces result. This follows from Newton's Third Law:

An action is always opposed by an equal reaction; or, the mutual actions of two bodies are always equal and act in opposite directions.

This means that natural forces come in opposing pairs. If substance of mass m_1 at position x_1 exerts force F on mass m_2 at position x_2, the latter exerts an exactly equal force $-F$ on the former. Examples would be an atmospheric boundary layer shearing over the ground surface, exerting force F_{ABL}, and opposed by equal and opposite force $-F_{GR}$; a fluid flowing down an inclined channel (Fig. 3.11); a lithospheric plate sliding over the asthenosphere exerting a force, call it F_L, which is opposed by an equal and opposite force from the asthenosphere of $-F_A$ (Fig. 3.12). In all such cases, the total rate of change of momentum with respect to time is zero, and momentum is said to be conserved.

3.3.5 Other forces and the nature of equilibrium

In subsequent sections, we will come across natural forces apart from those due to gravity including viscous, buoyant, pressure, radial, and rotational forces. Each, together with the gravitational and inertial forces described above (including contact forces due to friction), may contribute to the total force acting on any substance. For the moment we simply say that Newton's Second Law states that the

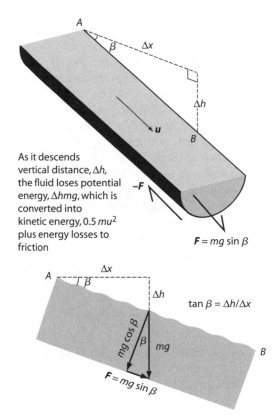

Fig. 3.11 A mass, m, of water, debris, or lava flowing at mean velocity, u, under influence of gravity down a channel sloping at angle, β, exerts a sharing force, $mg \sin \beta$, along the base.

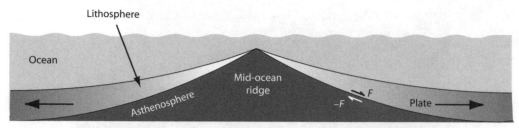

Fig. 3.12 A shear couple acts at the junction between lithosphere and asthenosphere.

sum, or resultant, of all these forces must equal the observed change of momentum of a substance, or $\Sigma F = ma$. In a steady flow of material, however fast or slow, or in a substance at rest, $\Sigma F = 0$ and for that very important case the arrayed forces must balance out to zero. Major progress in physical dynamics may be made in such cases.

3.3.6 Signs and orientations of forces

As a vector quantity, force acts in the same direction as the acceleration that produces it. Complications arise in turbulent flows where the direction of the force is constantly changing in time and space. In slowly deforming or static solid rock or ice (Sections 3.13–3.15), we can more easily speak of the orientation of forces with respect to the distortion they produce: *compressive* forces tending to push adjacent portions of rock together (which is what is happening under a surface load) or *extensional* forces doing the opposite. Across any plane, force vectors pointing toward each other are designated compressional and positive and those pointing away from each other, extensional and negative. Forces acting on a plane can also have any orientation. The two end members are *normal* and *shear* force orientations, the former normal to the plane in question and the latter parallel. As a vector, force may have any orientation and can always be resolved into components.

3.3.7 Energy, mechanical work done, and power

Energy, work done, and power are interrelated scalar quantities. When a mass of substance is displaced from one position to another during flow or deformation, *mechanical work* must have been done to achieve the displacement. This must be equal to the force required, F, times the distance moved, Δx, or $F \cdot \Delta x$. *Mechanical or flow work* of this kind comes in units of force times unit distance, or Nm, of dimensions $ML^2 T^{-2}$. A single unit of work done is termed a Joule, J. All objects, moving or stationary, may be said to be capable of doing mechanical work: they all possess *energy*. This energy must also have units $ML^2 T^{-2}$. A moving object or portion of substance has the energy proportional to its mass, m, times velocity squared, u^2. This is called *kinetic energy* (i.e. energy of movement) and may be shown (Cookie 5) equal to $0.5mu^2$. A stationary object or piece of substance has the energy of its weight force, mg, times distance, h, from the center of gravity to which it is being attracted. This is what Rankine originally called *potential energy* (i.e., energy of position), mgh. It is usual to define h with respect to some convenient reference level.

Energy may clearly be released at variable rate, either very slowly as during motion of a great lithospheric plate, or spectacularly rapidly as in an explosive volcanic eruption. The time rate of liberation of energy, in Js^{-1}, is termed *power*. It has dimensions ML^2T^{-3} and specified units called Watts, W. In terms of work, power is the rate at which work is done.

The interrelated concepts of force, energy, work, and power are perhaps easiest grasped by reference to fluid flowing down a sloping channel under the influence of gravity (Fig. 3.11). Fluid movement is created by gravity and the resulting applied force is opposed by the reactive friction force of the channel bed and banks. The flowing fluid has kinetic energy of motion that is provided by the fall in elevation of its surface as a loss of potential energy, though some of the latter is lost in friction. The available power of the river is the time rate of change of this potential energy to kinetic energy minus the frictional losses. The energy of flow is available to do mechanical work, artificially in turning a water wheel or naturally in sediment transport. See Section 3.12 for energy conservation in moving fluid.

3.4 Thermal energy and mechanical work

3.4.1 Thermal systems

We saw previously (Sections 1.5 and 2.2) that many Earth layers are defined by their temperature contrast with neighboring layers. The flow behavior of such thermal systems depends upon temperature-dependent properties such as density and viscosity. Air and shallow water change density and pressure because of variations in temperature caused by solar heating. Erupting volcanoes discharge lava which cools and changes viscosity during gradual descent. The venting of a "black smoker" is another obvious example. These are all physical systems in *thermal disequilibrium*; that is, their temperature is varying with time. By way of contrast, the main mass of lake or ocean waters may be said to be in *thermal equilibrium* where temperature changes are very slow over a time span large in relation to their speed of travel. Clearly, small volume thermal systems and/or those having thermal contact with other thermal systems are more likely to suffer changing temperature than large volume systems: we call these latter *thermal reservoirs*. The oceans are good thermal reservoirs, due to both their large volume and the high thermal capacity (Section 2.2) of water. The idea of thermal contact is a useful one with which to view the transfer of heat by conduction, convection, and radiation between thermal systems. As we shall see in more detail later (Sections 4.18–4.20), when two thermal systems directly touch, the heat is transferred by

conduction. In fluid systems, the heat can then be transported by convection. Heat transfer at a distance through the gases of the atmosphere occurs by radiation.

3.4.2 Thermal variables of state

These are pressure, p, temperature, T, volume, V, and mass, m; relationships between the variables are expressed as the equation of state. To illustrate this with an everyday example, if you compress air from a bicycle pump into a tire, the increase of pressure in the pump is accompanied by volume decrease and temperature increase. As we shall see in a later section, we are pushing the widely spaced air molecules together, causing them to collide more frequently, both with themselves and with the walls of the pump and its moving piston. Boyle first experimented on this interesting problem, though not with a bicycle pump, and showed that the increased pressure times the decreased volume was proportional to the increased temperature (Fig. 3.13). In other words, for a constant mass of air, the three variables vary in proportion as $pV \propto T$. When applied to dilute (low density) gases, the relationship becomes $pV = nRT$ the *Ideal Gas Law*, where the parameter n is the number of moles of gas present and R is the *universal gas constant*. This important equation has many applications; we use it to calculate the density of air (Box 3.1).

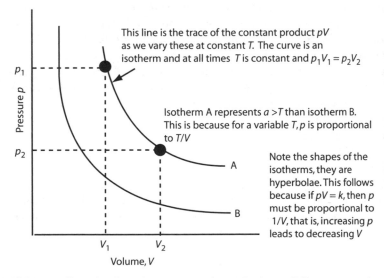

This line is the trace of the constant product pV as we vary these at constant T. The curve is an isotherm and at all times T is constant and $p_1V_1 = p_2V_2$

Isotherm A represents a $>T$ than isotherm B. This is because for a variable T, p is proportional to T/V

Note the shapes of the isotherms, they are hyperbolae. This follows because if $pV = k$, then p must be proportional to $1/V$, that is, increasing p leads to decreasing V

Boyle

Fig. 3.13 Boyle's Law tells us that for a given concentration and volume of dilute gas at a given or fixed temperature, the product pV is constant.

Box 3.1 For a constant volume of dilute gas, p and T are interdependent. We can write this as $p = RT$, where R is a proportionality constant known as the universal gas constant. R has a value of 8.314 J mol^{-1} K.

Combining this expression with Boyle's Law we obtain the ideal gas law, $pV = nRT$, where n is the number of moles of the gas.

TO FIND THE DENSITY OF AIR FROM FIRST PRINCIPLES

We use the ideal gas law for 1 mole of substance to determine the volume, V, of air at normal Earth surface temperature (in degrees kelvin, i.e. °C + 273) and pressure. $V = RT/p = 2.41 \cdot 10^{-2}$ m^3
We know that the molecular weight of 1 mole of air at standard surface pressure and 20°C T is 29.2 g. The density of air is thus $29.2/V = 1.21$ kg m^{-3}

3.4.3   Changing heat energy in Earth thermal systems

We may think of changes to thermal systems in two ways: *isothermal*, temperature constant, or *isobaric*, pressure constant. Rather than the absolute value of a substance's internal/thermal energy, E, we are much more interested in changes to that energy, ΔE. A thermal system can transfer energy by
• changing the temperature of an adjacent system it is in thermal contact with
• changing the phase (i.e. liquid, solid, gas) of an adjacent system
• doing mechanical work on its environment
Any change in temperature, ΔT, of a thermal system by the first two methods must be accompanied by a *flow of heat*, Q, the change, ΔQ, being proportional to the thermal capacity, c, for the particular substance making up the thermal system. Thus we have $\Delta Q = c\Delta T$, or, equal quantities of heat energy produce different changes of temperature in a substance if the thermal capacities are different. However, following on from the equation of state, the actual heat flow depends upon how c varies according to the path along which the change takes place, whether it is with volume or pressure kept constant.

An interesting example of heat flow of great relevance to Earth and environmental sciences occurs when a substance like water changes phase. Figure 3.14 shows a *phase diagram*, that is, a graph in *p–T* space in which experimental data for the phases of water are plotted. We may again speak of isobaric or isothermal changes and define *binding energy* as the energy required to change a mole of solid or liquid into a gas (Fig. 3.15). Importantly, there is no temperature change to the phases during the change of phase; rather a flow of heat energy arises from changing the molecular structure of the unstable phase. This flow of heat energy may pass from or to the ambient medium. Heat energy is required for fusion or evaporation, but is

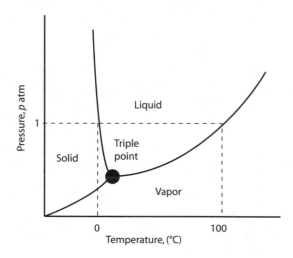

Fig. 3.14 Pressure–temperature ("phase") diagram to show stability of states of water.

given out during solidification or condensation. Such heat energy transfer is known as latent heat, L. The *latent heat of evaporation*, L_E, is much greater than the *latent heat of fusion*, L_F, because the energy required to change liquid molecules into widely spaced gaseous molecules is much greater than that to make solid molecules pack a little less tightly into liquid spacing (Fig. 3.15).

3.4.4   Mechanical equivalence of heat energy

Changes in temperature may also be brought about during the conversion of mechanical work into heat. Heat and mechanical work are interchangeable: the flow of heat, like work, is a transfer of energy. It follows that heat energy must take its rightful place alongside the other forms of energy, kinetic and potential, we encountered in Section 3.3. The production of heat energy by mechanical work begs the question, "How much energy from how much work?" Since Joule's classic experiments (Fig. 3.16) we can say that mechanical work, W, done on a thermal system produces a rise in temperature that corresponds to a particular flow of heat, ΔQ. If this occurs in a thermally isolated system, such as a calorimeter, where no heat can be lost or gained, then we can state that $\Delta Q = W$. Very accurate experiments have established that one heat flow unit, called a *calorie*, is equivalent to exactly 4.185 J of work. The equivalence of thermal and mechanical energy stamps itself more obviously on everyday life when we realize that food consumption releases energy in exactly the proportion indicated – the calorific value of foods and drinks stated on their packaging is given in both heat flow units and in energy units. The bottle of fruit juice Mike has just drunk, for example, has provided him with either 168 kJ of mechanical energy or 40 kcal of heat energy. In another example, a unit mass of Mississippi river water traveling at constant mean velocity loses potential energy to do work in descending from its source in the Rockies to the Gulf of Mexico (about 2000 km): the total temperature change expected (it could not be practically measured because of intervening energy changes) is 2°C.

3.4.5 Work done by thermal systems

Atmospheric dynamics depends upon work done on the ambient atmosphere during ascent and descent of air masses. The work is a consequence of the changing volume and density of the compressible gases that make up the atmosphere. Clearly, the net work done depends on the actual path taken, descent or ascent. Imagine that the volume changes during ascent or descent are recorded by a frictionless plunger (Fig. 3.17); the relationship we need is given by $\Delta W = p\Delta V$, where p is a function of volume and temperature. If ΔV is positive, that is, the volume of air is increased during ascent, then work has been done *by* the air on its surroundings to expand it (Fig. 3.18). If ΔV is negative, that is, the volume of air is decreased during descent, then work has been done *on* the air by its surroundings to compress it. This concept of path dependence applies to all forms of work done by both mechanical

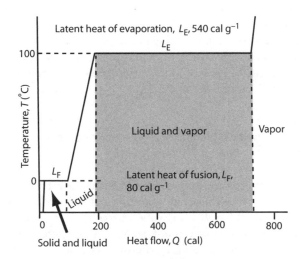

Fig. 3.15 Temperature–heat flow diagram for the phases of water.

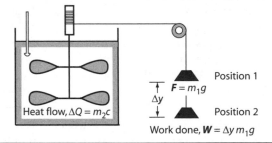

Rotating paddles turned by a falling mass, m_1, create a rise in temperature within the mass, m_2, of water of specific heat, c, in the thermally isolated calorimeter. Energy conservation means that the loss in potential energy of the mass (= the work done on the paddles) must be equivalent to the gain in thermal energy by heat flow into the water. 1 kg of water would require a descent of almost a kilometer to raise the temperature by 1°C

Fig. 3.16 Sketch to illustrate principles of Joule's apparatus.

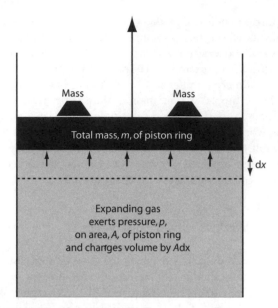

Fig. 3.17 Mechanical work done by expanding gas on its surroundings, illustrated here by a thought experiment with a cylinder and piston apparatus. Expanding gas exerts force, $F = pA$, and does work $W = pA\mathrm{d}x$. Generally, $\mathrm{d}W = p\mathrm{d}V$.

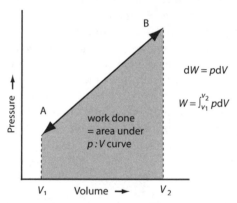

Fig. 3.18 To illustrate path dependence. Path A to B; from the integral the net work done in an expanding gas is positive and the gas does work on its surroundings. Path B to A; the net work done in a compressing gas is negative; the ambient medium has done work on the gas.

and thermal systems. *Isobaric* systems can do work because, by Boyle's Law, the constant pressure must be accompanied by a change in temperature. The resultant heat flow is then given by $\Delta Q = C_p \Delta T$, where C_p is the thermal capacity at constant pressure. Systems where volume is kept fixed while pressure changes are known as *isochoric*. Since there is no volume change no work can be done by the system, but despite this a heat flow exists of magnitude $\Delta Q = C_v \Delta T$, where C_v is the thermal capacity at constant volume.

An *adiabatic* transformation is one of the most important aspects of work done by thermal systems. Here the substance can do work but it is treated as being thermally isolated from its surrounding environment. In other words, no heat can flow from or into the substance, rather in the fashion of an imaginary super-efficient thermos flask. Thus $\Delta Q = 0$ for such systems. However, it is most important to realize that T can change within the adiabatic volume as it rises or falls, it is just that none of the heat energy can escape or be exchanged with the ambient environment. The process is best illustrated by reference to air masses once more, for air is a reasonably efficient thermal insulator. Our rising, expanding air mass must therefore increase in T as it does work against its surroundings, vice versa for descent. Another example is the adiabatic rise of deep mantle rock undergoing convection, a key solid Earth process; here the rising hot rock loses so little of the extra heat energy arising from decompression that it eventually melts to cause midocean ridge volcanism and plate creation (Sections 5.1 and 5.2).

3.4.6 Internal thermal energy, energy conservation, and the First Law of Thermodynamics

Molecules making up a thermal system have their own intrinsic energy, called internal thermal energy, denoted by the symbol U or E. This is easiest to comprehend with reference to an adiabatic transformation since here work is being done by the isolated thermal system. The system may be thought of as changing its internal thermal energy in proportion to the amount of this work, that is, $\Delta U = -\Delta W$. The minus sign indicates that W refers to the work being done by the system on its environment. In an expanding atmospheric air mass or gaseous volcanic column (Fig. 3.19), the system is doing work and losing internal energy, with the converse being true for contraction. Internal energy arises due to the motion of molecules as described by kinetic theory (Section 4.18) and is a function of the variables of state, p, V, and T. Just like potential energy but unlike work or heat flow, ΔU is path independent. In the special case of ideal gases, ΔU is only a function of T. The First Law of Thermodynamics recognizes the mechanical equivalence of heat energy. Any change in a substance's internal energy must be equal to work done plus any heat flow into the system. Thus, $\Delta U = -\Delta W + \Delta Q$. In reversible isochoric systems, the first term on the right-hand-side of the First Law is zero, since no work is done and $\Delta U = \Delta Q = C_v \Delta T$. In reversible isobaric systems, $\Delta U = -\Delta W + \Delta Q = -p\Delta V + C_p \Delta T$.

erupting Phillipines volcano

Fig. 3.19 In Nature the most spectacular demonstration of positive gaseous work occurs during volcanic eruption, like those illustrated here. The cylinder in the thought experiment is replaced by a subsurface magma chamber. Upward magma flux is accompanied by rapid degassing and blows off. Gaseous expansion may be slow or, as in this case, fast, causing magma eruption through a conduit to the surface. Here, work done includes the total force exerted on magma and crustal rock in the explosion.

3.5 Hydrostatic pressure

3.5.1 Pascal's result

The most important fact about the scalar quantity pressure, p, at any point in a stationary fluid is that it has the same magnitude in all directions, a result that was established by Pascal. The simplest argument for this is that if it were not the same in all directions, then a net force would exist which would cause motion – a paradox for a stationary fluid. For a formal proof of this striking property of static fluids, consider a diagonal half slice of a solid unit cubic volume, a stationary prismatic body, in air or completely immersed within stationary liquid (Fig. 3.20) of constant density. Imagine the prism has the same density as the fluid, so no buoyant forces act. We use the property of stationarity to demonstrate that there are no shearing forces acting on any of the planes AB, BC, and CA; if there were, then the larger face AB would have a larger force acting and the prism would rotate. But it does not. Our second result is a positive one; all forces acting on the prism must be normal, that is, in the geometrical sense, acting at right angles to each face. Now we can make rapid progress using simple vectorial geometry. First we shrink the prism so that it is infinitesimal. Then we resolve the forces acting. Call the force acting on each face $f_{AB}, f_{BC},$ and f_{CA}. From definition, BC = CA and from geometry AB > BC = CA. The tangent of angle ABC is given vectorially by $f_{CA} \cdot AC / f_{BC} \cdot BC$. Since the angle is 45°, $\tan \beta = 1$, AC = BC, and thus $f_{CA} = f_{BC}$. It is then easy, by resolving the cosine of angle ABC, to complete the calculation that $f_{AB} = f_{CA} = f_{BC}$, an exercise we leave to the reader.

Hydrostatic pressure arises from surface gravity forces. The units of p (mind your "p"s – do not confuse pressure, p, a scalar, with momentum, $\boldsymbol{p}$, a vector) are $\mathrm{N\,m^{-2}}$ or Pa and the dimensions are $\mathrm{ML^{-1}T^{-2}}$. We emphasize that pressure intensity is usually measured with respect to a difference in pressure between the particular fluid and some reference pressure, often taken as that of the atmosphere at the time of measurement. When dealing with pressures in the watery Earth, it is often sufficient to neglect variations in atmospheric pressure, but in meteorology such a course is not possible. For example, the column of atmosphere over us, rising over 100 km, exerts a mean pressure of some $1.01 \cdot 10^5$ Pa (1 atm) due to its weight; as weather fronts move past us this weight changes, typically up to ±4 percent (Fig. 3.21). The higher we climb on a mountain, the lesser pressure gets as the effective thickness of the column resting on us reduces. At 8.9 km altitude on Everest's crest, the pressure is only 0.31 atm.

It is worth pondering on the significance of Pascal's result: Even though static fluid pressure is caused by the downward action of gravity, the resulting stresses at a point are equal in all directions and give the pressure, p. If we define three orthogonal stresses, σ_x, σ_y, and σ_z as vector components (appendix), then we have the equality $p = \sigma_x = \sigma_y = \sigma_z$.

It is only when we realize that the gravity force is transmitted onto any surface by the random three-dimensional (3D)

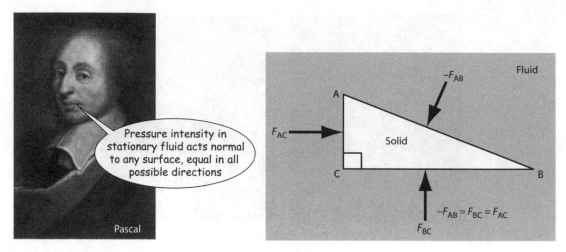

Fig. 3.20 Forces acting on a neutrally buoyant solid prism totally immersed in constant-density fluid.

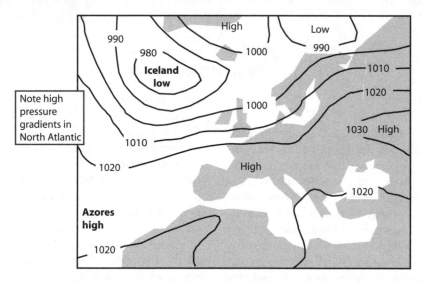

Fig. 3.21 Atmospheric pressure over North Atlantic, Europe, and North Africa, winter 2003: pressures varying over the area by an extreme 6%.

motion of molecules that we realize the solution to the paradox: gravity increases the frequency of molecular collisions in direct proportion to the quantity of fluid matter lying above. The concept of random molecular motions is the stuff of *kinetic theory* (see Section 4.18).

A final ponder on pressure in solid Earth. Although a mean *lithostatic pressure gradient* can be defined as the gradient of $\rho g z$, where ρ is rock density and z is depth, any analogy with hydrostatic pressure is misleading because of elastic behaviour in the upper crest and the existence of *tectonic stresses*. These exist generally in the solid Earth and cause the principal normal stresses defined above to be unequal. The mean pressure is then given by $p = 1/3$ $(\sigma_x + \sigma_y + \sigma_z)$ and it is convenient to then define by how

much individual principal stresses diverge from the mean stress by subtracting the latter from the former. This defines the differential normal stress (see Section 3.13.6), for example, $\sigma_x' = (\sigma_x - p)$.

3.5.2 Vertical gradient in hydrostatic pressure

Hydrostatic pressure is independent of direction, that is, it is a scalar property, just like temperature and density. However, gradients of pressure can certainly exist, giving rise to net forces. This is best appreciated by considering a definition diagram for another interesting thought experiment (Fig. 3.22). An imaginary small cylinder, open at both

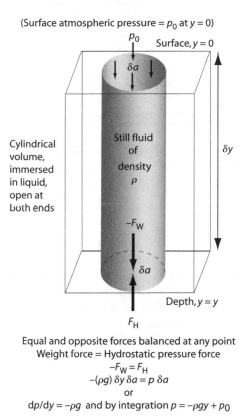

(Surface atmospheric pressure = p_0 at $y = 0$)

p_0

Surface, $y = 0$

δa

Cylindrical volume, immersed in liquid, open at both ends

Still fluid of density ρ

δy

$-F_W$

δa

Depth, $y = y$

F_H

Equal and opposite forces balanced at any point
Weight force = Hydrostatic pressure force
$$-F_W = F_H$$
$$-(\rho g)\, \delta y \, \delta a = p \, \delta a$$
or
$$dp/dy = -\rho g \text{ and by integration } p = -\rho g y + p_0$$

Fig. 3.22 Pressure gradient in stationary fluid.

ends, is immersed vertically within a stationary fluid of constant density. There is a weight force due to gravity acting positively downward on the base of the cylinder. We use Newton's Third Law to insist that an equal and opposite force, negative upward, must exist to balance this weight force. This is the *hydrostatic force*. When we balance the two forces and let the cylinder diminish to an infinitesimal point, we get an expression for the pressure gradient of universal significance. It reveals that the gravitational weight force at a point (remember this is equal in all directions) is given by the vertical spatial gradient of the hydrostatic pressure (Cookie 6). For incompressible liquids and solids at rest on and in the Earth, the effect of surface atmospheric pressure is commonly neglected. For the compressible atmosphere, a modification of the basic formula is necessary (Cookie 7).

3.5.3 Horizontal gradients in hydrostatic pressure

Thus far we have considered that in any fluid the pressure at all similar depths or heights must be equal. In order to arrive at this generalization for liquids we use the condition of invariant density via the assumption of incompressibility. For gases we assume that compressibility changes as a function of height alone; so for any given height above surface, density is similar. In both cases surfaces of equal pressure, isobars, parallel the horizontal fluid upper surface. This is known as the *barotropic condition*. However, density is free to vary independently of depth or height in Nature; we stressed in Section 2.3.3 that although density is a material property it depends on stated conditions of variables such as temperature or salinity. Because of such causes of density variations, pressure at similar heights in the atmosphere or depths in the ocean may differ laterally. This is known as the *baroclinic condition*.

Horizontal gradients in hydrostatic pressure act to cause fluid flow down the gradient from high to low pressure. In the oceans, lateral pressure gradients often arise due to slope of the ocean surface (Sections 4.1 and 6.4); in these cases the horizontal pressure gradient exists despite the fact that the barotropic condition exists, it is the slope of the isobars with respect to horizontal that matters. Such slopes may be caused by wind shear or variations in atmospheric pressure (Sections 6.2 and 6.4). Lateral gradients may also be due to vertical changes in the temperature gradient and therefore water density (e.g. Fig. 8.12). However, salinity contrasts also occur; these may either reinforce or diminish any temperature-driven density contrasts (see Section 6.4).

Horizontal pressure differences are commonly thermally driven in the atmosphere; adjacent parts may be differentially heated by variations in longwave radiation given off from the ground after solar heating or by differential shortwave radiative heating aloft. On a global scale this may be seen in the meridional contrast in surface temperature from equator to Pole, the lateral gradient giving rise to what is known as the *thermal wind* and ultimately responsible for the jet streams (Section 6.1). On a smaller scale, horizontal pressure gradients may be due to density differences arising from diurnal contrasts in reradiated heat flux from land and water surfaces, giving rise to the phenomena of land and sea breezes.

Horizontal lithostatic pressure gradients also exist in the outer 50 km or so of solid Earth, despite the tendency for pressures below that to be approximately equal (the concept of isostasy, see Section 3.6). This is because of lateral differences in rock composition and density above the "compensation" level. It is thought by some that the gradients are sufficient to cause slow lower crustal and upper mantle flow along the gradients, especially when the rock is weakened by water or elevated temperatures.

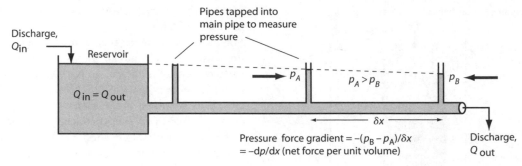

Fig. 3.23 Pressure gradient in moving fluid.

3.5.4 Horizontal gradients of static pressure in moving fluids

Rather confusingly, the static pressure condition also refers to moving fluid when the pressure is measured normal to the flow direction (Fig. 3.23); a downstream pressure gradient always exists. Pressure decreases downstream at a particular flow depth due to energy used up in overcoming viscous and turbulent frictional resistance. Using the definition diagram we can see that because of the downstream decrease of pressure there is a net positive force acting in the x-direction of $-\mathrm{d}p/\mathrm{d}x$ per unit fluid volume. We return to the energy consequences of pressure changes in moving fluids in Section 3.12.

3.6 Buoyancy force

We have seen that any mass, m, in a gravity field is acted upon by gravity equal to the force, mg. This is not quite the most general formulation of the situation for we cannot always ignore the density of the ambient medium. We must consider the magnitude of weight force acting upon a mass when the mass is immersed, that is, we must measure the weight force of a pear underwater (Fig. 3.24). In such cases both the mass of substance and gravity are constant at a point in space and the weight force must only depend upon the contrast in density between that of the mass and of the ambient medium. In situations arising in meteorology or oceanography, neighboring air or water masses may have densities that vary only slightly (Fig. 3.25) and the buoyancy must be taken into account. On the other hand, it is usual to neglect the tiny density differences between solid Earth material and air where the ratio between density and typical silicate minerals making up rocks is only $4 \cdot 10^{-4}$. In problems involving sedimentation of such particle through air we may ignore buoyancy, but not through water (Fig. 3.26).

Archimedes principle tells us that A is acted upon by an upthrust equal to the weight of ambient fluid displaced. This is because of the vertical gradient in hydrostatic pressure between the bottom and top of A. If the upthrust is less than the weight of the immersed or partially immersed substance, that is, density of A is greater than that of B, then descent will occur; vice versa for ascent. Generally when we mix two substances of contrasting density, ρ_A and ρ_B, any motion, or the lack of it, depends only upon the *sign* of the density contrast, $\rho_A - \rho_B = \Delta\rho$, and not in any way upon the magnitude of the masses involved. Three conditions are possible: neutral buoyancy ($\Delta\rho = 0$); negative buoyancy when positive $\Delta\rho$ gives rise to a net downward force causing descent of A; positive buoyancy when negative $\Delta\rho$ gives rise to a net upward force causing ascent of A. In each case the *buoyant force* per unit volume of substance, F_B, is given by the expression $\Delta\rho g$. The speed of any resultant motion due to this force depends upon other properties of the fluids involved such as absolute mass and viscosity. Examples of buoyancy in the oceans and ocean lithosphere are given in Fig. 3.27.

3.6.1 Archimedes and the buoyant force

Generally, for any solid or fluid mass, A, that rests within or partly within another ambient solid or liquid mass, B,

3.6.2 Reduced gravity

From the above discussion, you can appreciate that any mass partly or wholly immersed in an ambient medium of

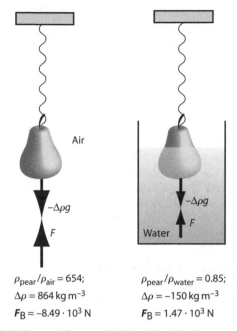

$\rho_{pear}/\rho_{air} = 654;$

$\Delta\rho = 864 \, \text{kg m}^{-3}$

$\boldsymbol{F}_B = -8.49 \cdot 10^3 \, \text{N}$

$\rho_{pear}/\rho_{water} = 0.85;$

$\Delta\rho = -150 \, \text{kg m}^{-3}$

$\boldsymbol{F}_B = 1.47 \cdot 10^3 \, \text{N}$

Fig. 3.24 Buoyant force and Archimedes principle.

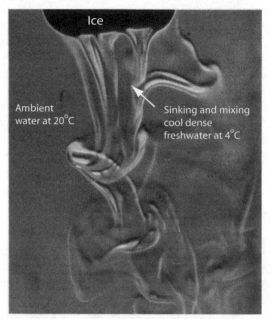

$\Delta\rho$ for ice : ambient water $= -80 \, \text{kg m}^{-3}$

$\Delta\rho$ for meltwater : ambient water $= +1.77 \, \text{kg m}^{-3}$

Fig. 3.25 Buoyancy in the water/ice system.

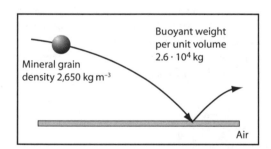
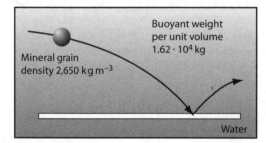

Fig. 3.26 Buoyant weights of mineral grains descending along transport paths in air and water.

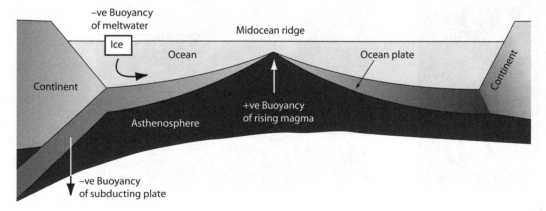

Fig. 3.27 Buoyancy in oceans and ocean plates.

differing density does not "feel" the same gravitational attraction as it would if the ambient medium were not there. For example, a surface ocean current of density ρ_1 may be said to "feel" *reduced gravity* because of the positive buoyancy exerted on it by underlying ambient water of slightly higher density, ρ_2. The expression for this reduced gravity, g', is $g' = g(\rho_2 - \rho_1)/\rho_2$. We noted earlier that for the case of mineral matter, density ρ_m, in atmosphere of density ρ_a, the effect is negligible, corresponding to the case $\rho_m \gg \rho_a$.

3.6.3 Natural reasons for buoyancy

We have to ask how buoyant forces arise naturally. The commonest cause in both atmosphere and ocean is density changes arising from temperature variations acting upon geographically separated air or water masses that then interact. For example, over the c.30°C variation in near-surface air or water temperature from Pole to equator, the density of air varies by c.11 percent and that of seawater by c.0.6 percent. The former is appreciable, and although the latter may seem trivial, it is sufficient to drive the entire oceanic circulation. It is helped of course by variations in salinity from near zero for polar ice meltwater to very saline low-latitude waters concentrated by evaporation, a maximum possible variation of some 4 percent. Density changes also arise when a bottom current picks up sufficient sediment so that its bulk density is greater than

that of the ambient lake or marine waters (Fig. 2.12); these are termed *turbidity currents* (Section 4.12).

Motion due to buoyancy forces in thermal fluids is called *convection* (Section 4.20). This acts to redistribute heat energy. There is a serious complication here because buoyant convective motion is accompanied by volume changes along pressure gradients that cause variations of density. The rising material expands, becomes less dense, and has to do work against its surroundings (Section 3.4): this requires thermal energy to be used up and so cooling occurs. This has little effect on the temperature of the ambient material if the adiabatic condition applies: the net rate of outward heat transfer is considered negligible.

3.6.4 Buoyancy in the solid Earth: Isostatic equilibrium

In the solid Earth, buoyancy forces are often due to density changes owing to compositional and structural changes in rock or molten silicate liquids. For example, the density of molten basalt liquid is some 10 percent less than that of the asthenospheric mantle and so upward movement of the melt occurs under mid-ocean ridges (Fig. 3.27). However we note that the density of magma is also sensitive to pressure changes in the upper 60 km or so of the Earth's mantle (Section 5.1).

In general, on a broad scale, the crust and mantle are found to be in hydrostatic equilibrium with the less dense

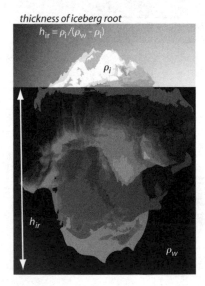
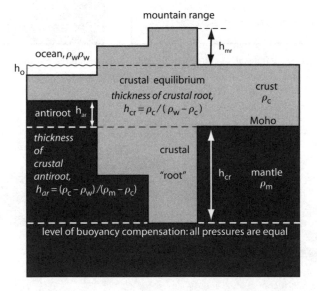

Fig. 3.28 Sketches to illustrate the Airy hypothesis for isostasy, analogos to the "floating iceberg" principle.

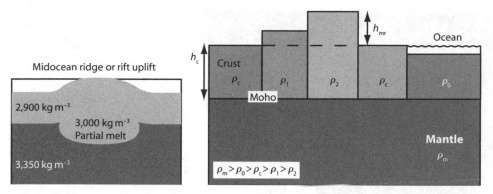

Fig. 3.29 Sketches to illustrate the Pratt hypothesis for isostasy. Here topography is supported by lateral density contrasts in the upper mantle (left) and crust (right).

crust either "floating" on the denser mantle or supported by a mantle of lower density. This equilibrium state is termed *isostasy*; it implies that below a certain depth the mean lithostatic pressure at any given depth is equal. As already noted (Section 3.5.3), above this depth a lateral gradient may exist in this pressure. In the Airy hypothesis, any substantial crustal topography is balanced by the presence of a corresponding crustal root of the same density; this is the floating iceberg scenario (Fig. 3.28). In the Pratt hypothesis, the crustal topography is due to lateral density contrasts in the upper mantle (at the ocean ridges) or in separate floating crustal blocks (Fig. 3.29). Sometimes the isostatic compensation due to an imposed load like an ice sheet takes the form of a downward flexure of the lithosphere, accompanied by radial outflow of viscous asthenosphere (Fig. 3.30). The reverse process occurs when the load is removed, as in the *isostatic rebound* that accompanies ice sheet melting.

An important exception to isostatic equilibrium occurs when we consider the whole denser lithosphere resting on the slightly less dense asthenosphere, a situation forced by the nature of the thermal boundary layer and the creation

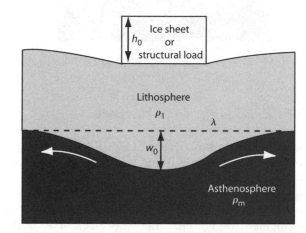

Fig. 3.30 Sketch to illustrate the Vening–Meinesz hypothesis for isostatic compensation by lithospheric bending and outward flow due to surface loading.

of lithospheric plate at the mid-ocean ridges (Sections 5.1 and 5.2). Lithospheric plates are denser than the asthenosphere and hence at the site of a subduction zone, a low-angle shear fracture is formed and the plate sinks due to negative buoyancy (Fig. 3.27).

3.7 Inward acceleration

In our previous treatment of acceleration (Section 3.2), we examined it as if it resulted solely from a change in the *magnitude* of velocity. In our discussion of speed and velocity (Section 2.4), we have seen that fluid travels at a certain speed or velocity in straight lines *or* in curved paths. We have introduced these approaches as relevant to *linear* or *angular* speed, velocity, or acceleration. Many physical environments on land, in the ocean, and

atmosphere allow motion in curved space, with substance moving from point to point along circular arcs, like the river bend illustrated in Fig. 3.31. In many cases, where the radius of the arc of curvature is very large relative to the path traveled, it is possible to ignore the effects of curvature and to still assume linear velocity. But in many flows the angular velocity of slow-moving flows gives rise to major effects which cannot be ignored.

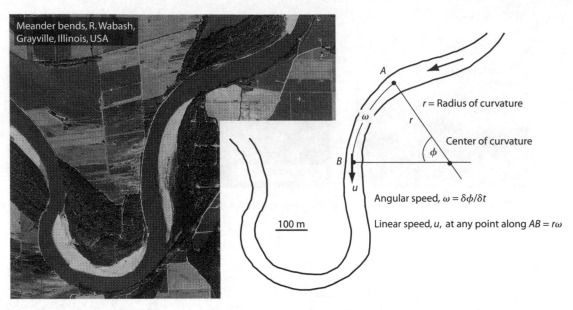

Fig. 3.31 The speed of flow in channel bends.

3.7.1 Radial acceleration in flow bends

Consider the flow bend shown in Fig. 3.31. Assume it to have a constant discharge and an unchanging morphology and identical cross-sectional area throughout, the latter a rather unlikely scenario in Nature, but a necessary restriction for our present purposes. From continuity for unchanging (steady) discharge, the magnitude of the velocity at any given depth is constant. Let us focus on surface velocity. Although there is no change in the length of the velocity vector as water flows around the bend, that is, the magnitude is unchanged, the velocity is in fact changing – in direction. This kind of spatial acceleration is termed a *radial acceleration* and it occurs in every curved flow.

3.7.2 Radial force

The curved flow of water is the result of a net force being set up. A similar phenomena that we are acquainted with is during motorized travel when we negotiate a sharp bend in the road slightly too fast, the car heaves outward on its suspension as the tires (hopefully) grip the road surface and set up a frictional force that opposes the acceleration. The existence of this radial force follows directly from Newton's Second Law, since, although the speed of motion, u, is steady, the direction of the motion is constantly changing, inward all the time, around the bend and

hence an inward angular acceleration is set up. This inward-acting acceleration acts centripetally toward the virtual center of radius of the bend. To demonstrate this, refer to the definition diagrams (Fig. 3.32). Water moves uniformly and steadily at speed u around the centerline at 90° to lines OA and OB drawn from position points A and B. In going from A to B over time δt the water changes direction and thus velocity by an amount $\delta u = u_B - u_A$ with an inward acceleration, $a = -\delta u/\delta t$. A little algebra gives the instantaneous acceleration inward along r as equal to $-u^2/r$. This result is one that every motorist knows instinctively: the centripetal acceleration increases more than linearly with velocity, but decreases with increasing radius of bend curvature. For the case of the River Wabash channel illustrated in Fig. 3.31, the upstream bend has a very large radius of curvature, c.2,350 m, compared with the downstream bend, c.575 m. For a typical surface flood velocity at channel centerline of $u = c$.1.5 m s^{-1}, the inward accelerations are $9.6 \cdot 10^{-4}$ and $4.5 \cdot 10^{-3}$ m s^{-2} respectively.

3.7.3 The radial force: Hydrostatic force imbalance gives spiral 3D flow

Although the computed inward accelerations illustrated from the River Wabash bends are small, they create a flow pattern of great interest. The mean centripetal acceleration must be caused by a centripetal force. From Newton's

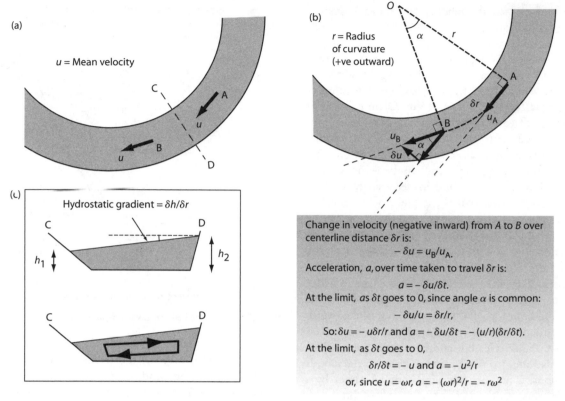

Fig. 3.32 (a) and (b) To define the radial acceleration acting in curved flow around channel bends; (c) Superelevation of water on the outside of any bend and sectional view of helical flow cell within any bend.

Third Law we know this will be opposed by an equal and opposite centrifugal (outward acting) force. This tends to push water outward to the outside of the bend, causing a linear water slope inward and therefore a constant lateral hydrostatic pressure gradient that balances the mean centrifugal force (Fig. 3.31). Although the *mean* radial force is hydrostatically balanced, the value of the radial force due to the faster flowing surface water (see discussion of boundary layer flow in Section 4.3) exceeds the hydrostatic contribution while that of the slow-moving deeper

water is less. This inequality drives a secondary circulation of water, outward at the surface and inward at the bottom (Fig. 3.32c), that spirals around the channel bend and is responsible for predictable areas of erosion and deposition as it progresses. The principle of this is familiar to us while stirring a cup of black or green tea with tea leaves in the bottom. Visible signs of the force balance involved are the inward motion to the center of tea leaves in the bottom of the cup as the flow spirals outward at the surface and down the sides of the cup toward the cup center point.

3.8   Rotation, vorticity, and Coriolis force

Earth's rotation usually has no obvious influence on motion, that is, motions closely bound to Earth's surface by friction, such as walking down the road, traveling by motorized transport, observing a river or lava flow, and so on. But experience tells us that rotary motion imparts its own angular momentum (Section 3.1) to any object, a fact never forgotten after attempted exit from, or walk onto, a rotating roundabout platform; in both cases a sharp lateral

push signifies an acceleration arising from a very real force (there *are* those who doubt the "realness" of the Coriolis force, referring to it as a "virtual" or "pseudo-force"). At a larger scale, the path of slow-moving ocean currents and air masses are significantly and systematically deflected by motion on rotating Earth. Such motions have come under the influence of the *Coriolis force*, a physical effect caused by gradients in angular momentum.

3.8.1 A mythological thought experiment to illustrate relative angular motion

King Aeolus governed the planetary wind system in Greek mythology; it was he who gave the bag of winds to Odysseus. He was ordered by his boss, Zeus, feasting as usual at headquarters high above Mount Olympus, to beat up a strong storm wind to punish a naughty minor goddess who had fled far to the East in modern day India. Aeolus, who was in Egypt close to the equator at the time observing a midsummer solstice, climbed the nearest mountain and pointed his wind-maker exactly East to release a great long wind that eventually reached and laid waste to the goddess's encampment by the River Indus. Zeus was pleased with the result and rewarded Aeolus with plenty of ambrosia. Some months later another naughty goddess fled north from Olympus in the direction of the frozen wastes of Scythia, for some reason (modern day Russia). Zeus again instructed Aeolus, now home from his Egyptian expedition, to let loose the punishing wind. Aeolus ascended Olympus, pointed his wind-makers exactly North and released another great long wind. However, this time, from his vantage point in the clouds above Olympus, Zeus sees the wind miss his target by a considerable margin, devastating a large area of forest well to the East. This happens over and over again. Zeus is highly displeased and calls an inquest into the sorry state of Aeolus's intercontinental wind punisher, vowing after the inquest to use Poseidon's earthquakes for the purpose in future.

3.8.2 The Aeolus postmortem: A logical conceptual analysis

Earth spins rapidly upon its axis of rotation; in other words it has vorticity. It has an angular velocity, Ω, of 7.292×10^{-5} rad s^{-1} about its spin axis that decreases equatorward as the sine of the angle of latitude. It also has a linear velocity at the equator of about 463 ms^{-1}; this decreases poleward in proportion to the cosine of latitude. Any large, slow-moving object (i.e. slow-moving with respect to the Earth's angular speed) not in direct frictional contact with a planetary surface and with a meridional or zonal motion is influenced by planetary rotation: a curved trajectory results *with respect to Earth-bound observers* (Fig. 3.33). The exceptions are purely zonal winds along the equator, by chance the first success of Aeolus. To Aeolus observing the wind from above Olympus (i.e. his reference axes were not on the fixed Earth surface), it seemed to travel in a straight line

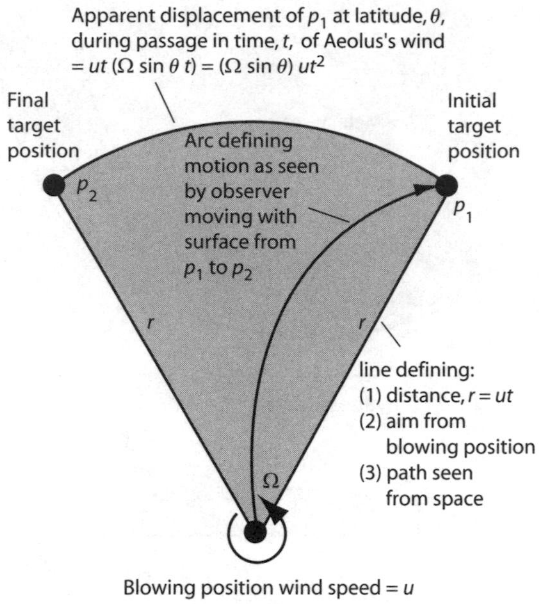

Apparent displacement of p_1 at latitude, θ, during passage in time, t, of Aeolus's wind $= ut\,(\Omega \sin \theta\, t) = (\Omega \sin \theta)\,ut^2$

Final target position p_2

Initial target position p_1

Arc defining motion as seen by observer moving with surface from p_1 to p_2

r

r

line defining:
(1) distance, $r = ut$
(2) aim from blowing position
(3) path seen from space

Ω

Blowing position wind speed $= u$

Fig. 3.33 The short-lived Aeolus wind-punisher.

(Fig. 3.33). However, to the terrified Scythians looking South the incoming wind seemed to them to be affected by a mysterious force moving it progressively to their left, that is, eastward, as it traveled northward.

We draw the following conclusions:
1 On a rotating sphere, fixed observers see radial deflections of moving bodies largely free from frictional constraints. Such deflections involve radial acceleration and a force must be responsible.
2 The magnitude of the deflection and the acceleration increases with increasing latitude.
3 The deflection *with respect to the direction of flow* is to the right in the Northern Hemisphere and to the left in the Southern.
4 To observers outside the rotating frame of reference (i.e. Gods) no deflection is visible.
5 For zonal motion at the equator there is no deflection.

3.8.3 Toward a physical explanation; First, shear vorticity

Streamline curvature in fluid flow signifies the occurrence of *vorticity*, ζ (eta) (Section 2.4). Clearly, fluid rotation can be in any direction and of any magnitude. Like in considerations of angular velocity (Section 2.4), the direction in question is defined with respect to that of a normal axis to the plane of rotation, both carefully specified with respect to three standard reference coordinates. Regarding signs

Definition of vertical component of
vorticity due to horizontal rotation

Fig. 3.34 Vorticity sign conventions and the negative vorticity evident from the flow of Coriolis's hair.

(Fig. 3.34), we define positive *cyclonic vorticity* with anticlockwise rotation viewed looking down on or into the vortical axis; vice versa for negative or *anticyclonic vorticity*. Looked at this way it is clear that vorticity, ζ is a vector quantity; it has both magnitude and direction with vertical, ζ_z, streamwise, ζ_x, and spanwise, ζ_y, components. Each of these components defines rotation in the plane orthogonal to itself, for example, streamwise vorticity involves rotations in the plane orthogonal to the streamwise direction and since x is the streamwise component the vorticity refers to rotation in the plane yz.

Now here is the tricky bit (Figs 3.35 and 3.36). In order for rotation to occur there must be a gradient of velocity acting upon a parcel of fluid; if there is no gradient there can be no vorticity. The velocity gradient sets up gradients of shearing stress and hence this kind of *shear vorticity* (also called *relative vorticity*) depends upon the magnitude of the gradient, not the absolute velocity of the flow itself. This is best imagined by spinning-up a small object, like a top, with one's fingers to create vertical vorticity (ignore the tendency for precession): a shear couple is required from you to turn the object into rotation. Better still for use in flowing fluids, you can make your own *vorticity top* from a wooden stick and two orthogonal fins (or you can just imagine the vorticity top in a thought experiment). Now, with respect to the plane normal to the vertical spin axis of the vorticity top, only two velocity gradients may exist in the xy plane that, between them,

Vertical paddle stick with fins

Anticlockwise (positive) vertical vorticity contribution 1

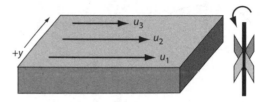

$u_1 > u_2 > u_3$ so gradient of u across direction $+y$ is negative, that is, $du/dy = -ve$

Anticlockwise (positive) vertical vorticity contribution 2

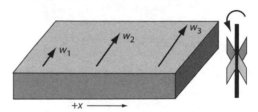

$w_1 < w_2 < w_3$ so gradient of w across direction $+x$ is positive, that is, $dw/dx = +ve$

Vertical component of vorticity, ζ_z
$$-\partial u/\partial y + \partial w/\partial x$$
is overall positive

Fig. 3.35 Shear vorticity and the Taylor vorticity top.

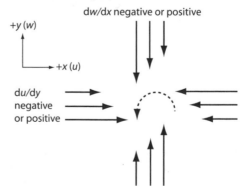

Fig. 3.36 Combination of velocity gradients that might produce overall positive vorticity.

Definition of vertical
component of vorticity due
to horizontal rotation

Fig. 3.37 Vorticity of curved flow.

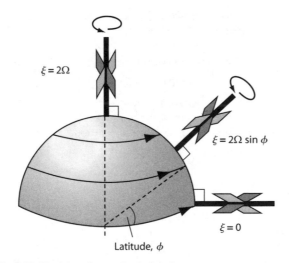

Fig. 3.38 Vorticity of a rotating hemisphere.

Angular speed of Earth surface is a function of latitude

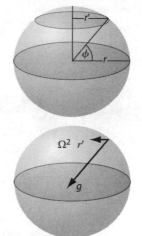

The small centripetal acceleration acting at the surface due to rotation

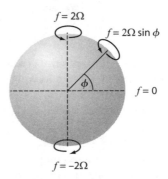

Since the rate of change of apparent displacement with respect to time is a definition of acceleration, the mean Coriolis acceleration is:

$$\frac{d^2}{dt^2}(\Omega \sin \phi)vt^2 = 2\Omega \sin \phi v$$

Fig. 3.39 Planetary vorticity.

can cause spin along the z-axis specified (Fig. 3.36): (1) a gradient of the horizontal streamwise velocity, u, in the spanwise direction, y, gives the gradient, $-\partial u/\partial y$, (2) a gradient of spanwise velocity, v, in the streamwise direction, x, gives $-\partial v/\partial x$. Either or both of these gradients

(see Fig. 3.36 for the various possibilities) contribute to the vertical vorticity, ζ_z, represented by the local spin of the horizontal flow about the vertical axis $(-\partial u/\partial y + \partial v/\partial x)$. It is possible of course that the velocity gradients could partially or wholly cancel each other out with resulting reduced or even zero vorticity; the signs in the expression take care of these possibilities. A similar argument holds for the other two reference planes enabling us to specify the total vorticity, ζ.

For curved flows we can make use of the coordinate system shown in Fig. 3.37, with s along the flow direction and n in the plane of rotation normal to s and positive inward toward the center of rotation. V is the local fluid speed and r is the radius of the curved flow. The vertical vorticity component, ζ_z, is now the sum of the shear $(-\partial V/\partial n)$ and the curvature (V/r) components, both of which are positive.

3.8.4 Toward a physical explanation; Second, solid vortical motions

Solid vorticity pertains to solid Earth rotation or to plate and crustal block rotation. It also applies to the rapidly rotating cores of tropical cyclones like hurricanes. It is best investigated initially as curved solid flow, as in the last example (Fig. 3.37), with a rotating disc or turntable setup. In the disc case, both velocity components are generally nonzero. Consider first the shear term, $(-\partial V/\partial n)$. In solid rotation, $V = \Omega r$, and the shear term is $(-\partial \Omega r/\partial n)$. Since V is increasing outward with n chosen positive inward, $-\partial r/\partial n = -1$, the term becomes simply Ω. The contribution, (V/r), due to curvature flow is also Ω, since $V = \Omega r$ by definition. Thus for solid body rotations we have the simple result that the shear and curvature components contribute equally to the total vorticity, and this is equal to 2Ω.

Now consider the vorticity, f, of a solid sphere like Earth. Viewed from the North Polar rotation axis (Figs 3.38 and 3.39) Earth spins anticlockwise, with each successive latitude band, ϕ, increasing in angular velocity poleward by $\Omega \sin \phi$. Since the vorticity of a solid sphere is twice the angular velocity, therefore for a given latitude, $f = 2\,\Omega \sin\phi$. With respect to local normal directions from the surface, we realize that only at the Pole does the vertical vorticity axis align exactly normal to the plane of the rotation. In fact, vertical vorticity, necessarily defined as parallel to the Earth's axis of rotation, must decrease to zero at the equator when the local normal to the surface is in the plane of the rotation. We commonly call Earth's vorticity, f, the *Coriolis parameter*. For southern latitudes ϕ is taken as negative and thus cyclonic vorticity is negative, vice versa for anticyclonic vorticity. The magnitude of the Coriolis parameter is quite small, of order $10^{-4}\,\mathrm{m\,s^{-2}}$ between latitudes 45° and 90°.

3.8.5 Finally: Absolute fluid vorticity on a rotating Earth

Any unbounded fluid, be it water or air, moving slowly over the Earth, must possess not only its own relative or shear vorticity, ζ, but also the Earth's vorticity, f. This is the absolute vorticity, ζ_A, given by the sum, $\zeta_A = \zeta + f$. In the slow-moving and slow-shearing oceans, $\zeta \ll f$. Just as we have to conserve angular momentum so we also have to conserve absolute vorticity. The poleward increase in absolute vorticity explains why the slow flows of ocean and atmosphere are turned by the Coriolis effect, the fluid motion is turned in the direction of angular velocity increase as extra angular momentum is obtained from the spinning Earth, that is, to the right in the Northern Hemisphere and to the left in the Southern Hemisphere. This, finally, is why earthquakes are better than winds for punishing transgressive minor goddesses.

3.9 Viscosity

3.9.1 Newtonian behavior

Viscosity, like density, is a *material property* of a substance, best illustrated by comparing the spreading rate of liquid poured from a tilted container over some flat solid surface or the ease with which a solid sphere sinks through the liquid. Viscosity thus controls the rate of deformation by an applied force, commonly a shearing stress. Alternatively, we can imagine that the property acts as a frictional brake on the rate of deformation itself, since to set up and maintain relative motion between adjacent fluid layers or between moving fluid and a solid boundary requires work to be done against viscosity. An analog model combining these aspects (the idea was first sketched as a thought experiment by Leonardo) is illustrated in Fig. 3.40.

Newton himself called viscosity (the term is a more modern one, due to Stokes) *defectus lubricitatis* or, in colloquial translation, "lack of slipperiness." While pondering on the nature of viscosity, Newton originally proposed that the simplest form of physical relationship that could explain the principles involved was if the work done by a shearing stress acting on unit area of substance (fluid in this case) caused a gradient in displacement that was linearly proportional to the viscosity (Fig. 3.41). He defined a coefficient of viscosity that we variously know as Newtonian, molecular, or dynamic viscosity, symbol μ (mu)

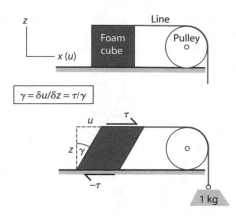

$$\gamma = \delta u / \delta z = \tau / \gamma$$

Fig. 3.40 Leonardo's implicit analog model for the action of viscosity in resisting an applied force. In this case the force is exerted on the top unit area of a foam cube. In continuous fluid deformation, as distinct from the finite displacement of solids, the displacement in x is the velocity, $\boldsymbol{u}$ (as shown).

For a given applied stress, shear strain is proportional to viscosity; it varies linearly and continuously with time and is irreversible

Shear stress and rate of strain are linearly related by the viscosity coefficient; zero stress gives zero strain and any finite stress gives strain

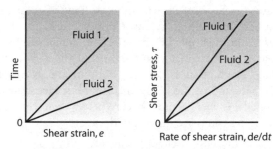

Defectus lubricatus is a material property of any fluid, with a constant value for the pure fluid appropriate only under specified conditions of T and P

Fig. 3.41 Newtonian fluids.

or η (eta). This is equal to the ratio between the applied shearing stress, $\boldsymbol{\tau}$ (tau), that causes deformation and the resulting *displacement gradient* or *rate of vertical strain*, $\mathrm{d}\boldsymbol{u}/\mathrm{d}z$. We call a fluid *Newtonian* when this ratio is finite and linear for all values (Fig. 3.41). We shall briefly examine the behavior of *non-Newtonian* substances in Section 3.15. From knowledge of the units involved in $\boldsymbol{\tau}$, and $\mathrm{d}\boldsymbol{u}/\mathrm{d}z$, check that the dimensions of viscosity are $ML^{-1}T^{-1}$, and the units, $N\,s\,m^{-2}$ or $Pa\,s$. Viscosity is sometimes quoted in units of *poises* (named in honor of

Poiseuille who did pioneer work on viscous flow): these are $10^{-1}\,Pa\,s$. Viscosity is a scalar quantity, possessing magnitude but not direction. The most succinct formal definition goes something like "the force needed to maintain unit velocity difference between unit areas of a substance that are unit distance apart."

The ratio of molecular viscosity to density, confusingly termed *kinematic viscosity*, is given the symbol, ν (nu) and has dimensions $m^2\,s^{-1}$, often quoted in Stokes (St), one stoke being $10^{-4}\,m^2\,s^{-1}$. Authors sometimes forget to specify which viscosity they are using, so always check carefully.

3.9.2 Controls on viscosity

As for density it is important to realize that Newtonian viscosity is a *material* property of *pure homogeneous* substances: the warning italic letters signifying caveats, exceptions, and potential sources of confusion;
- Specific conditions of T and P must be quoted when a value for viscosity is quoted. Some variations of molecular (dynamic) viscosity with temperature are given in Fig. 3.42.
- Natural materials are often impure, with added contaminants; particles may also be of variable chemical composition. For example, the viscosity of molten magma is highly dependent upon Si content (Section 5.1), and the viscosity of an aqueous suspension of silt or clay differs radically from that of pure water (Fig. 3.42).

3.9.3 Maxwell's view of viscosity as a transport coefficient

In fluid being sheared past a stationary interface, those molecules furthest from the interface have a greater forward (drift) momentum transferred to their random thermal motions as they are dragged along. Under steady conditions (i.e. shear is continuously applied) the combination of forward drift due to shear and random thermal molecular agitation (very much faster) must set up a continuous forward velocity gradient; molecules constantly diffuse drift momentum as they collide with slower moving molecules closer to the interface where momentum is dissipated as heat. We see clearly from this approach why Maxwell viewed molecular viscosity as a *momentum diffusion* transport coefficient, analogous to the transport of both conductive heat and mass (Section 4.18). Thermal effects thus have a great control on the value of viscosity. Although it is a little more difficult to imagine the viscous transport of momentum in a solid, we can nevertheless measure the angle of shear achieved by a

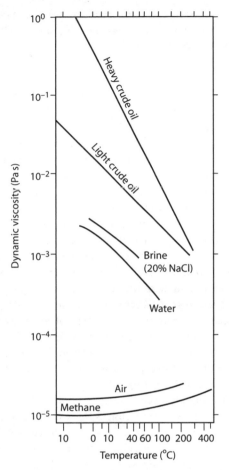

Fig. 3.42 The dynamic viscosity of some common pure substances as a function of temperature.

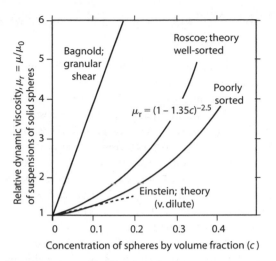

Fig. 3.43 The variation of relative dynamic viscosity (with respect to pure water at zero solids concentration, μ_0) with solid sphere concentration according to two theoretical models; Einstein is for vanishingly small c, Roscoe for finite c. The Bagnold curve is for experimental data on the behavior of spheres under shear when solid–solid reactions are induced by the shear and intragranular collisions are produced.

shear couple acting on an elastic solid in just the same way (Section 1.26; see Fig. 3.84).

It is simplest to grasp why $\mu_{solid} > \mu_{liquid} > \mu_{gas}$ from the point of view of molecular *kinetic theory* (Section 4.18) applied to the states of matter. Thus decreasing concentrations of molecules cause deformation or flow to be easier as the molecules are more widely spaced. Maxwell's view

of viscosity in terms of the diffusion of momentum by viscous forces is again essential. Thus any swinging pendulum put into motion and then left, once corrected for friction around the bearings, slows down (is damped) progressively; the time required for damping being inversely proportional to viscosity. As Einstein later explained in a relation between viscosity and diffusion, the damping is due to molecular collisions between fluid and the pendulum mass moving through it. This makes it easier to conceptualize the reason why solid suspensions have increased viscosity over pure fluid alone (Fig. 3.43). Maxwell and Einstein were able to show from similar molecular collisional arguments why experimentally determined viscosities of liquids were inversely proportional to temperature while the viscosity of gas is broadly independent of pressure.

3.10 Viscous force

In Section 2.4 on motion we neglected frictional effects arising from viscosity. Here we consider the simplest type of viscous fluid flow and ask how net forces might come about. The flows are steady Newtonian systems moving past an interface, most simply a stationary solid obstruction to the flow or another fluid of similar or contrasting material and kinematic properties. Such physical systems are clearly common in Nature.

3.10.1 Net force and the rate of change of velocity close to an interface

We can imagine that the further we go away from an interface the less likely it will be that the flow "feels" the influence of the surface; it will be increasingly retarded by its own constant internal property of viscosity. This is our introduction to the concept of a *boundary layer*, being that part of a flowing substance close to the boundaries to the flow where there is a spatial change in the flow velocity (Section 4.3). Such boundary layer gradients were first investigated systematically by Prandtl and von Karman in the early years of the twentieth century. At this stage we are not concerned with calculating or predicting the exact nature of the change in the rate of flow in a boundary layer, but are content to accept that the field and experimental evidence for such change is in no doubt. We shall look at the question in more detail in Sections 4.3–4.5.

We make use of thought experiments at this point: let velocity stay constant, increase, or decrease away from a flow boundary (Fig. 3.44). In the first case no viscous stress or net force exists. In the second and third cases viscous stresses exist. There are two further possibilities:

1 The velocity of flow may decrease linearly from any boundary so that the rate of change of velocity is constant. Here there can be no *net* force acting across the constant velocity gradient, du/dy. This is because there is no rate of change, d/dy, of the gradient, that is, $d^2u/dy^2 = 0$ and the applied Newtonian viscous stresses acting on both sides of an imaginary infinitesimal plane normal to the y-axis are equal and opposite.

2 The rate of change of velocity with distance may decrease away from the boundary (Fig. 3.45). This possibility is discussed next.

3.10.2 Net viscous force in a boundary layer

Careful measurements of flow velocity at increments up from the bed of a river or through the atmosphere demonstrate how the shape of a boundary layer is defined and that while the velocity slows down through the boundary layer toward the boundary itself, the velocity gradient actually increases (Section 4.3). If we now consider an imaginary infinitesimal plane in the xy plane of this boundary layer flow (Fig. 3.45) it is immediately apparent that the viscous stress, τ_{zx} acting on unit area will be greater on one side than the other, because the velocity gradient is itself changing in magnitude. We call this difference in stress the *gradient of the stress per unit area*, or $d\tau_{zx}/dy$. We have already come across the concept of stress gradients in our development of the simple expression that determines the force due to static pressure (Section 3.5). Since a stress is, by definition, force per unit area, any change in force across an area is the *net* force acting.

Since we already have Newton's relationship for viscous stress, $\tau_{zx} = \mu du/dz$ (Section 3.9), we can combine the previous expressions and write the net force per unit area as $d/dz\,(\mu du/dz)$, more concisely written as the constant molecular viscosity times the second differential of the velocity, $\mu d^2u/dz^2$ (Fig. 3.45). This is the second time we

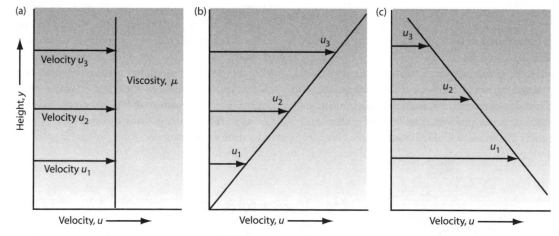

Fig. 3.44 By Newton's relationship, $\tau = \mu du/dy$, viscous frictional forces can only be present if there is a gradient of mean flow velocity in any flowing fluid. The three graphs are sketches of simple hypothetical velocity distributions. (a) has no gradient and therefore no viscous stresses; (b) has a positive linear velocity gradient, that is, velocity increasing at constant rate upward, and hence has viscous stresses of constant magnitude; (c) has a negative linear velocity gradient, that is, velocity decreasing at constant rate upward, and hence also has viscous stresses of constant magnitude.

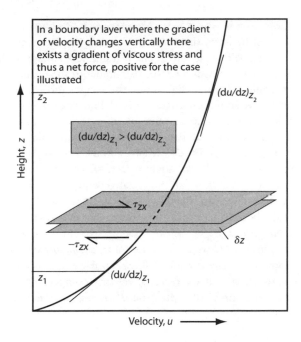

In a boundary layer where the gradient of velocity changes vertically there exists a gradient of viscous stress and thus a net force, positive for the case illustrated

z_2

$(du/dz)_{z_2}$

$(du/dz)_{z_1} > (du/dz)_{z_2}$

τ_{zx}

$-\tau_{zx}$

δz

z_1

$(du/dz)_{z_1}$

Height, z

Velocity, u

Fig. 3.45 To show definitions of velocity gradients and viscous shear stresses in a boundary layer whose velocity is changing in space across an imaginary infinitesimal shear plane, δz. Such boundary layers are very common in the natural world and the resulting net viscous force reflects the mathematical function of a second differential coefficient of velocity with respect to height, that is, $F_{\text{viscous}} = -\mathrm{d}\tau_{zx}/\mathrm{d}z = -\mu \mathrm{d}^2 u/\mathrm{d}z^2$.

have come across the concept of a second differential in this book, the first was for acceleration, as rate of change of velocity with time. Luckily this particularly second differential can be just as easily interpreted physically; it is the rate of change of velocity gradient with distance. In other words it is a spatial acceleration in the sense discussed in Section 3.2. So we have just derived Newton's Second Law again, force equals mass times acceleration, but this time in a physical way as the action of viscosity upon a gradient in velocity across unit area, that is, $F_{\text{viscous}} = \mu |\mathrm{d}^2 u/\mathrm{d}z^2|$.

3.10.3 The sign of the net force

But one thing is missing from our discussion above – the sign of the net force. Thinking physically again we would expect the viscosity to be opposing the rate of change of fluid motion, giving a negative sign to the term, that is, $F_{\text{viscous}} = -[\mu \mathrm{d}^2 u/\mathrm{d}z^2]$. For the particular case of the boundary layer we need to look again at the nature of velocity change; the velocity is decreasing less rapidly per given vertical axis increment the further away from the boundary we get. We will play a simple mathematical trick with this property of the boundary layer later in this book; for the moment we will not specify the exact nature of the change. Now, since the rate of change is negative, the net viscous force acting must be overall positive in all such cases.

3.11 Turbulent force

Turbulent flows of wind and water dominate Earth's surface. Much of the practical necessity for understanding turbulence originally came from the fields of hydraulic engineering and aeronautics. It is perhaps no coincidence that "modern" fluid dynamical analysis of turbulence started around the date of *Homo sapiens*' first few uncertain attempts at controlled flight. Eighty years later photographs of turbulent atmospheric flows on Earth were taken from the Moon, and using radar we can now image turbulent Venusian and Martian dust storms.

3.11.1 Steady in the mean

We know about the intensity of turbulence from experience, like the gusty buffeting inflicted by a strong wind.

The wind may be steady when averaged over many minutes, but varies in velocity on a timescale of a few seconds to tens of seconds; thus a slower period is followed by a period of acceleration to a stronger wind, the wind declines and the process starts over again. This is the essential nature of turbulence; seemingly irregular variations in flow velocity over time (Figs 3.46 and 3.47). If we investigate a scenario where we can keep the overall discharge of flow constant, such as in a laboratory channel, then we still have the fluctuating velocity but within a flow that is overall *steady in the mean*. Insertion of a sensitive flow-measuring device into such a turbulent flow for a period of time thus results in a fluctuating record of fluid velocity but with a statistical mean over time. By way of contrast, in steady laminar flow any local velocity is always constant.

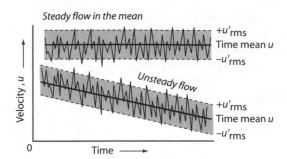

Fig. 3.46 Turbulent flow velocity time series in u, the streamwise velocity component.

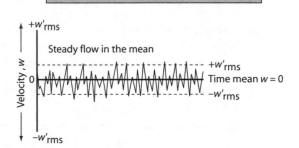

Fig. 3.47 Turbulent flow velocity time series in w, the vertical velocity component.

3.11.2 Fluctuations about the mean

Quite what to do about the physics of turbulent flow occupied the minds of some of the most original physicists of the latter quarter of the nineteenthcentury. Reynolds' finally solved the problem in 1895 using arguments for solution of the equations of motion (Newton's Second Law as applied to moving fluids; see Section 3.12). These were partly gained from experiments (Section 4.5) into the physical nature of such flows and from analogs with nascent kinetic molecular theory of heat and conservation of energy. The solution Reynolds' came up with was that both the magnitude of the mean flow *and* of its fluctuation must be considered: both contribute to the kinetic energy of a turbulent flow. To illustrate this, take the simplest case of steady 1D turbulent flow (Fig. 3.46); the arithmetic gets quite cumbersome for 3D flows (see Cookie 8). The instantaneous longitudinal x-component of velocity, u, is equal to the sum of the time-mean flow velocity, $\overline{u}$, and the instantaneous fluctuation from this mean, u'. In

symbols: $u = \overline{u} + u'$. Over a longer time period, the mean of u' must be zero, since u' is positive and negative about the mean at different times. The instantaneous magnitude of u' gives us a measure of the instantaneous magnitude of the turbulence. But what about the longer-term magnitude; can we somehow characterize the fluctuating system? Although the long-term value of u' is zero, the positive and negative values all canceling, there is a statistical trick, due originally to Maxwell, that we can use to compute the long-term value. If we square each successive instantaneous value over time, all the negative values become positive. The mean of these positive squares can then be found, whose square root then gives what is known as the *root-mean-square fluctuation*, $(\overline{u'^2})^{0.5}$ or in shorthand, u'_{rms}. This is how we express the mean *turbulent intensity* component of any turbulent flow. Similar expressions for the vertical, w, and spanwise, v, velocity components give us a measure of the *total turbulent intensity*, $q'_{rms} = (u'_{rms} + v'_{rms} + w'_{rms})$.

3.11.3 Steady eddies: Carriers of turbulent friction

Turbulent flows are very efficient at mixing fluid up (Fig. 3.48) – far more so than simple molecular diffusivity can achieve in laminar flow. Since mixing across and between different fluid layers involves accelerations, new forces are set up once turbulent motion begins. These are

Fig. 3.48 Turbulent air flow in a wind tunnel is visualized by smoke generated upflow close to the lower boundary. The top view shows the flow from above, the thin light streak along the central axis being the intense beam of light used to simultaneously illuminate the lower side view. Turbulent eddies are mixing lower speed fluid (the smoky part) upward and at the same time transporting faster fluid downward.

Fig. 3.49 Eddies provide a variable turbulent friction far greater in magnitude than viscous friction. Boussinesq added the turbulent friction as an "eddy viscosity" term, η, to Newton's viscous shear expression: $\tau = (\eta + \mu)\,\mathrm{d}u/\mathrm{d}y$.

additional to those molecular forces created by the action of the change of velocity gradient on dynamic viscosity (see Section 3.10). The extra mixing process resulting from turbulence was given the name *eddy viscosity*, symbol η, by Boussinesq in 1877 (Fig. 3.49). Although this was a useful illustrative concept, η is not a material constant like the dynamic viscosity, μ, for it varies in time and space for different flows (i.e. it is anisotropic) and must always be measured experimentally.

3.11.4   Reynolds' accelerations for turbulent flow

Now back to Reynolds': he proposed to take the Second Law and replace the total acceleration term involving mean velocity, $\bar{u}$, by a term also involving the turbulent velocity, $\bar{u} \pm u'$. After some manipulation (Box 3.2) although the arithmetic looks complicated, it is not (see Cookie 8). The total acceleration term for a steady, uniform turbulent flow becomes simply the spatial change in any velocity fluctuation. The result is staggering – despite the fact that a turbulent flow may be steady and uniform in the mean there exist time-mean accelerations due to gradients in space of the turbulent fluctuations. The acceleration gradients, when multiplied by mass per unit fluid volume, are conventionally expressed as *Reynolds' stresses*. Net forces produce the gradients *because* there is change of momentum due to the turbulence. Or, since we are discussing accelerations, we say the turbulent acceleration requires a net force to produce it. We shall return to this topic in Section 4.5; in fact we constantly think about it.

3.12   Overall forces of fluid motion

We have seen that in stationary fluids the *static forces* of hydrostatic pressure and buoyancy are due to gravity. These forces also exist in moving fluids but with additional *dynamic forces* present – viscous and inertial – due to gradients of velocity and accelerations affecting the flow. In order to understand the dynamics of such flows and to be able to calculate the resulting forces acting we need to understand the interactions between the dynamic and static forces that comprise ΣF, the total force. This will enable us to eventually solve some dynamic force equations, the equations of motion, for properties such as velocity, pressure, and energy. Such a development will inform Chapter 4 concerning the nature of physical environmental flows.

3.12.1 General momentum approach

To begin with, we make simple use of Newton's Second Law and consider the total force, ΣF, causing a change of momentum in a moving fluid, not inquiring into the various subdivisions of the force (Fig. 3.50). To do this we take the simplest steady flow of constant density, incompressible fluid moving through an imaginary conic streamtube orientated parallel with a downstream flow unaffected by radial or rotational forces. From the continuity equation (Section 2.5) the discharges into and out of the tube are constant but a deceleration must be taking place along the tube, hence momentum must be changing and a net force acting. The net downstream force acts over the entire streamtube and comprises both pressure forces normal to the walls and ends of the tube and shear forces parallel to the walls. The approach also allows us to calculate the force exerted by fluid impacting onto solid surfaces and around bends.

3.12.2 Momentum–gravity approach

In many cases, we need to know more about the components of the total force in order to find relevant and interesting properties of environmental flows, such as velocity and pressure distributions. One major problem in the early development of fluid dynamics was what to do

with Newton's discovery of viscosity and the existence of viscous stresses. This was because the origin and distribution of viscous forces was seen as an intractable problem. In a bold way, Euler, one of the pioneers of the subject, decided to ignore viscosity altogether, inventing *ideal* or *inviscid flow* (see Section 2.4; Cookie 9). In fact, viscous friction can be relatively unimportant away from solid boundaries to a flow (e.g. away from channel walls, river or sea bed, desert surface, etc.) and the inviscid approach yields relevant and highly important results. In the interests of clarity, we again develop the approach for the simplest possible case (Fig. 3.51), a steady and uniform flow through a cylindrical streamtube involving two forces, gravity and pressure, acting in a vectorially unresolved direction, s. The Second Law tells us that

$$\Sigma \boldsymbol{F} = \boldsymbol{F}(\text{pressure}) + \boldsymbol{F}(\text{gravity}) = \text{mass} \times \text{acceleration}$$

Since in this flow there is no acceleration:

$$\boldsymbol{F}(\text{pressure}) + \boldsymbol{F}(\text{gravity}) = 0 \text{ or } \boldsymbol{F}(\text{pressure}) = \boldsymbol{F}(\text{gravity})$$

The principles involved may be illustrated by a simple but dramatic experiment. A large reservoir feeds a length of horizontal tube which has a middle section of lesser diameter that leads smoothly and gradually to and from the larger diameter end sections. Vertical tubes are let out from the horizontal tube to measure the static pressures acting at the boundary. When the fluid is at rest, the outlet

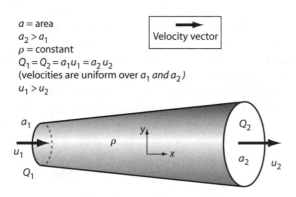

a = area
$a_2 > a_1$
ρ = constant
$Q_1 = Q_2 = a_1 u_1 = a_2 u_2$
(velocities are uniform over a_1 and a_2)
$u_1 > u_2$

Net force acting in x direction per unit time is:

F_x = x-momentum out – x-momentum in

$F_x = (\rho a_2 u_2) u_2 - (\rho a_1 u_1) u_1$

$F_x = \rho Q (u_2 - u_1)$ N

that is, product of mass flux times velocity change. For the case in point, f_x is overall negative, that is, force acts upstream

If all momentum is lost at a_2, the force of the water jet is $\rho Q u = \rho a_2 u_2^2$ N.

Fig. 3.50 General momentum approach.

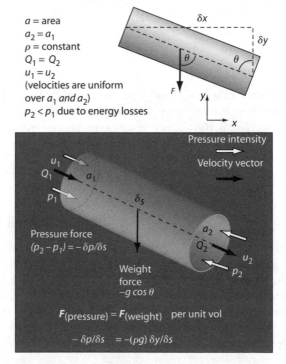

a = area
$a_2 = a_1$
ρ = constant
$Q_1 = Q_2$
$u_1 = u_2$
(velocities are uniform over a_1 and a_2)
$p_2 < p_1$ due to energy losses

Pressure intensity

Velocity vector

Pressure force
$(p_2 - p_1) = -\delta p / \delta s$

Weight force
$-g \cos \theta$

$\boldsymbol{F}(\text{pressure}) = \boldsymbol{F}(\text{weight})$ per unit vol

$-\delta p / \delta s = -(\rho g) \delta y / \delta s$

Fig. 3.51 Pressure–gravity approach for constant velocity.

valve being closed, the pressures in each vertical tube are equal. The outlet valve is now opened and constant water discharge (i.e. steady flow conditions) is let into the inlet end of the tube to freely pass through the whole tube. A dramatic change occurs in the pressure, which in the narrow bore section being much reduced compared with that measured in the upstream and downstream wider bore sections.

How do we explain this startling result? As the flow passes into the narrow part of the tube, continuity (Section 2.5) tells us that the flow must accelerate (remember that water is incompressible under the experimental conditions) and that this must be caused by a net force. Since there is no change in the mean gravity force, the tube centerline being horizontal throughout, this net force must come about by the action pressure in order that the force balance between inertia and pressure is maintained. We thus have

$$\Sigma F = F(\text{pressure}) + F(\text{gravity}) = \text{mass} \times \text{acceleration}$$

The result means that the frequency of intramolecular collisions responsible for pressure is decreased by the acceleration. Also, if forces are balanced then energy must also be balanced, the increase in flow kinetic energy due to the acceleration being balanced by a decrease in the flow energy due to pressure.

By generalizing the approaches above (Fig. 3.52), we arrive at *Bernoulli's equation* (Cookie 9).

3.12.3 Scope of application of Bernoulli's equation

The production of flow acceleration as a consequence of pressure change is a major feature of fluid dynamics which has major consequences (Fig. 3.53). Despite its simplicity in ignoring the effects of frictional forces exerted by flow boundaries, application of Bernoulli's equation has enabled increased understanding of flight (Fig. 3.54), wave generation, hydraulic jumps, and erosion by wind and water, to name but a few. Consider flow over a convexity on a free boundary, such as a protruding sediment grain or wingspan. The mean streamlines converge and then diverge. From the continuity equation the flow will

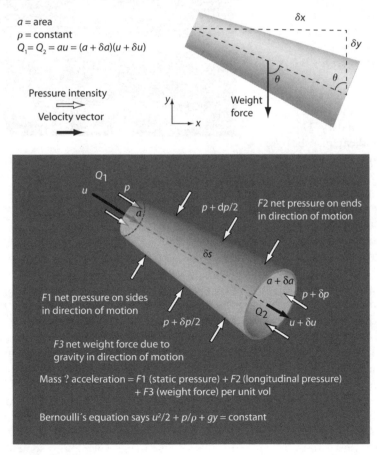

Fig. 3.52 Euler–Bernoulli energy approach for variable velocity. D. Bernoulli and Euler pioneered the application of Newtonian mechanics and the calculus to physical and engineering problems.

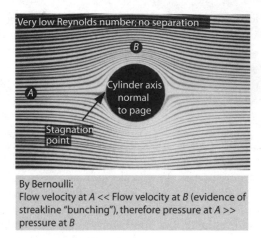

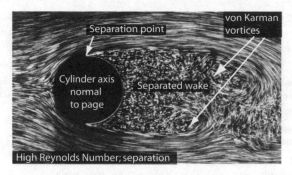

By Bernoulli:
Flow velocity at *A* << Flow velocity at *B* (evidence of streakline "bunching"), therefore pressure at *A* >> pressure at *B*

Flow pathlines visualize periodic von Karman vortices formed by shear at the unstable margins to the separated fluid. They tend to be shed alternately from one side to the other of the obstacle, diffusing gradually downstream after intense turbulent mixing

Fig. 3.53 Lateral pressure gradients cause flow separation around obstacles at high particle Reynolds number: an important consequence of the conservation of energy expressed in Bernoulli's equation.

speed up and then slow downstream. Bernoulli's equation states that the pressure should decrease in the accelerated flow section. This decrease of pressure produces a pressure gradient and a lift force that may reach sufficient magnitude to exceed the downward acting weight force and so cause upward movement. All flight and some forms of sediment transport depend upon this *Bernoulli effect* for the conservation of flow energy. When a convexity reaches a certain critical height, the pressure gradients $dp/dx < 0$, upstream, and $dp/dx > 0$, downstream, have the greatest effect on the lower-speed fluid near to the boundary. This fluid retarded by the adverse pressure gradient may be moved upstream at some critical point, a process known as *flow separation*. Flow separation creates severe pressure energy degradation and destroys the even pressure gradients necessary for lift (Fig. 3.54); a process known as *stall* results. Flow separation also occurs when a depression (negative step; Fig. 3.55) exists on a flow boundary; accentuated erosion results due to energy degradation in the separation and reattachment zones.

Another application of Bernoulli's equation occurs when fluid flow occurs within another ambient fluid. In such cases, with shear between the two fluids, the situation becomes unstable if some undulation or irregularity appears along the shear layer, for any acceleration of flow on the part of one fluid will tend to cause a pressure drop and an accentuation of the disturbance. Very soon a striking, more-or-less regular system of wavy vortices develops, rotating about approximately stationary axes parallel to the plane of shear. Such vortices are termed *Kelvin–Helmholtz instabilities* that are important mixing mechanisms across a vast variety of scales, from laboratory tube to the Gulf Stream (Sections 4.9 and 6.4).

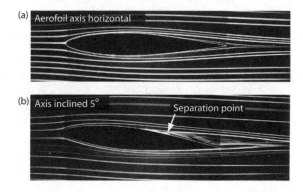

Fig. 3.54 In these symmetrical aerofoils, only a slight change (5° here) in the angle of incidence can cause flow separation.

3.12.4   Real-world flows of increased complexity

For real-world flows of hydraulic, oceanographic, and meteorological interest several additional terms are relevant, including those for friction (viscous and turbulent), buoyancy, radial, and rotational forces. We sample just a few of the various possibilities here, to give the reader an idea of the richness presented by Nature.

Frictionless oceanographic and meteorological flows: In the open oceans and atmosphere, away from constraining boundaries to flow, currents have traditionally been viewed as uninfluenced by viscous or turbulent frictional forces. This is because in such regions there was thought to be very little in the way of spatial gradient to the velocity flow field and therefore not much in the way of viscous or turbulent forcing. Clearly this somewhat unrealistic scenario is inapplicable in regions of fast ocean surface and bottom current systems, where dominant

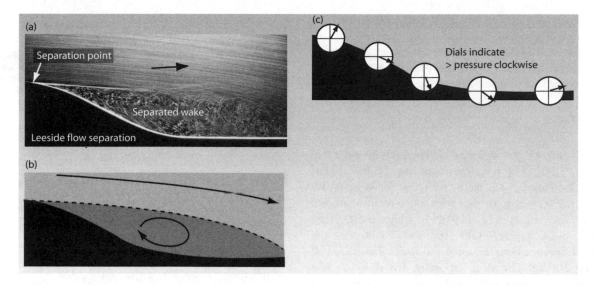

Fig. 3.55 (a) Many situations occur in nature where downstream slope gradients lead to a decrease of mean flow velocity. Desert and subaqueous dunes and ripples are obvious examples. By Bernoulli's equation we can predict that such decelerations should cause a rise in the fluid pressure or flow work. Given the right conditions this downstream increase in pressure can be sufficient to reverse the flow locally and cause flow separation: see sketches below; (b) Time-mean upstream recirculating flow in the leeside; (c) Pressure guages measure downstream increases in pressure over the steeper leeside slope. This negative pressure gradient causes upstream flow close to the boundary and hence flow separation.

turbulent mixing occurs. Nevertheless, the *geostrophic approximation* has enabled major progress in understanding the large-scale oceanic circulation substantially affected by the Coriolis force:

$$\Sigma F = F(\text{pressure}) + F(\text{gravity}) + F(\text{Coriolis})$$
$$= \text{mass} \times \text{acceleration}$$

In situations of steady flow with no acceleration, where there are no density changes and where gravity is balanced in the hydrostatic condition, this expression becomes an equality between the pressure and Coriolis forces:

$$F(\text{pressure}) = F(\text{Coriolis})$$

Viscous friction flows: Navier–Stokes approach: The incorporation of frictional resistance via viscous forces into the Euler–Bernoulli versions of the equations of motion was a major triumph in science, attributed jointly to Navier and Stokes. Refer back to Section 3.10 for an account of the derivation of the viscous stress and the net viscous force resulting in a flow boundary layer. The simplest form of the *Navier–Stokes equation* may then be written for an incompressible, nonrotating, straight-line flow, such as in a straight river channel or along a local wind, as

$$\Sigma F = F(\text{pressure}) + F(\text{gravity}) + F(\text{Viscous})$$
$$= \text{mass} \times \text{acceleration}$$

Turbulent friction flows: Reynolds' approach: As we have seen in the previous chapter, Reynolds' neatly deconstructed turbulent flow velocities into mean and fluctuating components. The latter are responsible for a very large increase in the resisting forces to fluid motions on account of the immense accelerations produced in the flow. Thus through most of the flow thickness these fluctuating turbulent forces dominate over viscous frictional forces. However, as we shall see subsequently, there still remain strong residual viscous resisting forces close to any flow boundary and so we keep the viscous contribution in the equation of motion for turbulent flows, written here for a simple case of straight channel flow:

$$\Sigma F = F(\text{pressure}) + F(\text{gravity})$$
$$+ F(\text{Viscous}) + F(\text{turbulent})$$
$$= \text{mass} \times \text{acceleration}$$

3.13 Solid stress

We have seen (Section 3.3) that vectorial force, F, is defined in classical Newtonian physics as an action which tends to alter the state of rest or uniform straight line velocity, u, of any object of a certain mass, m. Also, forces can be defined in terms of changes in momentum, $p = mu$ in time or space, $F = d(mu)/dt$. In relation to the solid deformation accompanying plate tectonics it is more likely that spatial changes in velocity are responsible, produced in or at the boundaries of the solid crust, whereas changes in mass due to thermal effects are less common. The fact that forces caused by changes in velocity or acceleration in space deform the solid crust is witnessed by structures formed in the rocky crust by them (Sections 4.14–4.16).

3.13.1 Stress

Stress, σ, is force F, per unit area, A, when acting over a surface (Fig. 3.56). The units for stress are the Pascal (Pa) which is Newton per square meter (Nm^{-2}). More useful units for the very large stresses relevant to tectonic studies are the Kilopascal, kPa (10^3 Pa), Megapascal, MPa (10^9 Pa), and Gigapascal, GPa (10^{12} Pa). To illustrate the physical difference between force and stress, imagine a ball weighing 200 kg. The most important part of solid stress is the force, the stress itself can be envisaged as the effect or intensity of the force applied to a mass of rock and how it is distributed or felt by the rock in every conceivable direction of space. The force exerted by the ball in Newtons (N) over any surface should be constant: $F = mg = 200$ kg $\cdot 9.8$ ms^{-2} = 1,960 N. In a simple example, consider the 200 kg ball resting on the top surface of a wide column, of area say 2 m^2 (Fig. 3.57a). The force is distributed over all the surface and the column will hold up depending on its material resistance. If the same force is applied over a surface of a smaller column, for example, 0.5 m^2, of the same material, the identical force of 1,960 N may lead the column to break (Fig. 3.57b). Remembering that $\sigma = F/A$ (Equation 1; Fig. 3.56) the resulting stresses for both columns are σ = 1,960 N/ 2 m^2 = 0.98 kPa and σ = 1,960 N/0.5 m^2 = 3.9 kPa, respectively.

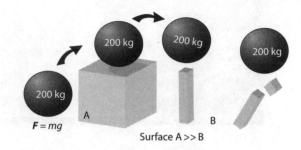

Fig. 3.57 Stress can be explained as the intensity of force over a surface. The constant force exerted by the rock ball gives a higher stress when it is put to rest over a column with a smaller surface area: producing fractures if the internal resistance is exceeded.

The resulting stress over the smaller surface is bigger even when the force has the same value in both cases. In other words, the force is felt with more intensity over the small surface and is thus able to produce deformation easier than over the wide surface. The single vector corresponding to the force per unit area which reflects the force intensity over a surface F/A is called a *traction* and is only one of the infinite components of the overall *stress system* (Fig. 3.58). Because forces can change magnitude or direction over a surface, we have to define different stress values for every infinitesimal part of the surface or at any point as dF/dA. If we consider a surface in a state of equilibrium, any traction has to be balanced by an equal and opposite traction. This pair of tractions constitutes the *surface stress* (Fig. 3.59b), which can be resolved into normal and shear components (Fig. 3.59c). It is important to remember that in a state of equilibrium the sum of the forces acting over a surface equals zero.

We have so far considered only how the stress value is distributed according to a single surface orientated perpendicular to the applied force. However, the orientation of the surface in relation to the applied force is essential in determining the resulting traction value and there should be an orientation and magnitude for every possible surface inclined at different angles. The stress tensor is composed of *all* the individual surface stresses acting over a given point. Stress analysis can be approached in two dimensions (2D) or three dimensions (3D): in 2D orientations are referred to an *xy* coordinate system, whereas in 3D an *xyz* system is used, in which *z* is conventionally taken as the vertical component. If all the tractions are equal in magnitude, as occurs in static fluids, the stress tensor has the shape of a circle in 2D and of a sphere in 3D; we have seen in our discussions of fluid stress (Section 1.16) that such a state is called *hydrostatic stress* (Fig. 3.58b). In solids, when the tractions usually have different magnitudes, the shape

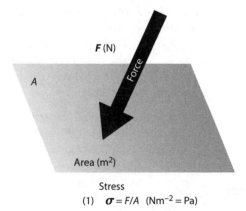

$$F\,(N)$$

Stress
(1) $\sigma = F/A$ (Nm^{-2} = Pa)

Stress is force per unit area, measured in Pascals (Pa) or Nm^{-2}

Fig. 3.56 Revision: force and stress.

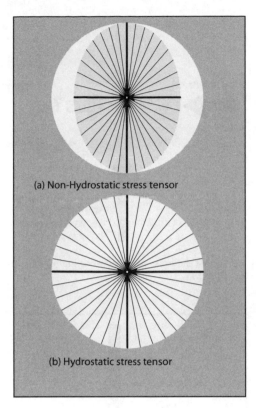

Fig. 3.58 The stress tensor defines an ellipse in 2D and an ellipsoid in 3D. In the particular case of hydrostatic stress, it defines a circle in 2D and a sphere in 3D.

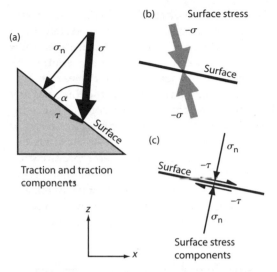

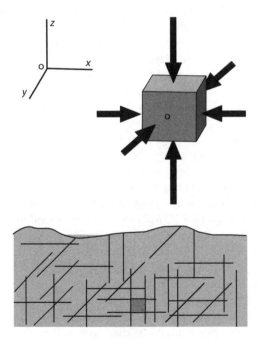

Fig. 3.59 (a) As forces, **tractions** (forces acting over a surface) can be resolved into a normal (σ_n) and a shear component (τ); (b) Surface stress, a pair of equal in magnitude and opposed in direction tractions at equilibrium; (c) Surface stress components resolved for (b).

Fig. 3.60 Forces acting over a small volume of rock in an outcrop. Many surfaces with different orientations can be defined. We can extract an imaginary cube and define the tractions acting on the surfaces.

of the stress tensor is still very regular but acquires the nature of an ellipse in 2D or an ellipsoid in 3D (Fig. 3.58a).

As in fluid forces, solid tractions can be resolved into stress components, one normal to the surface which is called *normal traction* $\boldsymbol{\sigma}_n$ and a component parallel to the surface named *shear traction*, $\boldsymbol{\tau}$ (Fig. 3.59a). It is usual to name these traction components *normal stress* and *shear stress* respectively. In the hydrostatic state of stress all tractions are normal components and are applied at 90° to all possible surfaces; there are no shear stresses in any direction. In order to have shear stress components, the tractions for different directions have to be unequal and they have to act over a surface orientated at a certain angle different to 90°.

In order to illustrate this point and to visualize how stresses are distributed in a rock, imagine a mass of stationary rock in which many different surfaces exist (Fig. 3.60). These surfaces are abundant in natural crustal rock and may comprise joints, faults, stratification planes, crystal lattices, and so on and even so we can also imagine other virtual surfaces which, even when not present at the moment, may potentially have been developed or can be developed at some time in the future. Forces acting at a point, actually over a infinitesimal volume of rock (as points are dimensionless and lack surfaces) will give the full stress tensor, with tractions or vectors acting over all the different surfaces with all possible orientations. From the mass of rock we can extract a tiny imaginary cube over which forces are acting and stresses are developed. If only gravity force is applied, our expectation from the fluid state is that the weight of the column of rock mass over the point-cube should give a state of hydrostatic stress at the point. In the case of solid rocks, this stress state is called *lithostatic stress* or confining pressure, and corresponds to the force per unit area produced by the weight of a column of rock due to the gravity force. However, there are *always* other forces at work in the Earth and the lithostatic stress state is a rare in rigid solids. In fact, the vertical stress, a compressive stress due to gravitational loading, affects the horizontal stresses as rocks tend to expand perpendicularly to applied compression. We will discuss this point further when we consider rheological properties of solid deformation (Section 3.15).

3.13.2 Magnitude of the stress tensor in the Earth

It is useful at the outset to have an idea of the magnitude of stress in the crust. For example, if we consider a mass of static granite, density $\rho = 2{,}700 \, \mathrm{kg\,m^{-3}}$, at a depth, h, of 1,000 m, the lithostatic load per unit area should be $\sigma = \rho g h$, which gives 26.5 MPa. This lithostatic load is responsible for the vertical compressional stress and should increase linearly with depth due to the increasing weight of the rock itself over progressively deeper surface areas. To try to understand how the stress tensor is constructed and which are the magnitude and orientation of the individual tractions acting upon different surfaces on a small volume of this granite rock we can analyze the stress in 2D with unequal values for the normal tractions acting parallel to the vertical z-axis (σ_{zo} and σ_{oz}), and the horizontal x-axis (σ_{xo} and σ_{ox}). Assuming that the point-cube is located at a depth of about 1.1 km in the crust, the value of the vertical traction will be roughly 30 MPa and will correspond to the vertical compressional loading in granite rocks. We now arbitrarily take the horizontal traction acting over the center of the cube to be less than the vertical, say 20 MPa. In order for this situation to exist some external force must operate.

Now let an inclined surface cut the cube so that a triangular prism is developed. Let this surface be defined by an angle with respect to the vertical axis z, for example 65° (Fig. 3.61). To simplify things we can call this inclined surface $A1$; $A2$ is then the surface parallel to the horizontal plane xy and $A3$ to the surface parallel to the vertical plane zy. Remember that all these surfaces have to be very small, of order $10^{-12} \, \mathrm{m}^2$ or less as the stress state is now only representative of a single point in the mass of rock (stress gradients will occur along any larger finite rock volume). We can now relate all the surfaces as a function of $A1$ using simple trigonometric relations (Equations 2 and 3; Fig. 3.61).

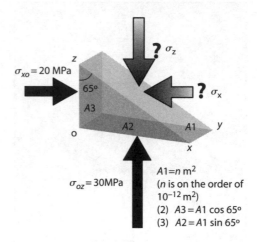

Fig. 3.61 3D view of the stress components acting over a triangular prism in two directions.

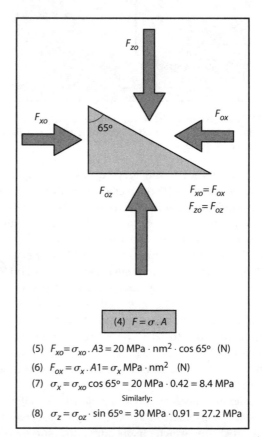

Fig. 3.62 Frontal view of the prism showing forces in equilibrium.

In any state of static equilibrium, forces have to be balanced so the prism does not move or rotate; the horizontal forces $\boldsymbol{F}_{ox}$ and $\boldsymbol{F}_{xo}$ have to be balanced so that $\boldsymbol{F}_{ox} = \boldsymbol{F}_{xo}$ and similarly the vertical forces so that $\boldsymbol{F}_{oz} = \boldsymbol{F}_{zo}$ (Fig. 3.62).

Knowing that $F = \sigma A$ (Equation 4; Fig. 3.62), we can substitute the values of the corresponding stress components and the values of the surfaces over which they are applied (with respect to $A1$, as developed in Fig. 3.61) and obtain the horizontal σ_x and vertical σ_z components of the traction, σ_{zx}, which we are looking for (Fig. 3.62). After simplifying (Equations 5–7; Fig. 3.62), the results are $\sigma_x = \sigma_{ox} \cos 65°$, and $\sigma_z = \sigma_{oz} \sin 65°$.

From the sketches of Figs 3.61 and 3.62, and the resulting equations, it is obvious that the value of the horizontal traction (σ_{xo}) acting over $A3$ has to be bigger than the value of the horizontal component of the traction σ_x acting on $A1$. Similarly, the vertical component σ_{oz} acting over $A2$ will have also a higher value than σ_z acting over $A1$, as the surface $A1$ is of larger area than $A2$ and $A3$. Calculated values are 27.2 MPa for σ_z and 8.4 MPa for σ_x. Having computed these values, σ_{zx} can be obtained from Pythagoras' theorem (Equation 9; Fig. 3.63). The resulting vector is a traction with a magnitude of 28.5 MPa, which forms an angle with the surface divergent from the normal to the surface (i.e. $\neq 90°$). The angle which defines σ_{zx} with respect to the surface can be easily determined as shown in Fig. 3.64 by simple trigonometric calculations (Equations 10 and 11), resulting in a value of 82°. To simplify things we divide the total angle, α, between the inclined surface and the traction σ_{zx}, into two angular segments, α_1 and α_2, so $\alpha = \alpha_1 + \alpha_2$. As the vertical component of the traction σ_z and the direction oz are parallel and they both intersect surface $A1$ (zx in the 2D sketch of Fig. 3.64), the angle α_1 equals 65°. α_2 can be calculated using any trigonometric relation as the three sides of the triangle (σ_{zx}, σ_z, σ_x) are known. In the example we used $\tan \alpha_2 = \sigma_x/\sigma_z$ (Equation 10; Fig. 3.64) but $\sin \alpha_2$ or $\cos \alpha_2$ could also be used.

Once the magnitude of this particular traction and the angle over the surface are known, the normal (σ_n) and shear (τ) stress components can be found (Equations 12 and 13; Fig. 3.64c). In the present example the normal stress can be calculated as $\sigma_n = 28.5 \sin 82° = 28.2$ MPa and the shear stress as $\tau = 28.5 \cos 82° = 3.96$ MPa.

Similar calculations can be made to obtain the magnitude and angle of other tractions acting over surfaces with different angles with respect to the vertical axis, using the initial values of 20 MPa and 30 MPa for the horizontal and vertical stress components. For any given orientation an inclined traction results. All tractions in this example will have a value bigger than 20 MPa and smaller than

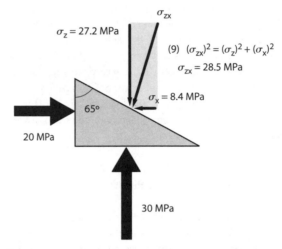

$\sigma_z = 27.2$ MPa

σ_{zx}

(9) $(\sigma_{zx})^2 = (\sigma_z)^2 + (\sigma_x)^2$

$\sigma_{zx} = 28.5$ MPa

$\sigma_x = 8.4$ MPa

65°

20 MPa

30 MPa

Fig. 3.63 Resolving the traction acting over the surface inclined at 65°, having found the values of the horizontal (x) and vertical (z) components.

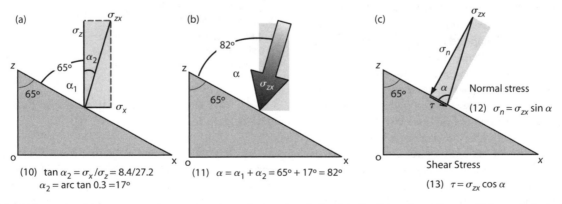

(a)

σ_{zx}

σ_z

α_2

z

65°

α_1

65°

σ_x

o x

(10) $\tan \alpha_2 = \sigma_x/\sigma_z = 8.4/27.2$
$\alpha_2 = \arctan 0.3 = 17°$

(b)

82°

α

z

σ_{zx}

65°

o x

(11) $\alpha = \alpha_1 + \alpha_2 = 65° + 17° = 82°$

(c)

σ_{zx}

σ_n

z

α

65°

τ

o x

Normal stress

(12) $\sigma_n = \sigma_{zx} \sin \alpha$

Shear Stress

(13) $\tau = \sigma_{zx} \cos \alpha$

Fig. 3.64 (a) The angle of the traction with respect to the surface can be calculated easily adding the angle α_1, which equals the inclined plane with respect to the vertical axis and α_2, which can be calculated by a simple trigonometric relation. (b) The final result: the thick arrow corresponds to the calculated traction σ_{zx} acting over the inclined surface. The angle with respect to the surface is 82°. (c) Calculation of the normal (σ_n) and shear (τ) components of a traction.

30 MPa. It is important to realize that all resulting angles of these tractions over the corresponding surfaces, except the two initial directions acting over *A2* and *A3*, are different from 90°. In other words, in a 2D study of the stress tensor, for all possible orientations there are only two directions in which the tractions are located at right angles over the surface. The rest are inclined with different angles over the inclined surfaces. This allows us to resolve for normal and shear stress components of the tractions.

As we have seen for the particular case of hydrostatic stress, all tractions have the same value in all directions, act perpendicular to any surfaces and there are no shear stresses. As an example, try substituting equal values of the vertical and horizontal components into the previous case, for example 25 MPa. Calculate the value of the traction acting over any inclined surface (65° or any other) and subsequently, the angle of the selected surface. The calculated tractions should have a constant value of 25 MPa and the angle over the corresponding surface should be of 90°. Thus all tractions are normal stresses and the shear stress values are always zero. Try different angles for the inclined surfaces, the result should always be the same.

3.13.3 Stress notation conventions

Normal stresses are compressive when the traction components converge at the surface and the arrows of the pair of tractions, defining the surface stress components point to each other; any two blocks of rock at either side of the surface are being pushed closer together (Fig. 3.65). Conversely, normal stresses are tensile when the surface stress traction components diverge from the surface, the two arrows point away from each other and the blocks of rock on each side are being pulled apart. Normal stresses are the components which tend to press or decompress the masses of rock over a surface. In structural geology, it is conventional that compressive stresses are considered positive and tensile stresses negative. Shear stress components promote sliding of rock masses at each side of the surface. Shear stresses are right-handed or clockwise when, facing a surface, the direction is to the right (a ball positioned between both arrows would rotate clockwise). Shear stresses are left-handed or anticlockwise when the arrow points left and a ball would rotate anticlockwise. Right-handed shear stresses are generally considered negative and left-handed positive, although sign conventions do not have general agreement. Our usage is the same as that used for vorticity (Section 3.8).

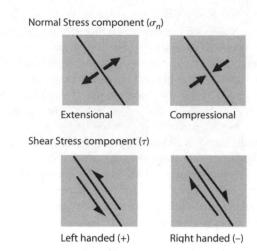

Fig. 3.65 Sign conventions for the normal and shear stress components of the stress.

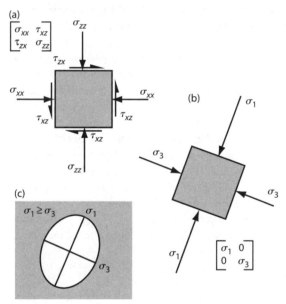

Fig. 3.66 (a) The stress state at a point centered on a cube in 2D is defined by two normal stresses and two shear stresses; (b) The main stress directions; (c) The stress ellipse.

3.13.4 The stress ellipse, the stress ellipsoid, and the principal stresses

The whole family of tractions acting over a certain point define an ellipse in 2D and an ellipsoid in 3D. In 2D, the longest and shortest tractions correspond to the axis of the ellipse which are located at 90° to each other and are called respectively σ_1 and σ_3 (Fig. 3.66). These surface stresses are called the *principal stresses* and their corresponding directions *principal stress directions*. The principal stresses are orientated perpendicularly to the surfaces over which they act; they are thus normal stresses and the shear stresses (τ) here are zero. The surfaces, perpendicular to the principal stresses, are called *principal planes of stress*. For any other direction the corresponding traction is inclined with respect to the surface and thus, has normal and shear stresses acting.

In 3D the stress tensor is an ellipsoid and the three axes are the principal stresses: σ_1, σ_2 and σ_3 (Fig. 3.67). σ_1 is the longest, major axis of the ellipsoid and represents the traction with the biggest magnitude, σ_2 has an intermediate value (which *does not* correspond to the mean stress value) and σ_3 is the smallest. Two of the stresses may have the same value: $\sigma_1 \geq \sigma_2 \geq \sigma_3$.

Stresses are described in 2D by two normal stresses and two shear stresses acting over two perpendicular surfaces (Fig. 3.66). The values may be represented in a matrix with four components. Rows in the matrix represent the surface stress components acting over a particular surface. Each component is referenced to a coordinate system x,z and named by two subscripts. The first subscript refers to the axis *perpendicular* to the surface over which the force is acting. The second refers to the *direction* of the traction component. All of them share the same first subscript and

increasing values from left to right. Columns show increasing values for the first subscript from top to bottom and the same value for the second subscript.

In 3D, the state of stress at a point over a surface is commonly represented by three mutually perpendicular components: a normal stress and two shear stresses. In the general case, the sum of the three components will result in an inclined traction over the surface. To fully define the state of stress at a point in a rock we need three normal stresses and six shear stresses defined in three reciprocally normal surfaces (Fig. 3.67a). With these nine components it is possible to define the *stress tensor*, which is represented in matrix form (see Appendix 1 for general information on tensors). Each component is referenced to a coordinate system x, y, z and, as in 2D, named by two subscripts. For example, the normal stress acting over the horizontal plane xy, perpendicular to the axis z, is named σ_{zz} and the two shear components τ_{zx} and τ_{zy} (Fig. 3.67a).

The principal diagonal, top-left to bottom-right is occupied by normal stresses and the remaining positions by shear stresses. It is important to note that in a state of equilibrium, where torques are not allowed, only six of the nine components are independent as the shear stresses acting over adjacent surfaces have to be balanced such as

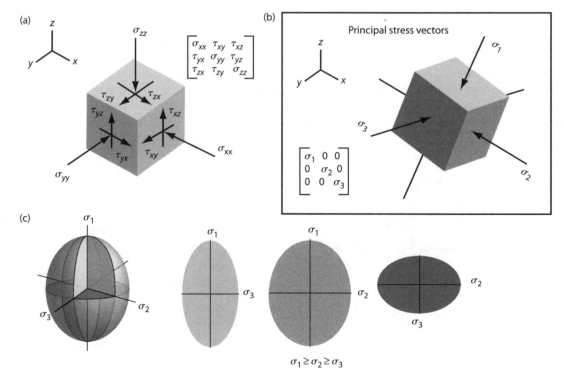

Fig. 3.67 (a) The stress state at a point centered in a small cube is defined by three normal stress components and six shear stress components. (b) The principal stress vectors are normal stresses with shear stress values of 0. (c) The stress ellipsoid. Right: a 3D representations; Left: 2D representations of the three principal planes of stress showing the corresponding main stresses and directions.

$\tau_{yx} = \tau_{xy}$, $\tau_{zx} = \tau_{xz}$, and $\tau_{yz} = \tau_{zy}$. The matrix can be defined also for a situation in which the axes are coincident with the directions of the principal normal stresses σ_1, σ_2, and σ_3 (Fig. 3.67b). In this particular case the principal stresses are located in the principal diagonal and all the shear components are zero, as the tractions are normal to the corresponding surfaces.

We explained earlier that the stress tensor has the shape of an ellipse in 2D and an ellipsoid in 3D. To derive the equation of the stress ellipse we go back to a prism of rock oriented with respect to a coordinate system x, y, z (Fig. 3.68). We will consider the projection of the prism in the vertical plane z, x to analyze the stress conditions in 2D. The two mutually perpendicular faces of the prism parallel to z and x will be principal planes of stress, so that σ_1 is perpendicular to the plane z, y (in 2D normal to the direction z) and σ_3 is perpendicular to the plane xy (normal to the direction of x). An inclined surface of area A forms an angle θ with respect to the plane zy (normal to σ_1). Over the surface A the traction σ will be resolved into two components: σ_z parallel to z and σ_x, parallel to x as depicted in Fig. 3.68.

3.13.5 Equations of equilibrium

These have to be defined for the vertical and horizontal forces acting upon the prism. It is important to remember that at equilibrium, where no movement of the prism is allowed, all force components in any given direction sum to zero. Accordingly, in order to establish these conditions for the prism in the state of equilibrium, forces acting over A in the principal stress directions σ_1 and σ_3 have to be balanced. The equilibrium equation for the horizontal forces (Equation 14; Fig. 3.68a) gives the balance between the force $\sigma_x A$ which is trying to move the prism left and the force $\sigma_1 A \cos\theta$ which is trying to push the prism right. From this relation we extract $\cos\theta = \sigma_x/\sigma_1$. Following a similar approach through Equation 15 (Fig. 3.68a), the vertical forces are balanced and we extract $\sin\theta = \sigma_z/\sigma_3$. The trigonometric relations between the directions of σ_z and σ_x and the angle θ are depicted in Fig. 3.68b. Since $\sin^2\theta + \cos^2\theta = 1$ we can substitute the values for $\cos\theta$ and $\sin\theta$ obtained earlier (Equations 16 and 17) which yields the equation of the stress ellipse (Equation 18; Fig. 3.68b), which is centered at the origin of the coordinate system. The major and minor axis of this ellipse are σ_1 and σ_3 respectively and are orientated in the x- and z-directions. Any other radius of the ellipse will be a traction σ or a stress vector with a magnitude between σ_1 and σ_3. A 3D analysis will result in equation 19 (Fig. 3.68b).

3.13.6 The fundamental stress equations

The fundamental stress equations are derived from the equations of equilibrium showing the stress balance over a prism of rock, relating the values of the main stresses σ_1 and σ_3 and the normal (σ_n) and shear (τ) stress components of a given traction (σ) acting over a surface A, which forms any angle θ with the normal to σ_1. These equations give the values of the normal and shear components of a traction for any plane knowing the principal stresses σ_1 and σ_3 and the angle θ that we choose.

As an example, we consider (Fig. 3.69) a triangular prism with two surfaces at right angles inclined with

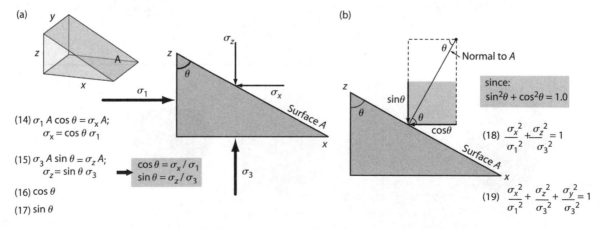

Fig 3.68 (a) Forces at equilibrium for the derivation of the equation of an ellipse with major and minor axis σ_1 and σ_3 parallel to the x- and z-axis; (b) Trigonometric relations for the directions of σ_x and σ_z related to the angle θ and after Pythagoras theorem, the resulting Equation 18 of the stress ellipse (for a 2D analysis). For a 3D analysis a similar approach can be applied resulting in Equation 19 of the stress ellipsoid.

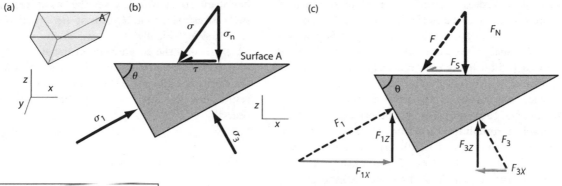

(a)

(b) Surface A

(c) Relations between the force components and the corresponding tractions

Forces at equilibrium

(20) $F_S + F_{3X} = F_{1X}$; $F_S = F_{1X} - F_{3X}$

(21) $F_N = F_{1Z} + F_{3Z}$

(22) $F_{1Z} = \sigma_1 \cos\theta\, A \cos\theta$
(23) $F_{3z} = \sigma_3 \sin\theta\, A \sin\theta$

(24) $F_{1x} = \sigma_1 \sin\theta\, A \cos\theta$
(25) $F_{3x} = \sigma_3 \cos\theta\, A \sin\theta$

(d)

Normal to A

Surface A

(e)

The fundamental stress equations

(26) $\sigma_n A = \sigma_1 \cos\theta \cdot A \cos\theta + \sigma_3 \sin\theta \cdot A \sin\theta$; $\sigma_n = \sigma_1 \cos^2\theta + \sigma_3 \sin^2\theta$

Considering the trigonometric identities:

$\cos^2\theta = \dfrac{1 + \cos 2\theta}{2}$ and $\sin^2\theta = \dfrac{1 - \cos 2\theta}{2}$ →   (27) $\sigma_n = \dfrac{\sigma_1 + \sigma_3}{2} + \dfrac{\sigma_1 - \sigma_3}{2} \cos 2\theta$

(28) $\tau A = \sigma_1 \sin\theta \cdot A \cos\theta - \sigma_3 \cos\theta \cdot A \sin\theta$; $\tau = (\sigma_1 - \sigma_3) \cos\theta \sin\theta$

Considering the trigonometric identity:

$\sin\theta \cos\theta = \dfrac{\sin 2\theta}{2}$ →   (29) $\tau = \dfrac{\sigma_1 - \sigma_3}{2} \sin 2\theta$

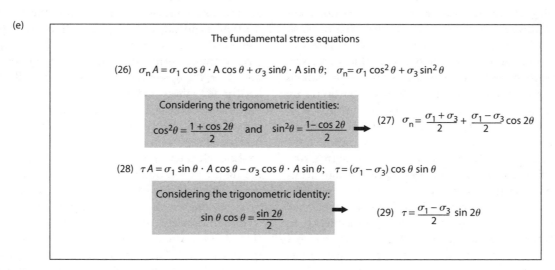

Fig. 3.69 Derivation for the principal stress equations in two dimensions: (a) A 3D view of a triangular prism showing the upper surface of area A over which a traction σ is acting. The coordinate system shows the orientation of the x-, y-, and z-axis. The surface A forms an angle θ with respect to the normal to the principal stress σ_1; (b) 2D diagram showing the prism projected in the zx plane, and the settings to state the relations between the normal (σ_n) and shear (τ) stress components of σ and the principal stresses σ_1 and $\sigma_3 \cdot \sigma_n$ will be parallel to the direction z and τ will be parallel to x; (c) Balance of forces over the prism, showing the force components parallel to z and x; (d) Diagram showing the forces ($F = \sigma A$) as a function of the stresses and all the surfaces as a function of the area A. (e) the fundamental stress equations.

respect to the coordinate system x, y, z, which will be principal planes of stress, and a horizontal surface (y, x) of area A (Fig. 3.69a). As in the previous case (Fig. 3.68), we will develop a 2D analysis projecting the prism in the plane z, x. A traction $\boldsymbol{\sigma}$ will be acting over the surface A (Fig. 3.69b) whose normal is inclined at an angle θ with respect to the principal stress σ_1. As usual, the traction $\boldsymbol{\sigma}$ can be resolved into the normal $(\boldsymbol{\sigma}_n)$ and shear $(\boldsymbol{\tau})$ components.

To establish equilibrium conditions the forces have to be balanced so the sum of all the components acting in every direction is zero. The force $\boldsymbol{F} = \boldsymbol{\sigma} A$ will be the force pushing over the surface A, and can be resolved into the normal force F_N $(F_N = \boldsymbol{\sigma}_N A)$ and the shear force F_S $(F_S = \boldsymbol{\tau} A)$. F_1 and F_3 are the forces acting in the directions of σ_1 and σ_3. As before, F_1 can be resolved into a vertical F_{1Z} and a horizontal F_{1X} component (for simplicity in writing the equations, we will use two subindexes in a force so the first subindex marks the main force to which the component belongs, and the second subindex the direction of application of the force (for instance F_{1X} is the force component of F_1 in the x, horizontal direction). Similarly, F_3 can be resolved into the horizontal F_{3X} and vertical F_{3Z} components. Starting with the horizontal forces, $F_S + F_{3X}$, both pushing the prism to the left has to be equal to F_{1X}, which pushes the prism to the right (Equation 20; Fig. 3.69d), so that $F_S = F_{1X} - F_{3X}$. Similarly, the vertical force component F_N, which is trying to move the prism down, has to be equal to the two forces F_{1Z} and F_{3Z} which try to move the prism up, so $F_N = F_{1Z} + F_{3Z}$ (Equation 21; Fig. 3.69c). The final conditions for the derivation (Fig. 3.69e) has the main stresses and all the traction components involved. All surfaces are as a function of A, so that the surface perpendicular to σ_1 is $A \cos \theta$ and the surface perpendicular to σ_3 is $A \sin \theta$. If the values of the vertical and horizontal force components given in Equations 22–25 as a function of the corresponding tractions are substituted in Equations 20 and 21 (in the form $F = \boldsymbol{\sigma} A$; for example, for F_{1Z}, $\sigma_1 \cos \theta$ is the vertical traction or stress component and $A \cos \theta$ is the surface on which the traction acts), we obtain Equation 27, for the normal $(\boldsymbol{\sigma}_n)$ and Equation 29 for the shear $(\boldsymbol{\tau})$ traction components. Considering the trigonometric identities specified in the gray boxes in Fig. 3.69e and simplifying, we obtain the fundamental stress equations for both components. In the fundamental stress equations the term $(\sigma_1 + \sigma_3)/2$ represents the mean normal stress taking into account all traction values and may be regarded as a circle in 2D or a sphere in 3D. The mean normal stress represents a value for the hydrostatic condition which will be responsible for changes in area or volume of rock bodies as compression or tension is applied. The term $(\sigma_1 - \sigma_3)/2$ is the

deviatoric stress, which gives the maximum value for the shear stress since $\sin 2\theta$ cannot be >1 in Equation 29 (Fig. 3.69e). Finally $(\sigma_1 - \sigma_3)$ is the differential stress, which reflects the eccentricity of the stress ellipse or ellipsoid. The bigger the differential stress the more eccentric the ellipse or ellipsoid. This part of the stress tensor will be responsible for distortions or changes in shape in the objects.

3.13.7 The Mohr circle

The Mohr circle, named after C. O. Mohr, is a useful tool to graphically represent the state of stress at a point in 2D or 3D; all possible values for normal and shear stresses are confined by the circle. The Mohr circle allows us to visualize and calculate quickly the normal and shear components of a traction over any surface forming any angle θ with respect to the normal of σ_1, if the magnitudes of σ_1 and σ_3 are known (Fig. 3.70). In other words, having established the values of σ_1 and σ_3 we can choose any surface and read directly from the diagram the values of the normal and shear stress components, which is much quicker than calculating the components from the respective equations or extracting them from the stress tensor by geometric measurements. This kind of diagram has many applications but is particularly useful in fracture analysis (Section 4.14). The Mohr diagram consists of a coordinate system, which represents magnitudes of the shear stress $(\boldsymbol{\tau})$ on the vertical axis and normal stress $(\boldsymbol{\sigma}_n)$ on the horizontal axis for any traction acting over a surface. From the origin, at zero, the right-hand-side of the $\boldsymbol{\sigma}_n$ axis will represent positive or compressive stresses, whereas the left side will be the negative or tensile normal stresses. Considering now the vertical axis, $\boldsymbol{\tau}$, the upper part from the origin will be positive, left-handed, shear and the lower part will be negative, right-handed, shear. In the general case, the state of stress is represented by a circle centered at any point along the horizontal axis, with coordinates $(\sigma_1 + \sigma_3)/2$ and $\tau = 0$, so the circle intercepts the axis at two points. The point at the right end will be the biggest normal stress σ_1 and the point at the left side will be the smaller normal stress σ_3.

As you can see, the Mohr circle provides a very intuitive way to visualize the fundamental stress equations (Fig. 3.70b). The diameter of the circle is the differential stress $(\sigma_1 - \sigma_3)$. The larger the circle the bigger will be the differential stress and the ability to cause distortions. The center of the circle will have the value of the mean normal stress $\sigma_n = (\sigma_1 + \sigma_3)/2$ (and $\tau = 0$). Any radius of the circle will have the value of the deviatoric stress $(\sigma_1 - \sigma_3)/2$. Note that the segment defined by the projection of the radius into the x-axis (from the center to σ_n) has the value $[(\sigma_1 - \sigma_3)/2] \cos 2\theta$, which added to the mean normal stress gives the equation for the normal stress component of a given traction (Equation 27; Fig. 3.69e). Following the same line of thought, it is easy to see that the distance τ in the diagram has the value $[(\sigma_1 - \sigma_3)/2] \sin 2\theta$, which is the equation for the shear stress component (Equation 29 in Fig. 3.69e).

To represent and calculate the shear and normal components of any traction acting over a surface inclined at angle θ, we have to draw a radius forming an angle 2θ starting at the zero value located at the point occupied by σ_1. In Fig. 3.71, two cubes projected in a plane (2D view) show

surfaces inclined at angle θ with respect to the normal to σ_1. The cube on the left (Fig. 3.71a) shows an anticlockwise angle, whereas the cube on the right (Fig. 3.71b) shows a clockwise angle. We will proceed up the circle if the angle is anticlockwise (from σ_1 to the normal of the surface), as in case (a) which corresponds to a left-handed shear stress, and down the circle, into the negative area, if the angle is clockwise corresponding in this case to a right-handed shear stress. From the points intercepting the circle at the end of the radius we can read the values at the coordinate axis.

A numerical example to illustrate the representation of stresses and how to use Mohr circles is illustrated in

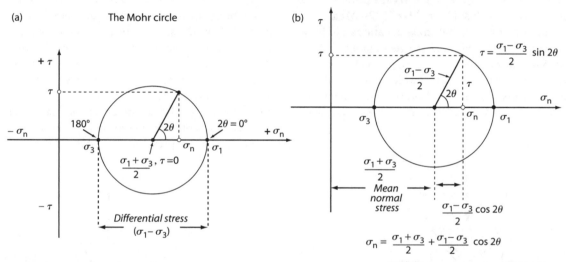

Fig. 3.70 (a) The Mohr circle is a tool to visualize and calculate stress components. It is represented as a circle positioned in a reference coordinate axis (σ_n, τ) and centered at a point whose values are the average normal stress $(\sigma_1 + \sigma_3)/2$ and $\tau = 0$. (b) Stress at a point and relationship to the fundmental stress equations.

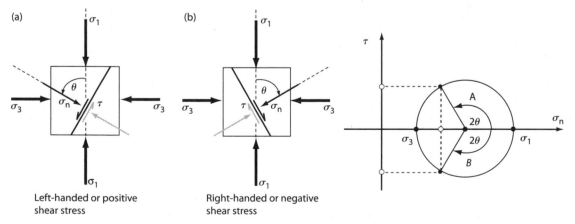

Fig. 3.71 Representation of clockwise and anticlockwise θ angles in the Mohr circle for positive and negative shear stresses.

Fig. 3.72. Initial data are $\sigma_1 = 40$ MPa, $\sigma_2 = 20$ MPa, and $\theta = 60°$. First the positions of σ_1 and σ_3 are marked in the x-axis at 40 and 20 MPa respectively; then both points are joined by a circle with extreme values 20 and 40 MPa, centered in the x-axis at the mean stress value $(\sigma_1 + \sigma_3)/2 = 30$ MPa. The value σ_n and τ of the surface orientated 60° with respect to the normal to σ_1 can be obtained by plotting the line measuring a double angle $(2\theta = 120°)$, in this case in the upper half of the circle, as the considered angle is positive (anticlockwise). Once the line is plotted, the point p (the intersection of the line with the circle), has coordinates (σ_n, τ) which can be obtained by reading the respective values at the coordinate axes. The values of σ_n and τ can also be calculated by using the fundamental stress equations explained earlier. Resulting values are $\sigma_n = 25$ MPa and $\tau = 8.7$ MPa. The differential stress can be easily calculated as $\sigma_1 - \sigma_3 = 20$ MPa.

Considering the fundamental stress equations and looking at the Mohr circles (Fig. 3.73) it is obvious that the maximum possible value for the shear stress is given by the value of the Mohr circle radius which has the value of the deviatoric stress $\sigma_1 - \sigma_3/2$. Two planes of maximum shear stress occur, each at 45° from σ_1 (θ is 45° too) since

in the equation $\tau = [(\sigma_1 - \sigma_3)/2]\sin 2\theta$, the value of $\sin 2\theta = 1$. In the numerical example (Fig. 3.72) the maximum value for τ corresponding to planes at $2\theta = 90°$ will be 10 MPa.

3.13.8 States of stress

Different states of stress are possible depending on the values of the nine tensor components (three normal tractions and six shear tractions). Some of those will be discussed briefly; they are represented by distinctive Mohr circles (Fig. 3.74). Definitions of the states of stress are relevant for rheological studies and fracture mechanics (Sections 3.15 and 4.14).

Hydrostatic stress or *hydrostatic pressure* (Section 3.5) is the state of stress characteristic of fluids in which all the tractions have the same value in all directions of space and so it is not possible to define directions for the principal stress axis as $\sigma_1 = \sigma_2 = \sigma_3 =$ pressure (Fig. 3.74a). All tractions are compressive and normal stresses. Hydrostatic stress is characterized by the absence of shear stress and consequently, in the Mohr circle this particular kind of stress is represented by a point in the x-axis corresponding to the value of the pressure. The *lithostatic stress*, applied to

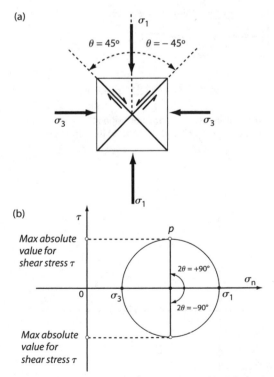

Fig. 3.72 The planes of maximum hear stress are those located at an angle $\theta = 45°$ (and also are 45° apart from the main principal stress σ_1).

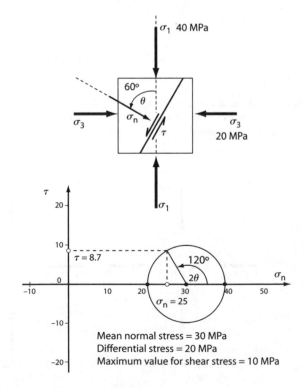

Fig. 3.73 A numerical example for the use of the Mohr circle.

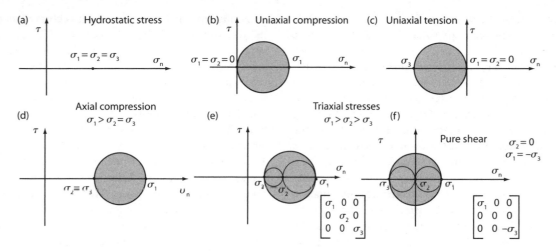

Fig. 3.74 Mohr circle diagrams for different states of stress.(modified from Twiss and Moors, 1992).

rocks in which all the tractions have the same value as the vertical load, should similarly be represented by a point.

Uniaxial compression occurs when a body of rock is compressed in a unique direction and unconfined laterally. This situation is not common in Nature but is used broadly in experiments in the laboratory to test mechanical properties in rocks and other materials applied not only to structural geology but also engineering problems. The principal stresses in uniaxial compression are $\sigma_1 > \sigma_2 = \sigma_3 = 0$ and are represented by a Mohr circle tangent to the τ-axis and in the right, positive side (Fig. 3.74b).

Uniaxial tension is produced by pulling a rock body in one direction. As for uniaxial compression, it is a favorite experimental condition in rock mechanics. The main stress axes have values $\sigma_1 = \sigma_2 = 0 > \sigma_3$. The Mohr circle will be tangent to the τ-axis but will be located in the negative normal stress field at the left side (Fig. 3.74c). Notice that both the uniaxial compression and extension allows the definition of a differential stress and so there are shear stresses $\tau \neq 0$ for different directions.

Axial stresses, both extensional and compressive, are those in which there is a confining pressure and an applied stress of different value in one direction. An *axial compression* is defined by $\sigma_1 > \sigma_2 = \sigma_3 > 0$, where $\sigma_2 = \sigma_3 > 0$ is the value of the confining pressure; whereas an axial extension or extensional stress is characterized by $\sigma_1 = \sigma_2 > \sigma_3 > 0$ (Fig. 3.74d). Both axial stresses are plotted in the right-hand-side of the Mohr diagram as all values for the normal stresses are positive.

Triaxial stresses are those in which all three principal axes have different values either positive or negative: $\sigma_1 > \sigma_2 > \sigma_3$. Triaxial stresses are represented by three circles, the bigger one confined between σ_1 and σ_3 and the other smaller ones included in the bigger circle, one confined between σ_1 and σ_2 and the other between σ_2 and σ_3 (Fig. 3.74e). A special example of triaxial stress is *pure shear stress* in which all main stresses are different, but $\sigma_2 = 0$ and $\sigma_1 = -\sigma_3$ (Fig. 3.74f); note that the surfaces corresponding to the maximum value of shear stress are those whose value for the normal stress is zero.

3.14 Solid strain

3.14.1 Kinematics of solid deformation

Deformation kinematics is the study of the reconstruction of movement that takes place during rock deformation at all scales of observation. Kinematics is not concerned with magnitude and orientation of stresses in terms of dynamics, it just describes the displacements that the rock suffers as stresses are applied to them. First of all, it is important to distinguish between the concepts of rigid and nonrigid

deformation. When stresses are applied to a volume of rock the outcome depends pretty much not only on the magnitude of the stress but also on the nature of the rock and external circumstances such as temperature, confining pressure by loading, fluid pressure in the pores, and so on. If a certain threshold of inner resistance of the rock or rock strength is attained, deformation occurs, otherwise the rock remains unaltered. If the rock is affected by applied stress, several alternative situations can happen. There are some rocks that are absolutely reluctant to change their shape or volume, preferring to shatter into pieces before experiencing any change. This is the case for a *rigid* rock body. Other rocks may respond to stress by changing their internal structure by shape or volume change. The resulting alternatives depend upon the differential stress: hydrostatic stress potentially causes changes in volume and stress ellipses cause changes in shape, the more eccentric the ellipse the more accentuated the change. In such cases we consider the deformation to be *nonrigid*. In kinematics four basic movements or displacements are defined: translation, rotation, distortion, and dilation. Any combination of the four displacements can be produced. The first two displacements characterize rigid deformation whereas the latter two correspond to nonrigid deformation. Rigid deformation does not cause the object to change its internal or external configuration but to move around, whereas nonrigid deformation causes the object to change its internal structure so that different, regularly spaced, points defined in the object will change position with respect to each other in a way that the spacing does not keep the original proportions and relative positions. This kind of deformation is called *strain*: strained bodies change shape or volume due to a nonrigid deformation.

3.14.2   Rigid deformation

To define the movements produced during deformation, displacement vectors must be defined in a coordinate frame (Fig. 3.75). Rigid deformation causes the rocks to move linearly or change position (translation) or orientation (rotation) but the internal structure, volume, or shape of the object is not altered and so any selected points in the object remain in the same position with respect to each other (Fig. 3.76a).

During *translation*, displacement vectors of all points in the rock are parallel (Figs 3.76a and 3.77b) and have the same magnitude and orientation. To describe displacement vectors in a translation, three parameters are used: vector magnitude, which reflects the transport or displaced distance; direction of movement in an orientated plane; and finally, sense of movement or transport of the rock or body. A real-world example of a translation is the vertical displacement of rock blocks on a flat surface fault(Section 4.15). To define the movement of this block we can measure the total amount of displacement as 5 m, the orientation of the displacement as a plane of strike 160° E and dip 60° and sense of displacement, for instance toward the SE. On a major scale, the linear displacement of the continent India toward the North during the Cenozoic (Fig. 3.78) can be approximated by a translation (although strictly speaking a rotation over a spherical surface).

Rotation is a rigid displacement involving turning of an object, that is, its orientation changes, about a rotation axis (Fig. 3.77c); it is a form of vorticity (Section 3.8). Examples of solid rotations are the movement of blocks on *listric* (arcuate) faults (Fig. 3.79) or the rotation of the

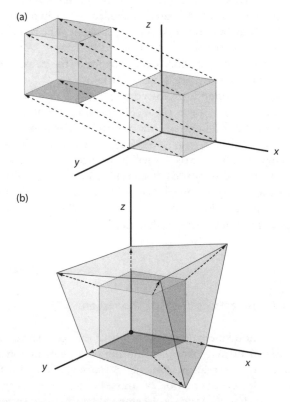

Fig. 3.75 Displacement vector of a point framed on a coordinate system.

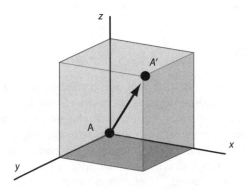

Fig. 3.76 Examples of rigid deformation (a) and nonrigid deformation (b) showing the displacement vectors of some reference points (the corners of the cube).

Iberian Peninsula during the Mesozoic, when the opening of the Bay of Biscay took place. Displacement vectors have different magnitudes, being zero in the rotation axis and systematically larger away from the rotation axis. Vector orientations are different in a predictable way. To fully describe movement by rotation, the orientation of the rotation axis has to be defined the magnitude of the rotation in degrees, and the sense of rotation as clockwise

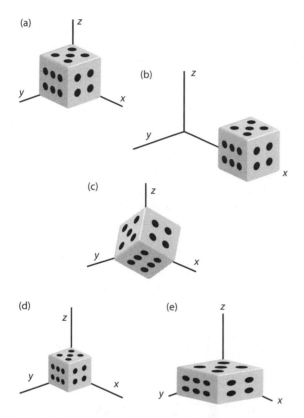

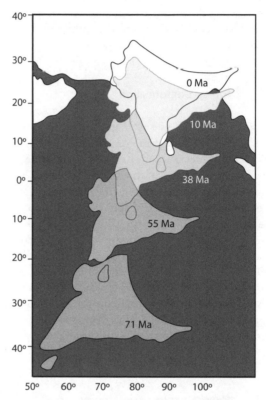

Fig. 3.77 The four basic displacements described in kinematics. (a) Original nondeformed object, a dice; (b) translation (linear movement); (c) rotation (change in orientation); (d) dilation (change in volume); (e) distortion (change in shape).

Fig. 3.78 Displacements at plate tectonics scale. The linear translation or drift of India toward the north.

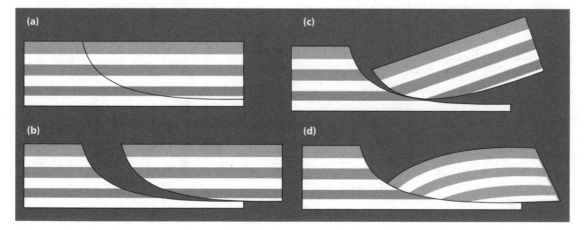

Fig. 3.79 Several examples of displacements applied to kinematic models of listric faults (Sectoin 4.15). (a) Original pre-fault position; (b) rigid translation of a block; (c) rigid rotation of a block; (d) Nonrigid distortion or inner deformation of the block to form a roll-over.

or anticlockwise. This last measurement depends on the observation point. Imagine looking at a spinning wheel from one end of the rotational axis; if this movement is clockwise, when we turn to the other side of the wheel and look at it, the rotational movement will be anticlockwise. To avoid such indetermination it is generally agreed that the observer will look at the axis in the sense of plunge, that is, looking down the axis. In the case of a horizontal rotation axis, the position of the observer has to be specified.

3.14.3   Nonrigid deformation

The evaluation of changes in shape and volume of objects or rock bodies is called *strain analysis*. This technique is a useful tool in kinematics and can be used if the original shape and size of objects in the rock are known. *Distortion* is a nonrigid deformation (Figs 3.77e) which causes the objects or rocks to change shape, preserving the original volume. Displacement vectors can have various orientations and magnitudes. *Dilation* is a displacement which produces a change in volume (Fig. 3.77d). The object can be enlarged or contracted so that the original regularly spaced points in the object are separated or get closer but they still preserve the original proportions so that no change in the shape of the object is produced. The displacement vectors converge or diverge radially from a point in a regular way.

3.14.4   Homogeneous strain analysis

Strain (generically represented as ε) can be *homogeneous* or *inhomogeneous* (Fig. 3.80) and, as with stresses, can be analyzed in 2D or 3D. Homogeneous strain is constant along the whole object. This means that all small portions of the deformed body have the same deformation proportions as the whole body. Homogeneous strain satisfies two conditions: (i) originally straight lines in the unstrained object remain straight after deformation, which applies also in 3D, where originally plane surfaces remain plane after deformation; and (ii) parallel lines or surfaces in the original object remain parallel after deformation. A consequence is that in 2D any circular object is transformed into a perfect ellipse and in 3D any sphere is converted into a perfect ellipsoid. Deformation is inhomogeneous when there are variable gradients of displacement through the object and so straight lines are changed into curved lines and originally parallel lines converge or diverge after being strained. In Fig. 3.80 a nondeformed square object,

depicting two mutually perpendicular black lines and an inner circle as decoration, is represented. In (b) the square has been deformed by flattening (pure shear), in (c) by shearing (simple shear), and in (d) by volume loss or dilation. Note that (b), (c), and (d) have suffered a homogeneous strain, as the original straight parallel lines remain parallel and straight after deformation and also the circle has been transformed into a perfect ellipse, whereas (e) and (f) suffered an inhomogeneous strain as the original straight lines became curved as in (e) or originally parallel lines converge or diverge in the deformed state as in (f). Note also that the circle has not become an ellipse but shows an irregular shape in (e) and (f). A good example of inhomogeneous strain is the generation of *folds* (Section 4.16) as the originally straight lines of rock layers become curved. In Nature when tectonic deformation takes place, nonhomogeneous deformations are most likely to occur and strain analysis cannot be used to predict deformation following simple mathematical rules. Nonetheless, inhomogeneous deformed terrains can be analyzed separately by dividing them into discrete homogeneous domains. Then, the whole deformation can be evaluated and *strain gradients* assessed.

In homogeneous strain different parameters are used to state the differences in the length of lines and angular changes between lines. To determine changes in length of straight lines, two of the most commonly used parameters

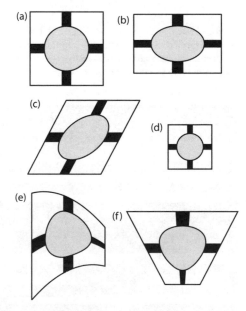

Fig. 3.80 Homogeneous and inhomogeneous strain. (a) is the object before deformation; (b), (c), and (d) show homogeneous strain; and (e) and (f) inhomogeneous strain.

Changes in length of lines

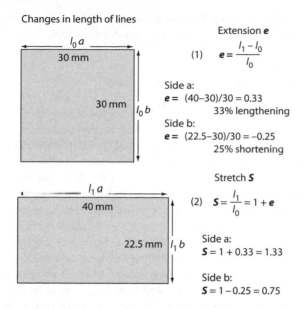

Extension **e**

$$(1) \quad \boldsymbol{e} = \frac{l_1 - l_0}{l_0}$$

Side a:
e = (40–30)/30 = 0.33
33% lengthening

Side b:
e = (22.5–30)/30 = –0.25
25% shortening

Stretch **S**

$$(2) \quad \boldsymbol{S} = \frac{l_1}{l_0} = 1 + \boldsymbol{e}$$

Side a:
S = 1 + 0.33 = 1.33

Side b:
S = 1 – 0.25 = 0.75

Fig. 3.81 Measuring the changes in the length of lines *a* and *b* in a square transformed into a rectangle.

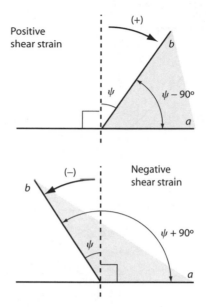

Fig. 3.82 Notation for negative and positive shear angles. The dashed line represents the original nondeformed state of line *b* perpendicular to line *a*. Clockwise rotation is considered positive and anticlockwise rotation negative.

are extension (*e*) and stretch (*S*). *Extension* is a nondimensional parameter defined by subtracting the original nondeformed length (l_0) from the final deformed length (l_1), and normalizing it by dividing the result by the original length (Fig. 3.81) so it becomes a proportion. All lines which are longer than the original after deformation (as line *a* in Fig. 3.81) have positive values of *e*, and all lines suffering shortening (like line *b* in Fig. 3.81) have negative values of *e*. Values of *e* range between −1 for maximum shortening and +∞ for maximum stretching, zero being the value before deformation. Maximum shortening will give a final deformed length l_1 equal to zero (a very theoretical situation unlikely to happen, but even so it will be the minimum possible value), and so $(0 - l_0)/l_0 = -l_0/l_0 = -1$. Maximum possible stretching (still very theoretical) will give a value for l_1 of +∞ and so the limit value for the extension *e* is $(\infty - l_0)/l_0 = +\infty$. Extension can also be given in percentages multiplying *e* by 100. *Stretch, S*, is also a nondimensional parameter used to measure shortening or lengthening of lines contained in an object. *S* is the ratio between the length of the line after deformation l_1 and the original length of the line l_0 before deformation (Fig. 3.81). The stretch can be also obtained by adding 1 to the extension *e* (since $e = (l_1 - l_0/l_0) = (l_1/l_0 - l_0/l_0) = (l_1/l_0 - 1)$ and finally $e + 1 = l_1/l_0 = S$). The value of *S* for a nonstrained body is 1, as in this case l_0 equals l_1. Limiting values for *S* are 0 for maximum shortening to +∞ for maximum stretching. Maximum shortening will happen when the final deformed length l_1 became 0 as

$0/l_0 = 0$, although this situation is not likely to happen. Maximum stretching will be produced when the deformed length l_1 is +∞ as $\infty/l_0 = +\infty$. The square of the stretch is also used to measure linear strain, when it is called *quadratic elongation* or *quadratic extension* (λ).

Angular changes between lines can be determined if the object contains two mutually perpendicular lines before deformation. When a line rotates and makes an angle different to 90°, the difference in angle from the original perpendicular position to the deformed position is called *angular shear*, ψ. The tangent of the angle ψ represents the angular deformation which is called *shear strain*, γ (Fig. 3.82). Positive and negative angular shear has to be defined to discriminate sense of rotation from the original nondeformed state. Defining clockwise and anticlockwise sense of rotation for reference has the same problem as for rigid rotations described previously, the observation point has to be defined. There is not a general agreement of which sense is the positive or the negative and both choices can be found in the literature. Another way of defining the sign is to consider that when the resulting deformed angle is bigger than 90° (90° + ψ) the shear is considered negative, and when the total deformed angle is smaller than 90° (90° − ψ) the shear will be defined as positive. Two examples of how to measure the shear strain are shown in Figs 3.83b and c. The object before deformation has a square shape (Fig. 3.83a). One of the examples

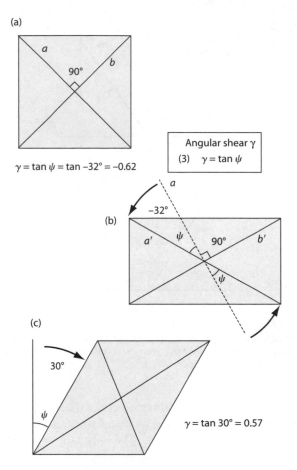

(a)

$$\gamma = \tan \psi = \tan -32° = -0.62$$

Angular shear γ

(3) $\gamma = \tan \psi$

(b)

(c)

$$\gamma = \tan 30° = 0.57$$

Fig. 3.83 Examples of measuring the angular shear in a square object (a) deformed into a rectangle by pure shear (b) and a rhomboid (c) by simple shear.

shows the object deformed by homogeneous flattening (Fig. 3.83b), the sides of the square remain perpendicular to each other, but notice that both diagonals of the square (*a* and *b* in Fig. 3.83) initially at 90° have experienced deformation by shear strain moving to the positions *a'* and *b'* in the deformed objects. To determine the angular shear, the original perpendicular situation of both lines has to be reconstructed and then the angle ψ can be measured. In this case the line *a* has suffered a negative shear with respect to *b*. The line *a* perpendicular to *b'* has been plotted and the angle between *a* and *a'* defines the angular shear. The shear strain is calculated by the tangent of the angle ψ. The same procedure can be followed to calculate the strain angle between both lines plotting a line normal to *a'*. Note that in this case the shear will be positive as the angle between *b* and *a'* is smaller than 90°. In the second example (Fig. 3.83c) the square has been deformed by

simple shear into a rhomboid, both the sides and the diagonals of the square have experienced shear strain.

3.14.5 Pure shear and simple shear

Pure shear and simple shear are examples of homogeneous strain where a distortion is produced while maintaining the original area (2D) or volume (3D) of the object. Both types of strain give parallelograms from original cubes. *Pure shear* or homogeneous flattening is a distortion which converts an original reference square object into a rectangle when pressed from two opposite sides. The shortening produced is compensated by a perpendicular lengthening (Fig. 3.84a; see also Figs 3.81 and 3.83b). Any line in the object orientated in the flattening direction or normal to it does not suffer angular shear strain, whereas any pair of perpendicular lines in the object inclined respect to these directions suffer shear strain (like the diagonals or the rectangle in Fig. 3.83b or the two normal to each other radii in the circle in Fig. 3.84a).

Simple shear is another kind of distortion that transforms the initial shape of a square object into a rhomboid, so that all the displacement vectors are parallel to each other and also to two of the mutually parallel sides of the rhomboid. All vectors will be pointing in one direction, known as *shear direction*. All discrete surfaces which slide with respect to each other in the shear direction are named *shear planes*, as will happen in a deck of cards lying on a table when the upper card is pushed with the hand (Fig. 3.84c). The two sides of the rhomboid normal to the displacement vectors will suffer a rotation defining an angular shear ψ and will also suffer extension, whereas the sides parallel to the shear planes will not rotate and will remain unaltered in length as the cards do when we displace them parallel to the table. Note the difference with respect to the rectangle formed by pure shear whose sides do not suffer shear strain. Note also that any circle represented inside the square is transformed into an ellipse in both simple and pure shear. To measure strain, fossils or other objects of regular shape and size can be used. If the original proportions and lengths of different parts in the body of a particular species are known (Fig. 3.85a), it is possible to determine linear strain for the rocks in which they are contained. Figure 3.85 shows an example of homogeneous deformation in trilobites (fossil arthropods) deformed by simple shear (Fig. 3.85b) and pure shear (Fig. 3.85c). Note how two originally perpendicular lines in the specimen, in this case the cephalon (head) and the bilateral symmetry axis of the body, can be used to measure the shear angle and to calculate shear strain.

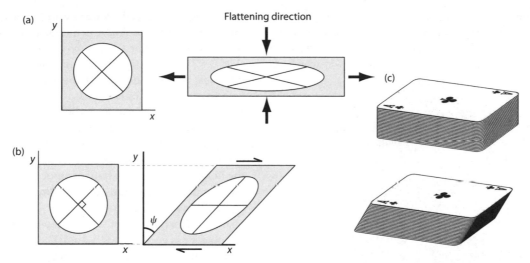

Fig. 3.84 Pure shear (a) and simple shear (b) are two examples of homogeneous strain. Both consist of distortions (no area or volume changes are produced; (c) Simple shear has been classically compared to the shearing of a new card deck whose cards slide with respect to each other when pushed (or sheared) by hand in one direction.

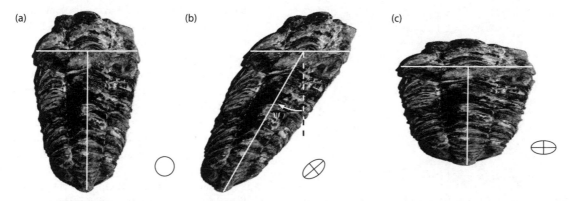

Fig. 3.85 Homogeneous deformation in fossil trilobites: (a) nondeformed specimen; (b) deformed by simple shear. Note how two originally perpendicular lines such as the cephalon base and the bilateral symmetry axis can be used to measure the shear angle and calculate shear strain (c) deformed by pure shear. If the original size and proportions of three species is known, linear strain can be established.

3.14.6 The strain ellipse and ellipsoid

We have seen earlier (Figs 3.80 and 3.84) that when homogeneous deformation occurs any circle is transformed into a perfectly regular ellipse. This ellipse describes the change in length for any direction in the object after strain; it is called the strain ellipse. For instance, the major axis of the ellipse, which is named S_1 (or e_1), is the direction of maximum lengthening and so the circle is mostly enlarged in this direction. Any other lines having different positions on the strained objects which are parallel to the major axis of the ellipse suffer the maximum stretch or extension. Similarly the minor axis of the ellipse, which is known as S_3 (or e_3) is the direction where the lines have been shortened most, and so the values of the extension e and the stretch S are minimum. The axis of the strain ellipse S_1 and S_3 are

known as the *principal axis of the strain ellipse* and are mutually perpendicular. The strain ellipse records not only the directions of maximum and minimum stretch or extension but also the magnitudes and proportions of both parameters in any direction. To understand the values of the axis of the strain ellipse imagine the homogeneous deformation of a circle having a radius of magnitude 1, which will be the value of l_0 (Fig. 3.86a). Now, if we apply the simple equation of the stretch S (Equation 2; Fig. 3.81) whereas for a given direction, the stretch e is the difference in length between the radius of the ellipse and the initial undeformed circle of radius 1 it is easy to see that the major axis of the ellipse will have the value of S_1 and the minor axis the value of S_3. An important property of the strain axes is that they are mutually perpendicular lines which were also perpendicular before strain. Thus the directions of maximum and minimum extension or stretch correspond to directions that do not experience (at that point) shear strain (note the analogy with the stress ellipse in which the principal stress axis are directions in which no

shear stress is produced). Shear strain can be determined in the ellipse by two originally perpendicular lines, radii R of the circle and the line tangent to a radius at the perimeter (Fig. 3.87). In (a), before deformation, the tangent line to the circle is perpendicular to the radius R. In (b), after deformation, the lines are no longer normal to each other and so an angular shear ψ can be measured and the shear strain calculated, as explained earlier.

In strain analysis two different kinds of ellipse can be defined, (i) the *instantaneous strain ellipse* which defines the homogeneous strain state of an object in a small increment of deformation and (ii) the *finite strain ellipse* which represents the final deformation state or the sum of all the phases and increments of instantaneous deformations that the object has gone through. In 3D a regular ellipsoid will develop with three principal axes of the strain ellipsoid, namely S_1, S_2, and S_3, being $S_1 \geq S_2 \geq S_3$.

Now that we have introduced the concept of the strain ellipse we can return to the previous examples of homogeneous deformation and have a look at the behavior of the strain axes. In the example of Fig. 3.84 the familiar square is depicted again showing an inner circle (Fig. 3.88). Two mutually perpendicular radius of the circle have been marked as decoration. Note that a pure shear strain has been produced in four different steps. The circle has become an ellipse that, as the radius of the circle has a value of 1, will represent the strain ellipsoid, with two principal axes S_1 and S_3. Note that when a pure shear is produced the orientation of the principal strain axis remains the same through all steps in deformation and so it is called *coaxial strain* (Fig. 3.88). This means that the directions of maximum and minimum extension are preserved with successive stages of flattening. A very different situation happens when simple shear occurs (Fig. 3.89): the axes of the strain ellipsoid rotate in the shear direction

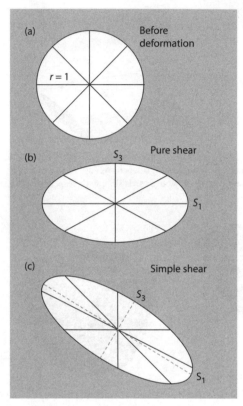

Fig. 3.86 The stress ellipse in 2D strain analysis reflects the state of strain of an object and represents the homogeneous deformation of a circle of radius = 1 transformed into an ellipsoid. As I_0 is 1, $S_1 = I_1/1 = I_1$ which represents the stretch S of the long axis. Similarly $S_3 = I_1$ giving the stretch S of the short axis.

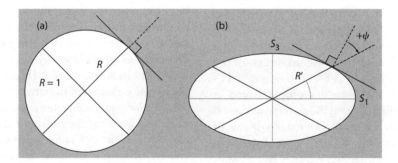

Fig. 3.87 Shear strain in the strain ellipse. In (a), before deformation, the tangent line to the circle is perpendicular to the radius R. In (b), after deformation, both lines are not normal to each other, the angular shear ψ can be obtained and the shear strain calculated by tracing a normal line to the tangent to the circle at the point where R' intercepts the circle, and measuring the angle ψ. The shear strain can be calculated as $y = \tan \psi$.

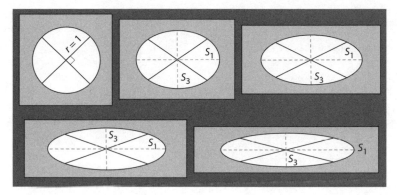

Fig. 3.88 Pure shear is considered to be a coaxial strain since the orientation of the axes of the strain ellipse S_1 and S_3 remain with the same orientation through progressively more deformed situations.

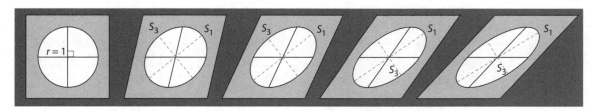

Fig. 3.89 Simple shear can be described as a noncoaxial strain as the orientation of the principal strain axis of the strain ellipse S_1 and S_3 rotates with progressive steps on deformation.

and so the strain is *noncoaxial*. The orientation of the axes is not maintained, which means that the directions of maximum and minimum extension rotate progressively with time.

3.14.7 The fundamental strain equations and the Mohr circles for strain

For any strained body the shear strain and the stretch can be calculated for any line forming an angle ϕ with respect to the principal strain axis S_1 if the orientation and values of S_1 and S_3 are known. As in the case of stress analysis the approach can be taken in 2D or 3D. Although it is important to remember that the physical meanings of strain and stress are completely different, the equations have the same mathematical form (Fig. 3.90) and can be derived using a similar approach. The fundamental strain equations allow the calculation of changes in length of lines, defined by means of the *reciprocal quadratic elongation* ($\lambda' = 1/\lambda$), of any line forming an angle ϕ with respect to the direction of maximum stretch S_1. To illustrate the use and significance of the Mohr circles for strain, an original circle of radius R can be used (as in Fig. 3.87).

After deformation, the circle has suffered strain and developed into a perfect ellipse by homogeneous flattening (Fig. 3.90b). The original radius R of the circle, with length l_0, has been elongated and will correspond to the radius R' of the ellipse of length l_1. Comparing both lengths, the extension, e (Equation 1; Fig. 3.81) or the stretch, S (Equation 2; Fig. 3.81), can be easily calculated. The reciprocal quadratic elongation can be directly obtained as $\lambda' = (l_0/l_1)^2$. The angular deformation can be measured by plotting the tangent to the ellipse at the point p, where the radius intercepts the ellipse perimeter, then plotting the normal to the tangent, and measuring the angle with respect to the radius R' (Fig. 3.90b).

The Mohr circle strain diagram is a useful tool to graphically represent and calculate strain parameters, following a similar procedure that was used to calculate stress components. In this case the ratio between the shear strain and the quadratic elongation (γ/λ) is represented on the vertical axis and the reciprocal quadratic elongation (λ') on the horizontal axis (Fig. 3.90c). The γ/λ ratio is an index of the relative importance of the angular deformation *versus* the linear elongation. When the ratio is very small, changes in length dominate, in fact when the ratio equals zero, there is no shear strain, which coincides with the directions of the principal strain axis. In homogeneous strain of pure

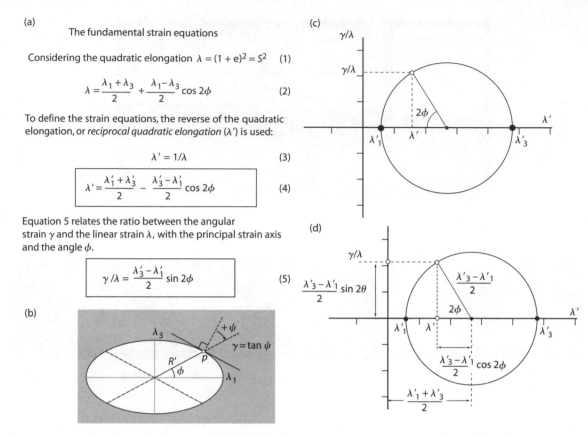

(a)

The fundamental strain equations

Considering the quadratic elongation $\lambda = (1 + e)^2 = S^2$ (1)

$$\lambda = \frac{\lambda_1 + \lambda_3}{2} + \frac{\lambda_1 - \lambda_3}{2} \cos 2\phi \quad (2)$$

To define the strain equations, the reverse of the quadratic elongation, or *reciprocal quadratic elongation* (λ') is used:

$$\lambda' = 1/\lambda \quad (3)$$

$$\lambda' = \frac{\lambda_1' + \lambda_3'}{2} - \frac{\lambda_3' - \lambda_1'}{2} \cos 2\phi \quad (4)$$

Equation 5 relates the ratio between the angular strain γ and the linear strain λ, with the principal strain axis and the angle ϕ.

$$\gamma / \lambda = \frac{\lambda_3' - \lambda_1'}{2} \sin 2\phi \quad (5)$$

(b)

(c)

(d)

Fig. 3.90 (a) The fundamental strain equations. The Mohr circles for strain display graphically the relations between the γ/λ ratio and the reciprocal quadratic elongation λ'. The γ/λ ratio reflects the relative importance of angular deformation versus linear deformation; (b) strained circle into an ellipse; (c) the Mohr circle strain diagram; and (d) Relation between the Mohr circle and the fundamental strain equations.

distortion, where there is no change in volume or area, there are two directions that suffer no finite stretch, where the value of $\lambda = 1$. Finally two directions of maximum shear strain are present, corresponding to the lines forming an angle $\phi = 45°$ with respect to S_1. The Mohr circle for strain has an obvious relation to the fundamental strain equations (Equations 4 and 5; Fig. 3.90a) as shown in Fig. 3.90d.

To plot the circle in the coordinate axes, the reciprocal values λ_1' and λ_3' of λ_1 and λ_3 are first calculated and represented along the horizontal axis. The circle will have a diameter $\lambda_3' - \lambda_1'$ and the center will have coordinates $(\lambda_1' + \lambda_3')/2, 0$. Note that as the expressions on the x-axis are the reciprocal quadratic elongations, the maximum

value, at the right end of the circle, corresponds to λ_3' and the minimum, at the left end, to λ_1'. Once the circle is plotted, it is possible to calculate the values of γ/λ and λ' (Fig. 3.90d) for any line forming an angle ϕ respect to the direction of the major principal strain axis S_1. The line is plotted from the center of the circle at the angle 2ϕ subtended from λ_1' into the upper half of the circle if the angle is positive or into the lower half if it is negative. The coordinates of the point of intersection between the line and the circle have the values $\gamma/\lambda, \lambda'$. Through λ' the value of λ and then that of S can be calculated. Knowing λ it is also possible to calculate γ and finally the angle of shear strain ψ.

3.15 Rheology

3.15.1 Rheological models

The reaction of rock bodies and other materials to applied stresses can only be observed and studied through laboratory

experiments: the study of strain–stress relations or how the rocks or other materials respond to stress under certain conditions is the concern of *rheology*. Different kinds of experiments are possible, generally undertaken on centimeter-scale cylindrical rock samples. Both tensional (the sample is generally pulled along the long axis) and compressive (sample is pushed down the long axis) stresses can be applied, both in laterally confined (axial or triaxial tests) or unconfined conditions (uniaxial texts). Experiments involving the application of a constant load to a rock sample and observing changes in strain with time are called *creep tests*. Experimental results are analyzed graphically by plotting stress, (σ), against strain, (ε), or strain rate (dε/dt), the latter obtained by dividing the strain by time (Fig. 3.91). Simple mathematical models can be developed for different regimes of rheological behavior. Stress is usually represented as the differential stress ($\sigma_1 - \sigma_3$). Other important variables are lithology, temperature, confining pressure, and the presence of fluids in the interstitial pores causing *pore fluid pressures*. There are three different pure rheological behavioral regimes: *elastic, plastic,* and *viscous* (Fig. 3.91). Elastic and plastic are characteristic of solids whereas viscous behavior is characteristic of fluids. Solids under certain conditions, for example, under the effect of permanent stresses, can behave in a viscous way. Elastic, plastic, and viscous are end members of a more complex suite of behaviors. Several combinations are possible, such as *visco-elastic, elastic–plastic,* and so on.

3.15.2 Elastic model

Elastic deformation is characterized by a linear relationship in stress–strain space. This means that the relation between the applied stress and the strain produced is proportional (Fig. 3.91a). An instantaneous applied stress is followed instantly by a certain level of strain. The larger the stress the larger the strain, up to a point at which the rock can be distorted no further and it breaks. This limit is called the elastic boundary and represents the maximum stress that the rock can suffer before fracturing. If the stress is released before reaching the elastic limit such that no fractures are produced, elastic deformation disappears. In other words, elastic strained bodies recover their original shape when forces are no longer applied. The classical analog model is a spring (Fig. 3.92a). The spring at repose represents the nondeformed elastic object. When a load is

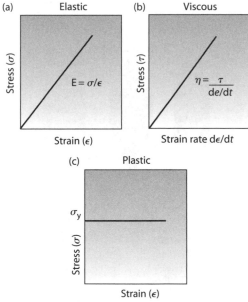

Fig. 3.91 Strain/stress diagrams for different rheological behaviors. (a) Elastic solids show linear relations. The slope of the straight line is the Young's modulus; (b) viscous behavior is characteristic of fluids. Fluids deform continuously at a constant rate for a certain stress value. The slope of the line is the viscosity (η); (c) plastics will not deform under a critical stress value or yield stress (σ_y).

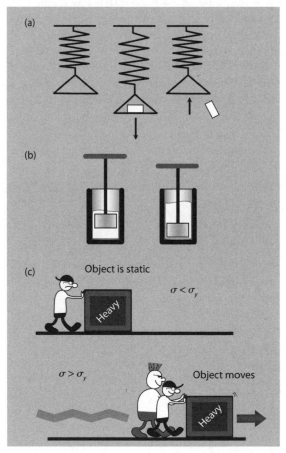

Fig. 3.92 Classical analogical models for (a) elastic behavior, is compared to a spring; (b) viscous behavior is compared to a hydraulic piston or dashpots; and (c) plastic behavior, like moving a load by a flat surface with an initial resistance to slide.

added to the spring in one of the extremes (as a dynamometer) or it is pulled by one of the edges, it will stretch by the action of the applied force. The bigger the load, or the more the spring is pulled on the extremes, the longer it becomes by stretching. When the spring is released or liberated from the load in one of the extremes the spring returns to the original length.

Elasticity in rocks is defined by several parameters; the most commonly used being *Young's modulus* (E) and the *Poisson coefficient* (ν). Young's modulus is a measure of the resistance to elastic deformation which is reflected in the linear relation between the stress (σ) and the strain (ε): $E = \sigma/\varepsilon$ (Fig. 3.93a). This linear relation, which was observed initially by Hooke in the mid-seventeenth century by applying tensile stresses to a rod and measuring the extension, is commonly known as Hooke's Law. Considering that all parameters used to measure strain (stretch, extension, or quadratic elongation) are dimensionless, the Young's modulus is measured in stress units ($N\,m^{-2}$, MPa) and has negative values of the order of -10^4 or -10^5. The reason why the values are negative is because the applied stress is extensional and hence has a negative value, and the strain produced is a lengthening, which is conventionally considered positive. Not all rocks follow Hooke's Law; some deviations occur but they are small enough so a characteristic value of E can be defined for most rock types (Fig. 3.93a). A high absolute value for the Young's modulus means that the level of strain produced is small for the amount of stress applied, whereas low values indicate higher deformation levels for a certain amount of stress. Rigid solids produce high Young's modulus values as they are very reluctant to change shape or volume. Rigid materials experience *brittle* deformation when their mechanical resistance is exceeded by the applied stress level at the elastic boundary.

When applying uniaxial compressional tests to rock samples, vertical shortening may be accompanied by some horizontal expansion. The Poisson coefficient (ν) shows the relation between the lateral dilation or barreling of a rock sample and the longitudinal shortening produced by loading: thus $\nu = \varepsilon_{\text{lateral}}/\varepsilon_{\text{longitudinal}}$ and it can be seen that Poisson's coefficient is dimensionless (Fig. 3.93b). When stresses are applied, if there is no volume loss, the sample has to thicken sideways to account for the vertical shortening. Typically, the sample should develop a barrel form (nonhomogeneous deformation) or increase its surface area as it expands laterally. For perfect, incompressible, isotropic, and homogeneous materials which compensate the shortening by lateral dilation without volume loss, the Poisson's coefficient is 0.5; although values for natural materials are generally smaller (Fig. 3.93b). In very rigid rock bodies, the lateral expansion may be very limited or not occur at all; in this case there is a volume loss and

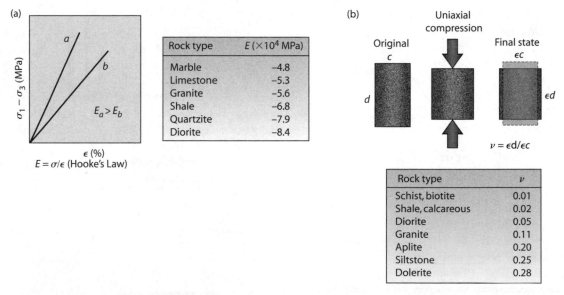

Fig. 3.93 Elastic parameters. (a) The Young's modulus describes the slope of the stress/strain straight line, being a measure of the rock resistance to elastic deformation. Line *a* has a higher value of Young's modulus (E_a) being more rigid than line *b* (Young modulus E_b) (i.e. it is less strained for the same stress values); (b) Poisson's coefficient relates the proportion in which the rock deforms laterally when it is compressed vertically. Comparing the original and final lengths before and after deformation strain ε can be calculated and the Poisson's ratio established.

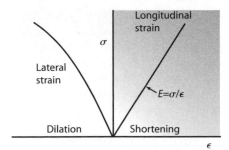

Fig. 3.94 Longitudinal and lateral strain experienced by a rock sample when an uniaxial compression is applied. The relation between both strains may not be linear as in this case, and the Poisson's ratio is not constant, it varies slightly for different stress values.

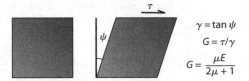

Fig. 3.95 Shear or rigidity modulus (G) and its relation to Young's modulus (E) and Poisson's number (m).

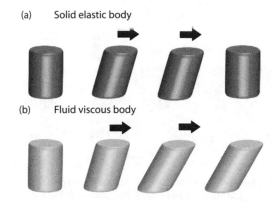

Fig. 3.96 (a) Solid elastic bodies are strained proportionally to the applied forces. If the intensity of the force is maintained there is not a further increase in strain. When the force is released the object recovers the initial shape; (b) The viscous fluid will be deformed when a shear force is exerted, but even when the intensity of the force is maintained, an increment in deformation will occur, defining a strain rate. That is why strain rate is used in rheological plots instead of strain as in solids. The fluid body will remain deformed permanently once the force is removed.

elastic stresses have to be accumulated somehow. Rock samples will fragment at the elastic limit after experiencing very little lateral strain when the Poisson's ratio is very small (close to zero). The reciprocal to the Poisson's coefficient is called the *Poisson's number* $m = 1/\nu$. This number is also constant for any material, and so the relation between the longitudinal and lateral strains have a linear relation. Nonetheless, as in the case of Young's modulus there may be slight variations in the linear trend of Poisson's coefficient (Fig. 3.94). It is important to remember that experiments to establish elasticity relationships under unconfined uniaxial stress conditions allow the rock samples to expand laterally. In the crust, any cube of rock that we can define is not only subject to a vertical load due to gravity but also due to adjacent cubes of rock in every direction and is not free to expand laterally; in such cases complex stress/strain relations can develop.

Other elastic parameters are the *rigidity modulus* (G) and the *bulk modulus* (K). The rigidity modulus or *shear modulus* is the ratio between the shear stress (τ) and the shear strain (γ) in a cube of isotropic material subjected to simple shear: $G = \tau/\gamma$ (Fig. 3.95). G is another measure of the resistance to deformation by shear stress, in a way equivalent to the viscosity in fluids. The bulk modulus (K) relates the change in hydrostatic pressure (P) in a block of isotropic material and the change in volume (V) that it experiences consequently: $K = \mathrm{d}P/\mathrm{d}V$. The reverse to the bulk modulus is the *compressibility* ($1/K$).

3.15.3 Viscous model

Viscous deformation occurs in fluids (Sections 3.9 and 3.10); fluids have no shear strength and will flow when shear stresses, even infinitesimal, are applied. One of the chief differences between an elastic solid and a viscous fluid is that when a shear stress is applied to a piece of elastic material it causes an increment of strain proportional to the stress, if the same level of stress is maintained no further deformation is achieved (Fig. 3.96a). In fluids when a shear stress (τ) is applied the material suffers certain amount of strain but the fluid keeps deforming with time even when the stress is maintained with the same value (Fig. 3.96b). In this case a level of stress gives way to a strain rate ($\mathrm{d}\varepsilon/\mathrm{d}t$), not a simple increment of strain as in the elastic solids. Higher stress values will give way to higher strain rates, so the fluid will deform at more speed. As in elastic materials there is no initial resistance to deformation even when stresses acting are very small, but the deformations are permanent in the viscous fluid case (Fig. 3.97a,b).

As we have seen earlier (Sections 3.9 and 3.10) the parameter relating stress to strain rate is the *coefficient of dynamic viscosity* or simply *viscosity* (η): $\eta = \tau/(\mathrm{d}\varepsilon/\mathrm{d}t)$, which is

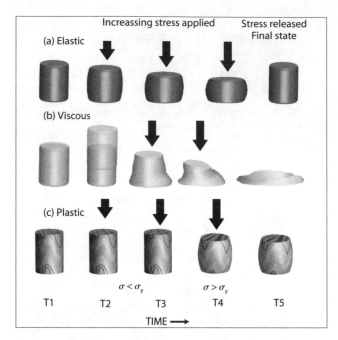

Fig. 3.97 Strain of different materials with time (stages $T1$ to $T5$) applying increasing levels of stress: (a) Elastic solids show discrete strain increments with increasing stress levels (linear relation); strain is reversible once the stress is removed ($T5$); (b) Viscous fluids flow faster (higher strain rates) with increasing stress; the deformation is permanent once the stress is released; (c) Plastic solids will not deform until a critical threshold or yield stress is overpassed (at $T4$ in this case). Deformation is nonreversible (at $T5$).

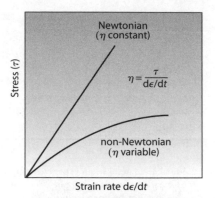

Fluid	η (Pa)
Water (30º)	$0.8 \cdot 10^{-3}$
Oil	0.08
Basalt lava	10^2
Rhyolite lava	10^8
Salt	10^{16}
Asthenosphere	10^{22}

Fig. 3.98 Viscosity is the resistance of a fluid to deform or flow: it is the slope of the curve stress/strain rate. Fluids showing linear relations (constant viscosity) are Newtonian. Fluids with nonlinear relation (η variable) are non-Newtonian. The table shows the values of viscosity (η) for some viscous materials.

measured in Pascals. Fluids that show a linear relation between the stress and the strain rate, and so have a constant viscosity, are called Newtonian. Fluids, whose viscosity changes with the level of stress are called non-Newtonian (Fig. 3.98). Viscous behavior is generally compared to a piston or a dashpot containing some hydraulic fluid (Fig. 3.92b). The fluid is pressed by the piston (creating a stress or loading) and the fluid moves up and down a cylinder, producing permanent deformation; the quicker the piston moves the more rapid the fluid deforms or flows up and down. The viscosity can be described as the resistance of the fluid to movement. High viscosity fluids are more difficult to displace by the piston up and down the cylinder. For non-Newtonian fluids (Fig. 3.98) as the piston is pushed more and more strongly in equal increments of added stress the rate of movement or strain rate rapidly increases in a non-linear fashion.

3.15.4 Plastic model

Plastic deformation is characteristic of materials which do not deform immediately when a stress is applied. A certain level of stress is required to start deformation, as the material has an initial resistance to deformation. This stress value is called yield stress $\boldsymbol{\sigma_y}$ (Fig. 3.91c). After the yield stress is reached the body of material will be deformed a big deal instantaneously, and the deformation will be permanent and without a loss of internal coherence. So, two important differences with respect to elastic behavior are that the strain is not directly proportional to the stress, as there is an initial resistance, and that the strain is not reversible as in elastic behavior (Fig. 3.97). An analogical model for plastic deformation is that of a heavy load resting on the floor (Fig. 3.92c). If the force used to slide the load along a surface is not big enough, the load will not budge. This would depend on the frictional resistance exerted by the surface. Once the frictional resistance, and so the yield stress, is exceeded, the load will slide easily and the movement can be maintained indefinitely as long as the force is sustained at the same level over the critical threshold or yield stress. The load will not go back on its own! So the deformation is not reversible (Fig. 3.92c).

3.15.5 Combined rheological models

Elastic, viscous, and plastic models correspond to simple mathematical relationships which apply to materials under ideal conditions; they are considered homogeneous (the rock has the same composition in all its volume) and isotropic (the rock has the same physical properties in all directions). Rocks are rarely completely homogeneous or isotropic due to their granular/crystalline nature and because of the presence of defects and irregularities in the crystalline structure, as well as layers, foliations, fractures, and so on. Nevertheless, although such aberrations would be important in small samples, on a large scale, when large volumes are being considered, rocks can be sometimes regarded as homogeneous. Usually, however, natural rheological behavior corresponds to a combination of two or even three different simple models, such as elastic–plastic, visco-elastic, visco-plastic, or elastic–visco-plastic. Also materials can respond to stress differently depending on the time of application (as in instantaneous loads *versus* long-term loads).

A well-known example of a combined rheological model is the elastic–plastic (Prandtl material) (Fig. 3.99); it shows an initial elastic field of behavior where the strain is recoverable, but once a yield stress (σ_y) value is reached the material behaves in a plastic way. The analogical model is a spring (elastic) attached to a heavy load (plastic) moving over a rough surface (Fig. 3.99b). The spring will deform instantly whereas the load remains in place until the yield stress is reached, then the load will move; after releasing the force, the spring will recover the original shape but the longitudinal translation is not recoverable. Elastic–plastic materials thus recover part of the strain (initial elastic) but partly remain under permanent strain (plastic). Remember that in a pure elastic material, permanent strain does not occur and after the elastic limit is reached the rock breaks (*b*, Fig. 3.99c; line I) whereas in a Prandtl material there is a nonreversible strain (*c*, Fig. 3.99c, line II). Once the plastic limit is reached, the material can then break but only after suffering some permanent barreling (*d*, Fig. 3.99c, line II).

Visco-elastic models correspond to solids (called *Maxwell materials*) which have no initial resistance to

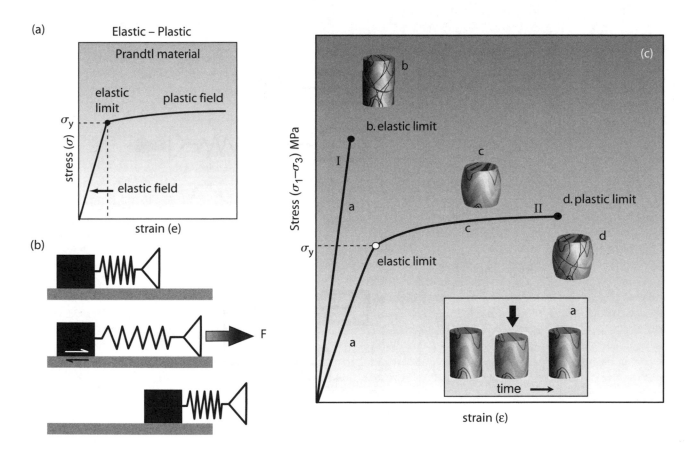

Fig. 3.99 (a) Elastic–plastic material shows an initial elastic field characterized by recoverable deformation strain followed by a plastic field in which the strain is permanent. The boundary between both fields is the elastic limit located at the yield stress value (σ_y); (b) The analog model is a load attached to a spring; (c) Part of the strain is recovered (the length of the spring) and part is not (the displacement of the load).

strain as in both elastic and viscous models (Fig. 3.100a). Part of the strain will recover following an elastic behavior but part will remain permanently deformed. Maxwell solids behave elastically when the stresses are short lived, like a ball of silicon putty that bounces elastically on the floor when thrown with some force; but will accumulate permanent deformations at a constant rate if the stress or load (like the proper weight of the material) is applied for a longer time. Visco-elastic models can be represented by a spring attached longitudinally to a dashpot (Fig. 3.100b). The spring will provide the recoverable strain whereas the dashpot will supply the nonrecoverable strain when a pulling force is applied parallel to the system.

Visco-plastic materials (called *Bingham plastics*) only behave like viscous fluids after reaching a yield stress, the strain rate subsequently being proportional to the stress; initially the material does not respond to the applied stress as for plastic solids (Fig. 3.100c). The analogy will be in this case a dashpot attached in parallel to a load sliding on a surface with an initial resistance to movement; once the load is in motion it behaves in viscous fashion.

3.15.6 Ductile and brittle deformation

From the different rheological models discussed above it can be concluded that there are several kinds of deformation. First, strain produced when loads are applied can be reversible; this is characteristic of elastic behavior as in the elastic curves or elastic–plastic materials (*a*, Fig. 3.99c) when small stress increments are applied. Deformations can also be nonreversible, which means that once the load is released the rock will be deformed permanently. Deformation is said to be *ductile* when rocks or other solids are strained permanently without fracturing, which happens in plastic or elastic–plastic materials once the elastic limit or yield strength (stress value which separates the elastic and plastic fields) is reached (as *c* in Fig. 3.99c).

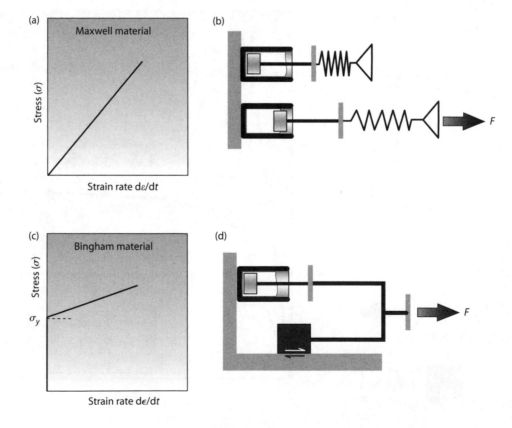

Fig. 3.100 (a) Visco-elastic or Maxwell materials have a recoverable strain part belonging to the elastic component and a permanent strain due to the viscous behavior like a spring attached to a dashpot (b); (c) visco-plastic or Bingham materials behave in a viscous way but after reaching a critical stress value or yield stress (σ_y) like a dashpot linked to a load moving on a rough surface (d).

Nonetheless, *ductile* is a general, descriptive term that does not involve a specific rheological behavior or strain mechanism. It is *not* a synonymous term for *plastic*, which is a very well-defined and particular rheological behavior. Strains produced during plastic deformations are larger in magnitude than those produced in the elastic field and are generally formed by dislocations of the crystalline lattices and/or diffusive processes. Ductile deformations are also called *ductile flows* as the material deforms or flows in a solid state (as a glacier sliding downslope does, Section 6.7.5). Examples of ductile deformation in rocks are the formation of folds and salt diapirs. Rocks have a limited ability to change their shape or volume, which also depends on such external parameters as the temperature, confining pressure, and so on.

Brittle deformation happens when the internal strength of rocks is exceeded by stresses; they bust, so internal cohesion is lost in well-defined surfaces or fractures. Brittle deformation can occur after the elastic limit is exceeded not only in pure elastic bodies (*b*, Fig. 3.99c) but also when the stresses reach the plastic limit after some ductile deformation has taken place. Such samples will be permanently deformed and also fractured (*d*, Fig. 3.99c).

3.15.7 Parameters controlling rock deformation

Lithology (rock type) is a variable which may cause diverse modes of stress–strain behavior. Different rocks or substances may need different rheological models with which to describe their deformation. *Competency* is a qualitative term used to describe rocks in terms of their inner strength or capacity for deformation. Rocks which deform easily and generally in a ductile way are described as *incompetent*, such as salts, shale, mudstone, or marble. Strong or *competent* rocks are those which are more difficult to deform, such as quartzite, granite, quartz sandstones, or fresh basalts. Competent rocks are stiffer and deform generally in a brittle way. Nevertheless, competency depends not only on lithology but also on temperature, confining pressure, pore pressure, strain rate, time of application of the stress, etc. To compare competencies of different kinds of rocks, experiments must take place at equal temperatures and confining pressures.

Temperature has particularly important effects in rheological behavior (Fig. 3.101). Comparing several experiments on samples of the same lithology under the same conditions of confining pressure, it is possible to compare stress–strain relations at different temperatures. At higher temperatures, rocks behave in a more ductile way, so competence is reduced and fractures are more difficult to produce. For rocks that are elastic at low temperatures a

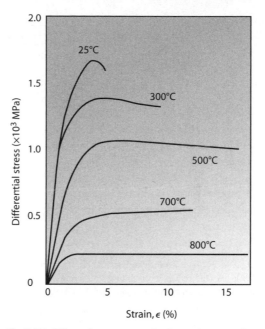

Fig. 3.101 Effect of temperature in the strain–stress diagram for basalts under the same confining pressure (5 kbars).

plastic field can develop. In elastic–plastic materials, temperature lowers the elastic limit, which is thus reached at lower stress levels. Rocks may also behave in a viscous way at high temperatures if the applied stresses are long lasting.

Confining pressure (lithostatic or hydrostatic pressure acting on all sides of a rock volume) can be simulated in laboratory experiments by introducing some fluid that exerts a certain amount of pressure in the sample (triaxial tests) in addition to that provided by the compressive load, and by isolating the sample in a constraining metal jacket to discriminate and separate the effects of the pore pressure in the rock. Experiments carried out on samples of the same lithology and at the same temperature show that higher confining pressures increase the yield strength in a rock, and also the plastic field, so fracturing, if it happens, occurs after more intense straining (Fig. 3.102). This means that rocks became more ductile at higher levels of confining pressure.

When there is fluid trapped in the rock pores, it exerts an additional hydrostatic pressure which has the effect of counteracting the confining pressure by the same value of the fluid pressure in the pores. The state of stress is lowered and an *effective stress tensor* can be defined by subtracting the values of the fluid stresses from those of the solid normal stresses (Fig. 3.103). The Mohr circle moves toward lower values by an amount equal to the *pore pressure* (p_f) sustained by the fluid. Thus, when fluids are present in the pores the effect is the same as lowering the confining

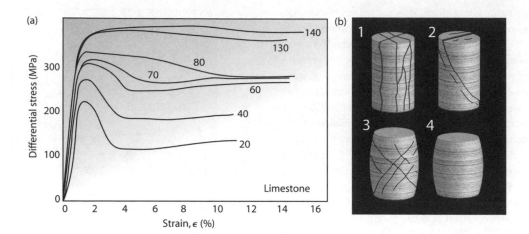

Fig. 3.102 (a) Strain–stress diagram showing several curves corresponding to limestone samples of the same composition at different confining pressures (in MPa); (b) Differences in confining pressure give way to different fracturing or deformation modes. Confining pressure from samples (from 0.1 to 35 MPa in the fractured samples and 100 MPa for the ductile flow).

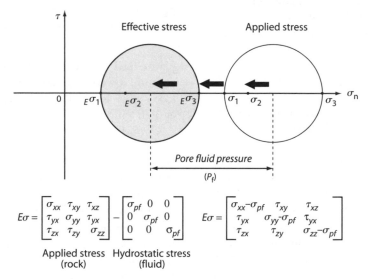

Fig. 3.103 When there is some pressurized fluid in the rock pores, part of the stress is absorbed. The state of stress is lowered and an effective stress tensor can be defined subtracting the values of the normal stresses from those of the fluid. The Mohr circle moves toward lower values by an amount equal to the pore pressure (P_f) sustained by the fluid.

pressure in the rocks, so that ductility decreases and fractures are produced more easily. Being hydrostatic in nature, the effectiveness of the normal stresses is lowered but the shear stresses remain unaltered. The control of pore pressure in the rocks is of key importance in fracture formation and will be discussed in some more detail in Section 4.14.

Other important factors are the time of application of the stresses: the instantaneous or long-term application of a certain level of stress may cause different rheological behaviors, like the case of the silicon putty discussed earlier. Rock strength decreases when the stresses are applied for long times under small differential stresses (creep experiments). Also in relation to time, the rates of loading (velocity of increased loading in the experiments) also have important implications for the production of strain. In a single experiment, the rate of strain is generally maintained constant but the rates of strain can be changed from one experiment to another. When changes in strain are produced rapidly (high loading rates) the rock samples become ductile and break at higher stress levels.

Further reading

P.M. Fishbane *et al.*'s *Physics for Scientists and Engineers: Extended Version* (Prentice-Hall, 1993) is again invaluable. Many good things of oceanographic interest can be found in the exceptionally clear work of S. Pond and G.L. Pickard – *Introductory Dynamical Oceanography* (Pergamon, 1983), while R. McIlveen's *Fundamentals of Weather and Climate* (Stanley Thornes, 1998) is good on the atmospheric side. A more advanced text is D.J. Furbish's *Fluid Physics in Geology* (Oxford, 1997). G.V. Middeton and P.R. Wilcox's *Mechanics in the Earth and Environmental Sciences* has a broad appeal at intermediate level and is very thorough. The best introduction to solid stress and strain is in G.H. Davies and S.J. Reynolds's *Structural Geology of Rocks and Regions* (Wiley, 1996); R.J. Twise and E.M. Moores's *Structural Geology* (1992) and J.G. Ramsay and M. Huber's *The Techniques of Modern Structural Geology*, vol. 1: *Strain Analysis* (Academic Press, 1993) are classics on structural geology for advanced studies on solid stress. W.D. Means's *Stress and Strain* (Springer-Verlag, 1976) takes a careful and rigorous course through the basics of the subject.

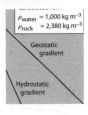

$\rho_{water} = 1,000 \text{ kg m}^{-3}$
$\rho_{rock} = 2,380 \text{ kg m}^{-3}$

Geostatic gradient

Hydrostatic gradient

4 Flow, deformation, and transport

4.1 The origin of large-scale fluid flow

Earth is a busy planet: what are the origins of all this motion? Generally, we know the answer from Newton's First Law that objects will move uniformly or remain stationary unless some external force is applied. The uniform motion of fluids must therefore involve a balance of forces in whatever fluid we are dealing with. In order to try to predict the magnitude of the motion we must solve the equations of motion that we discussed previously (Section 3.12). Bulk flow (in the continuum sense, ignoring random molecular movement) involves motion of discrete fluid masses from place to place; the masses must therefore transport energy: mechanical energy as fluid momentum and thermal energy as fluid heat. There will also be energy transfers between the two processes, via the principle of the mechanical equivalent of heat energy and the First Law of Thermodynamics (Section 2.2, conservation of energy). For the moment we shall ignore the transport of heat energy (see Sections 4.18–4.20) since radiation and conduction introduce the very molecular-scale motions that we wish to ignore for initial simplicity and generality of approach.

4.1.1 Very general questions

1 How does fluid flow originate on, above, and within the Earth? For example, atmospheric winds and ocean currents originate somewhere and flow from place to place for certain reasons. This raises the question of "start-up," or the beginnings of action and reaction.
2 If fluid flow occurs from place A to place B, what happens to the fluid that was previously at place A? For example, the arrival of an air mass must displace the air mass previously present. This introduces the concept of an ambient medium within which all flows must occur.
3 How does moving fluid interact with stationary or moving ambient fluid? For example, does the flow mix at all

with the ambient medium? If so, at what rate? How does the interaction look physically?
4 What is the origin and role of variation in flow velocity with time (unsteadiness problem)? It is to be expected that accelerations will be very much greater in the atmosphere than in the oceans and of negligible account in the solid earth (discounting volcanic eruptions and earthquakes). Why is this?

4.1.2 Horizontal pressure gradients and flow

Static pressure at a point in a fluid is equal in all directions (Section 3.5) and equals the local pressure due to the weight of fluid above. Notwithstanding the universal truth of Pascal's law, we saw in Section 3.5.3 that horizontal gradients in fluid pressure occur in both water and air. These cause flow at all scales when a suitable gradient exists. The simplest case to consider is flow from a fluid reservoir from orifices at different levels (Fig. 4.1). Here the flow occurs across the increasingly large pressure gradient with depth between hydrostatic reservoir pressure and the adjacent atmosphere.

The gradient of pressure in moving water (Fig. 4.1) is termed the *hydraulic gradient*, and the flow of subsurface water leads to the principle of artesian flow and the basis of our understanding of groundwater flow through the operation of Darcy's law (developed from the Bernoulli approach in Sections 4.13 and 6.7). The flow of a liquid down a sloping surface channel is also down the hydraulic gradient.

Similar principles inform our understanding of the slow flow of water through the upper part of the Earth's crust. Here, pressures may also be hydrostatic, despite the fluid held in rock being present in void space between solid rock particles and crystals (Fig. 4.2); this occurs when the rocks are porous to the extent that all adjacent pores communicate, as is commonly the case in sands or gravels. Severe lateral and vertical gradients arise when pores are closed by compaction, as in clayey rock; the hydrostatic condition now changes to the *geostatic condition* when pore pressures are greater due to the increased weight of overlying rock compared to a column of pore water (Fig. 4.3).

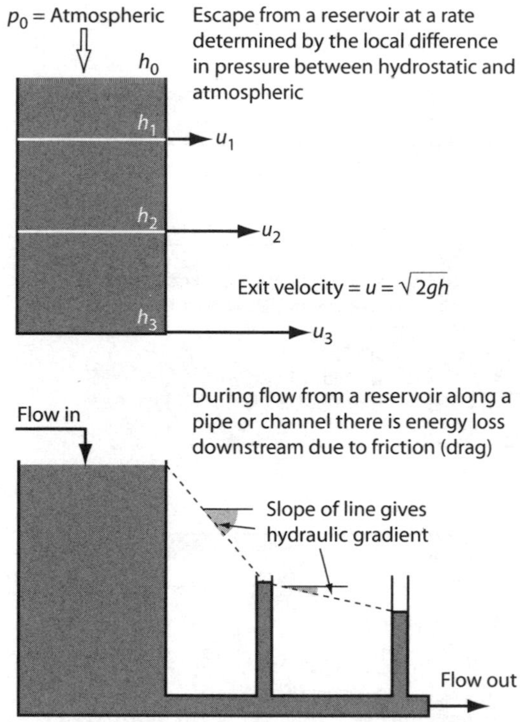

Interlayering of porous and nonporous rock then leads to high local pressure gradients down which subsurface fluids may move. In passage down an oil or gas exploration well, pressure may jump quickly from a hydrostatic trend toward

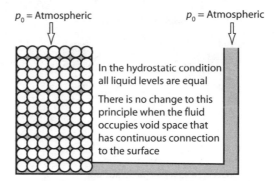

In the hydrostatic condition all liquid levels are equal

There is no change to this principle when the fluid occupies void space that has continuous connection to the surface

Fig. 4.2 The hydrostatic condition is equally valid for liquid in reservoirs or porous rock.

p_0 = Atmospheric

h_0

h_1 → u_1

h_2 → u_2

h_3 → u_3

Escape from a reservoir at a rate determined by the local difference in pressure between hydrostatic and atmospheric

Exit velocity = $u = \sqrt{2gh}$

Flow in

During flow from a reservoir along a pipe or channel there is energy loss downstream due to friction (drag)

Slope of line gives hydraulic gradient

Flow out

Fig. 4.1 Flows induced by hydrostatic pressure.

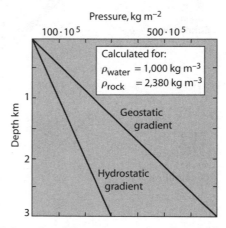

Pressure, kg m^{-2}

$100 \cdot 10^5$ $500 \cdot 10^5$

Depth km

Calculated for:
ρ_{water} = 1,000 kg m^{-3}
ρ_{rock} = 2,380 kg m^{-3}

Geostatic gradient

Hydrostatic gradient

Fig. 4.3 Hydrostatic and geostatic pressure gradients in the Earth's crust.

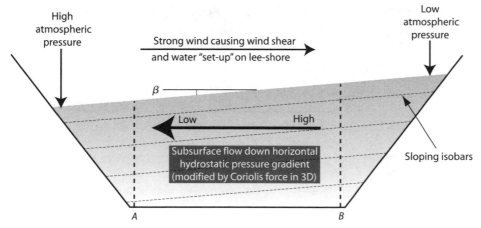

High atmospheric pressure

Low atmospheric pressure

Strong wind causing wind shear and water "set-up" on lee-shore

β

Low   High

Subsurface flow down horizontal hydrostatic pressure gradient (modified by Coriolis force in 3D)

Sloping isobars

A   B

Fig. 4.4 Barotropic flow due to a horizontal gradient in hydrostatic pressure caused and maintained by atmospheric dynamics. The spatial gradients in atmospheric pressure and wind shear may act together or separately. In both cases hydrostatic pressures above B are greater than hydrostatic pressures at all equivalent heights above A, by a constant gradient given by the water surface slope.

lithostatic, causing potentially disastrous consequences for the drill rig and possible "blowout." The regional hydraulic gradient drives the direction of migration of subsurface fluids like water and hydrocarbon. Pressures in partially molten rocks of the Earth's upper crust in crustal magma chambers (Section 5.1) may also vary between hydrostatic and geostatic values, with obvious implications for the forces occurring during volcanic eruptions.

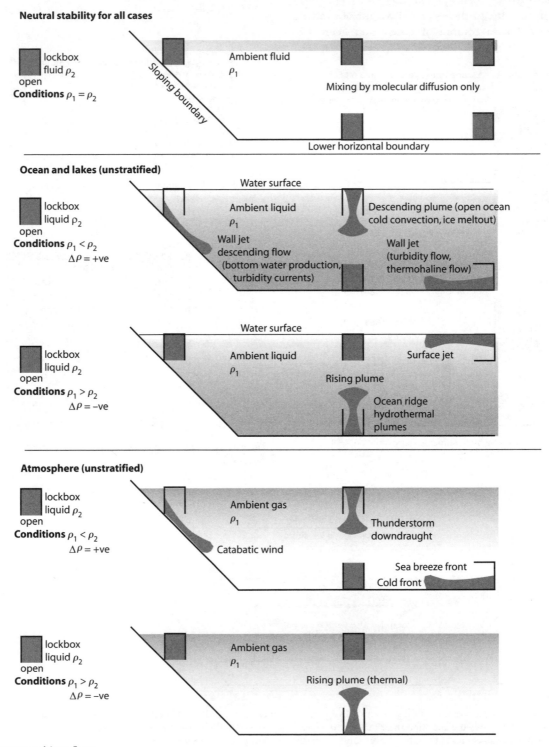

Fig. 4.5 Buoyancy-driven flows.

Table 4.1 Nomenclature and possible types of density currents.

	Gas +ve $\Delta\rho$ (e.g. cooler air)	Gas neutral	Gas −ve $\Delta\rho$ (e.g. warmer air)	Liquid +ve $\Delta\rho$ (e.g. Cooler/more saline/ suspensions of solids)	Liquid neutral	Liquid −ve $\Delta\rho$ (Warmer/ less saline)
Ambient gas	Sinking plume Bottom-spreading and undercutting current	Neutral stability No flow	Rising plume Interface-spreading jet	River flow downslope	NA	NA
Ambient liquid	NA	NA	Degassing bubbles in magma or lava	Sinking plume Bottom-spreading and undercutting wall jet	Neutral stability	Rising plume Spreading jet

4.1.3 Flow in the atmosphere and oceans

The atmosphere and oceans are in a constant state of flux, both experiencing "weather"; that is, the velocity of the ocean waters and atmosphere is unsteady with respect to either magnitude or direction over timescales of minutes to months. Here we briefly note that their longer-term *average* flow approximates to the *geostrophic* condition (see also Section 3.12). This is when pressure gradients are balanced by the Coriolis force alone, with no other forces involved: the fluid is assumed *ideal*, that is, inviscid. In terms of the relevant equations of motion, we have F(pressure) = F(Coriolis).

In the atmosphere, the pressure variations that cause geostrophic flow are up to 6 percent and are caused by lateral variations in air density between regional pressure cells like the Iceland Low or the Azores High in the northern hemisphere (see Fig. 3.21). Water density also varies with depth in the oceans but in the well-mixed surface layers of the open oceans this density variation is not so important. Regional ocean pressure gradients are set up due to variations in the elevation of the mean sea surface (Fig. 4.4), ignoring short-term topography due to storms, waves, and tides. The slopes involved are very small, up to 3 m over distances of a thousand kilometers or so, that is, gradients of $c.3 \cdot 10^{-6}$. These tiny gradients, in conjunction with the Coriolis force (not considered in Fig. 27.4), are quite sufficient to drive the entire average surface oceanic circulation (discussed in Sections 6.2 and 6.4).

4.1.4 Buoyancy/density flow

Many flows that take place in, on, and above the solid Earth occur because density contrasts, $\Delta\rho$, give rise to buoyancy forces (Section 3.6). The resulting flows are termed density or gravity currents. These may act between different parts of the same general state of matter (e.g. air, water, magma) or between different states of matter (e.g. water in and under air, gases in magma). We may illustrate the various possibilities for water and air by means of thought experiments with gravity lockboxes (Fig. 4.5). The lockbox is of unit volume with any side that can be opened instantaneously so that the contained fluid, air, or water, may be smoothly introduced within ambient masses of similar or different fluid. In all cases the gas phase has a lower density than the liquid phase. For simplicity, we examine the gravity lock in two dimension only, opening the locks in the top, bottom, or side as appropriate. The sketches show the expected flow direction as each box is opened; the types of flows possible are summarized in Table 4.1.

4.2 Fluid flow types

There is something immensely satisfying in discovering the efforts of pioneering scientists to reduce apparently complicated natural phenomena to simple essentials governed by some overall guiding principle. One such contribution that stands out in the area of fluid flow was that by Reynolds. Before Reynolds' contribution was published in 1883, it was generally recognized from observations in natural rivers, from experiments on flow in pipes (by Darcy), and from work on capillary flow in very narrow tubes (by Pouisseille, simulating the flow of blood in veins and arteries), that fluid flow could exhibit two basic kinds of behavior while in motion and that two flow "laws" must exist to explain the forces involved. In Reynolds' elegant words, "either the elements of the fluid follow one another along lines of motion which lead in the most direct manner to their destination, or they eddy about in sinuous paths the most indirect possible." In a series of careful experiments (Fig. 4.6), Reynolds visualized these flow types by

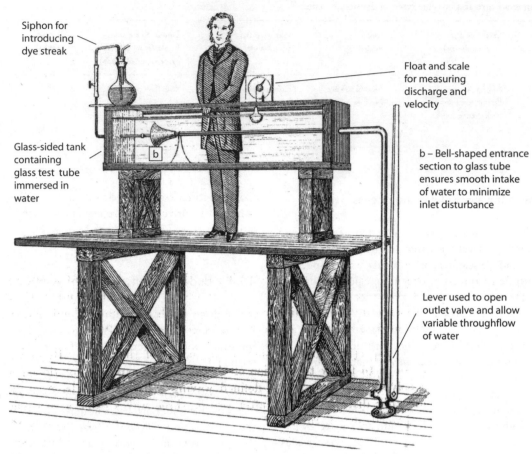

Siphon for introducing dye streak

Float and scale for measuring discharge and velocity

Glass-sided tank containing glass test tube immersed in water

b – Bell-shaped entrance section to glass tube ensures smooth intake of water to minimize inlet disturbance

Lever used to open outlet valve and allow variable throughflow of water

b

Fig. 4.6 Reynolds' apparatus as presented in his 1883 paper.

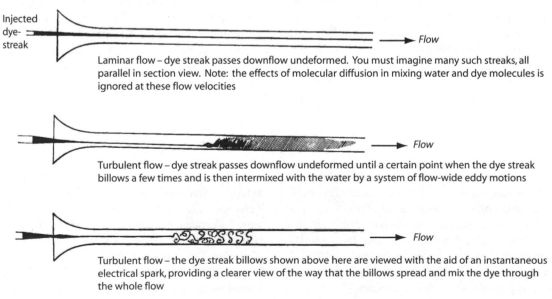

Injected dye-streak

Flow

Laminar flow – dye streak passes downflow undeformed. You must imagine many such streaks, all parallel in section view. Note: the effects of molecular diffusion in mixing water and dye molecules is ignored at these flow velocities

Flow

Turbulent flow – dye streak passes downflow undeformed until a certain point when the dye streak billows a few times and is then intermixed with the water by a system of flow-wide eddy motions

Flow

Turbulent flow – the dye streak billows shown above here are viewed with the aid of an instantaneous electrical spark, providing a clearer view of the way that the billows spread and mix the dye through the whole flow

Fig. 4.7 Details of flow patterns as sketched by Reynolds.

carefully introducing a dye streak into a steady flow of water through a transparent tube (Figs 4.7 and 4.8). At low flow velocities the dye streak extended down the tube as a straight line. This was his "direct" flow, what we nowadays call laminar flow. With increased velocity the dye streak was dispersed in eddies, eventually coloring the whole flow. This was Reynolds' "sinuous" motion, now known as turbulent flow. It was Reynolds' great contribution, first, to recognize the fundamental difference in the two flow types and, second, to investigate the dynamic significance of these. The latter process was not completed until he published another landmark paper in 1895 on turbulent stresses (see Section 3.11); more on these in Section 4.5.

4.2.1 Energy loss and flow type: Reynolds critical experiments

Concerning the forces involved, it was previously known that "The resistance is generally proportional to the square of the velocity, and when this is not the case it takes a simpler form and is proportional to the velocity." Reynolds approached the force problem both theoretically (or "philosophically" as he put it) and practically, in best physical tradition. His philosophical analysis was "that the general character of the motion of fluids in contact with solid surfaces depends on the relation between a physical constant of the fluid, and the product of the linear dimensions of the space occupied by the fluid, and the velocity." Designing the apparatus reproduced in Fig. 4.6, he

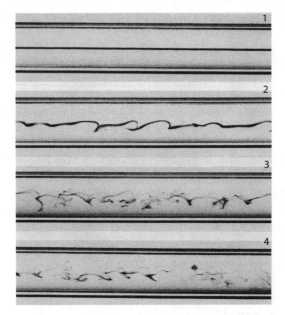

Fig. 4.8 Photographic records of laminar to turbulent transition in a pipe flow.

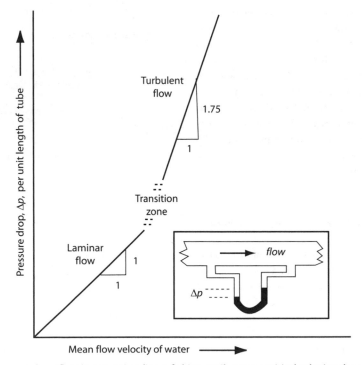

Fig. 4.9 The rate of pressure decrease downflow increases in a linear fashion until at some critical velocity, the rate of loss markedly increases as about the 1.8 power of velocity.

measured the pressure drop over a length of smooth pipe through which water was passed at various speeds. As we have seen in Sections 3.12 and 4.1, pressure drop is due to energy losses to heat as fluid moves, accompanied by conversion of potential to kinetic energy. Pressure loss per unit pipe length increased with velocity in a straightforward linear fashion, but at a certain point the losses began to increase more quickly, as the 1.8 power of the velocity (Fig. 4.9). Measurements confirmed Reynolds' intuition that this implied change in force balance was accompanied by the previously observed change in flow pattern from "direct" to "sinuous."

4.2.2 A general statement establishing universal flow types

Reynolds repeated the pipe experiments with different pipe diameters (the pipes always being smooth) and with the water at different temperatures so that viscosity, the "physical constant" noted in the quote above, could be varied. He showed that the critical velocity for the onset of turbulence was not the same for each experiment and that the change from laminar to turbulent flow occurred at a fixed value of a quantity of variables that has become known as the *Reynolds number* (*Re*) in honor of its discoverer. We may think of *Re* as a ratio of two forces acting in a fluid (Fig. 4.10). Viscous forces resist deformation: the greater the molecular viscosity, the greater the resistance. Inertial forces cause fluid acceleration. *Re* may be derived from first principles in this way as shown in Fig. 4.10. When viscous forces dominate, as say in the flow of liquefied mud or lava, then *Re* is small and the flow is laminar. When inertial forces dominate, as in atmospheric flow and most water flows, then *Re* will be large and the flow turbulent. For flows in pipes and channels, the critical value of *Re* for the laminar-turbulent transition usually lies between 500 and 2,000, but this depends upon entrance conditions and can be very much larger.

4.2.3 Turbulence is a property of a flow, not a fluid

We should be careful in any identification of laminar flow with high viscosity liquids alone. As Reynolds himself took great pains to emphasize, a flow state is dependent upon four parameters of flow, not just one. Thus a very low density or very low velocity of flow has the same reducing effect on *Re* as a very high viscosity. As Shapiro, the author of a classic introductory text in fluid mechanics, states "it is more meaningful to speak of a very viscous situation than a very viscous fluid."

4.2.4 More on scaling and the fundamental character of the Reynolds number

The origin of the *Re* criterion as a fundamental indicator of flow type must be sought ultimately in the equations of motion (Sections 3.2 and 3.12) and in the interplay between forces trying to destabilize laminar flow and those trying to control the deformation (Fig. 4.10). Turbulent acceleration is the advective kind in which a nonlinear form of the velocity change is implied by terms like $u(\partial u/\partial x)$ and $v(\partial u/\partial y)$, that is, the terms involve squares of the velocity that grow rapidly as velocity increases. Let us simplify the approach by assuming that to first order the overall term varies as u^2/l. The viscous friction force (Section 3.10) is proportional to the rate of change of velocity gradient that is, $(\partial^2 u/\partial z^2)v$ times the coefficient of kinematic viscosity. To first order this can be written as vu/l^2. The ratio of the acceleration term to the viscous one gives an idea of

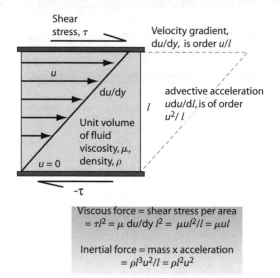

Viscous force = shear stress per area
$$= \tau l^2 = \mu \, du/dy \, l^2 = \mu u l^2/l = \mu u l$$

Inertial force = mass × acceleration
$$= \rho l^3 u^2/l = \rho l^2 u^2$$

Reynolds, number, Re = inertial force/viscous force

$$= \rho l^2 u^2/\mu l u$$
$$= \rho u l/\mu = l u/\nu$$

The birth of eddies depends on some definite value of lu/ν

Reynolds

Fig. 4.10 Simple derivation of Reynolds number.

the likely control on the nature of the flow, in this case $(u^2/l)/(\nu u/l^2)$ or ul/ν, the Reynolds' number. Flows should be dynamically similar if they have the same magnitude of the ratio of acceleration to viscous resistance. This is useful when we conduct experiments on flows, for if the model Re is of same magnitude as the prototype then the nature of the flow should also be similar; the flows are said to be *dynamically scaled*. There is no hard and fast magnitude for the value of Re at which the transition to turbulence takes place, it can be as low as 500 or as great as 10,000 depending upon the conditions that trigger the turbulent instability in the apparatus. The vast majority of atmospheric, fluvial, and oceanographic flows are highly turbulent: the effects of viscous friction are only important close to solid boundaries or in very small flows.

Table 4.2 Some representative Re for natural flows.

Flow	Velocity, u (ms^{-1})	Density, ρ (kgm^{-3})	Length scale, (m)	Molecular viscosity, μ (Pas^{-1})	Re
Basalt magma in fissure flow at 1,400°C	1	2,700	20	350	154
Stream	2	1,000	1	10^{-3}	2,000
River	2	1,000	5	10^{-3}	10,000
Ocean current	0.25	1,028	1,000	10^{-3}	$2.57 \cdot 10^5$
Sea breeze	2	1.293	500	$18.3 \cdot 10^{-6}$	$7.1 \cdot 10^6$
Wind storm	20	1.293	2,000	$18.3 \cdot 10^{-6}$	$2.8 \cdot 10^9$
Jet stream	50	0.52	5,000	$c.18.3 \cdot 10^{-6}$	$7.1 \cdot 10^9$

4.3 Fluid boundary layers

Natural boundaries are multitudinous and may comprise other fluids or solids. For example, atmospheric flows interact with land, sea, and other atmospheric flows. Ocean currents interact with each other and with the ocean floor. The solid boundary itself frequently comprises loose grains sticking partly into the flow, like on the gravel bed of a river or the sandy desert surface. Or it may be rough on a larger scale, with crops, trees, or mountains which an atmospheric flow must interact with as it passes overhead. The boundary may be at the same temperature as the fluid flowing over it or might show a marked temperature contrast. Thus a wind blowing off a high mountain or a glacier, a firn wind like the Mistral of the French Alps, has a very distinct chill to it which diminishes as heat is gained from the surrounding land surface over which it blows. Desert winds like the famous simmoom of the Sahara do the opposite. The boundary may be stationary, or it may be moving at some relative velocity. A solid boundary may be porous like a soil surface or it may even be soluble to the fluid flow, like the surface of a limestone cave.

4.3.1 A simple experiment

We start our explanation with a simple experiment (Fig. 4.11). We observe water in a very deep flow moving slowly to the right over the upper part of a solid lower boundary formed by a flat plate. A fine vertical wire of tellurium to the left of the frame of view is subject to an electrical impulse which induces formation of colloidal-sized H_2 bubbles along the length of the wire. After the line of bubbles has drifted with the flow for a few seconds it is photographed using a high-speed camera. The photograph

spectacularly reveals the form of the boundary layer. What do we make of it, remembering that the bubbles were all produced at the same time along the whole length of the wire by a millisecond electrical impulse?

4.3.2 Qualitative analysis

1 The line of bubbles defines a sharp curve that is well defined: the bubbles have not intermixed appreciably during their brief lifetime in the flow. This means that we can neglect molecular diffusion (Section 4.18) when we are considering fluid flow.

2 The displaced line of bubbles defines a smooth curve whose distance from the wire increases away from the solid boundary. Since distance from the wire is proportional to velocity, the latter must increase similarly. There is no abrupt change of velocity with height; the variation is entirely smooth and in the same direction. The bubble line defines a *velocity profile* adjacent to the solid boundary. If we draw a few arrows from the wire horizontally to the bubble curve then these lengths will be proportional to velocity. The arrows define a *velocity field*. Physically, the *boundary layer* is a zone in the velocity field where there is a *velocity gradient*, that is, $du_x/dz \, 0$.

3 The curve of the line of bubbles is concave-upward, diminishing in slope upward away from the boundary. Thus the rate of flow velocity increase per unit of height, the gradient, must decrease upward. You may recollect that spatial gradients in velocity cause viscous and inertial forces (Section 3.10). The viscous retardation gradually dies out away from the wall until at some point in the flow, the *free stream*, there is no velocity gradient and hence no

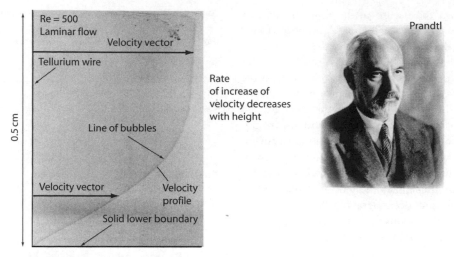

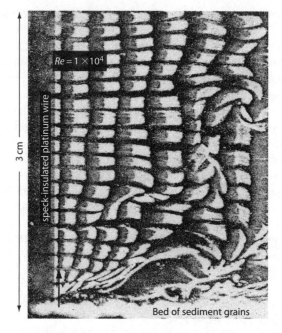

Prandtl

Fig. 4.11 Flow visualization of laminar flow boundary layer by a cloud of H_2 bubbles released by continuous hydrolysis.

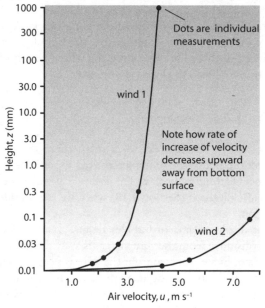

Fig. 4.12 Measurements of wind speed with height above the floor of a wind tunnel to illustrate boundary layers.

Fig. 4.13 Instantaneous photo of strain markers in a turbulent shear flow of water to show heterogeneous strain in a boundary layer. Water flows left to right past a speck-insulated vertical platinum wire; pulsed voltage across the wire gives hydrolysis and production of initially square blocks of hydrogen bubbles. Blocks are released 0.2 s apart. Compare this with the smoothly varying gradient of velocity in the laminar flow case in Fig. 4.11. Note the progressive deformation of individual bubble block strain markers from left to right and the very high strains and strain rates close to the lower flow boundary over a roughened surface of sand grains.

force. It follows that there must be important localization of stresses close to the boundary.

4 The fluid molecules immediately adjacent to the solid boundary surface have not moved at all. It is a characteristic of all moving fluid that there is no "slip," that is, no mean drift, downstream at a solid boundary.

4.3.3 Boundary layer concept

The theory of the boundary layer was first proposed by Prandtl in 1904. The concept simplifies the study of many

fluid dynamic problems because any natural or experimental flow may be considered to comprise two parts: (1) the boundary layer itself, in which the velocity gradient is large enough to produce appreciable viscous and turbulent frictional forces and (2) the free stream fluid outside the boundary layer where viscous forces are negligible. The audience of applied mathematicians who listened to the young Prandtl give his 10 min paper in Heidelburg in 1904 witnessed the birth of a concept that was to revolutionize the study of fluid mechanics. The first major challenge was to apply the boundary layer concept to the study of natural and experimental turbulent flows (Figs 4.12 and 4.13).

4.4   Laminar flow

Laminar flow is rare in the atmosphere and hydrosphere, but the unseen laminar flow of mantle and lower crust turns over a far greater annual discharge of material, not to mention subsurface movement of molten magma and surface flows of high viscosity substances, like mud, debris, and lava. As we have seen, laminar flow is characterized by individual particles of fluid following paths which are parallel and do not cross the paths of neighboring particles; therefore, no mixing occurs. All forces set up within and at the boundaries to flow are due to molecular viscosity and three-dimensional (3D) flow patterns essentially conform to the shape of the vessel through which the fluid happens to be passing. In a wide channel, for example, the flow may be imagined to comprise a multitude of parallel laminae while in a pipe-like conduit the fluid layers comprise a series of coaxial tubes (Fig. 4.14). In all cases the Reynolds number is small and thus viscous forces predominate over inertia forces and prevent 3D particle mixing. In a steady laminar flow any instantaneous measurement of velocity at a point will be exactly the same at that point every time.

4.4.1   The continuum approach to fluid flow

But, steady on, you might exclaim! What is all this talk of nonmixing and "particles"? Surely, fluid *molecules* are the only "particles" present in a laminar flow, or any other flow come to that, and these are whizzing around randomly all the while. These cogent points are why we continue to disregard molecular scale processes until we return to the subject of heat conduction. When dealing with bulk flow we must treat the fluid as if the molecules did not exist and concentrate on the behavior of small fluid volumes ("particles"). Molecular diffusion would surely destroy the color bands in the photo opposite line of bubbles in Fig. 4.11, given enough time, but we are dealing with bulk flow velocities that are rapid compared to the diffusion time of the fluid molecules. This approach to fluid flow is termed the *continuum approach*. However, random molecular movements still occur and are influenced by the fluid velocity: witness the interrelationships between flow velocity and fluid pressure inherent in Bernoulli's equation (Section 3.12).

4.4.2   Viscous shear across a boundary layer

For each and every moving layer in a laminar flow there must exist a shear stress due to the displacement of one layer over its neighbors. Newton's relation for this in the case of flow over a plane solid boundary orientated in the zx plane (Sections 3.9 and 3.10) is $\tau_{zx} = \mu \partial u / \partial z$. This says that "the coefficient of molecular viscosity, μ, induces a shear stress, τ_{zx} (subsequently referred to as *the* shear stress, τ), in any fluid substance when the substance is placed in a gradient of velocity." We must emphasize however that this relation is only true when momentum is transported by molecular transfer.

How does the shear stress vary across the laminar flow boundary layer? Since viscosity is constant for the experimental conditions, the viscous shear stress depends only on the gradient of velocity which, as we have seen (Section 3.10), decreases away from the solid flow boundary. So, the stress must also die away across the boundary layer in direct proportion to the velocity gradient. The greatest value of stress, τ_0, will occur at the solid boundary itself (Fig. 4.15). In the free stream, where the velocity gradient has disappeared, or at least greatly diminished, there is no or little viscous stress. In such areas of flow, remote from boundary layers, the flow is said to be *inviscid* or *ideal* (Section 2.4) and the property of viscosity can be neglected entirely.

In the area of the boundary layer between the solid boundary and the free stream the simplest assumption concerning the falloff of stress with distance (Fig. 4.15) would be to assume that it changes at a constant rate

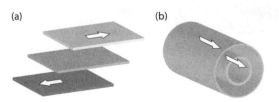

(a)   (b)

Fig. 4.14 3D laminar flow: (a) Couette flow: shearing laminae in a wide channel, (b) Poiseuille flow: shearing concentric cylinders in a pipe or conduit.

throughout the boundary layer thickness, δ. A simple expression for this is given by $\tau = \tau_0(1 - z/\delta)$; as z goes to δ at the outer edge of the boundary layer, τ vanishes. As z goes to zero at the solid boundary, τ goes to τ_0.

In fact, the linear assumption is untrue in detail for laminar flows, though curiously enough it is true for the very thin innermost layer of turbulent flow (Section 4.5). Making use of the Navier–Stokes expressions for viscous force balance, Reynolds originally deduced that the profile of velocity across the laminar boundary layer for a Newtonian fluid actually has the shape of a parabola

(Fig. 4.16; see derivation in Cookie 10). A laminar, non-Newtonian fluid flow has a characteristic plug-like profile in the middle of the flow (Fig. 4.16) where there is virtually no velocity gradient and hence no internal shear.

What does all this mean in practice? Any laminar flow will exert greatest stress, and therefore greatest strain, across the boundary layer. At some point in the flow

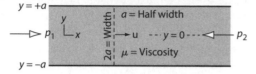

(a) Definitions for derivation of velocity profiles

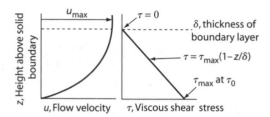

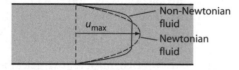

(b) Computed velocity profiles

Fig. 4.15 Velocity and shear stress distribution (to first order only) in laminar flow.

Fig. 4.16 Laminar flow boundary layers between parallel walls (Coutte flow) for Newtonian and non-Newtonian fluids.

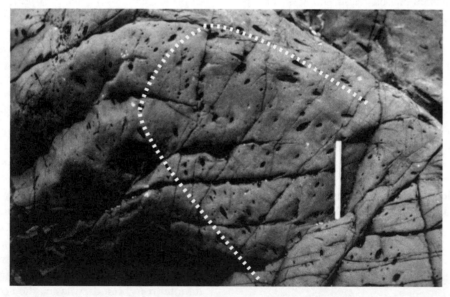

Fig. 4.17 A natural boundary layer "frozen in time." This Namibian dyke intrusion (*Coward's dyke*) was once molten silicate magma flowing as a viscous fluid along a crack-like conduit opening up in cool brittle "country-rock." Gas exsolved from solution to form bubbles. These were deformed (strained) into ellipses (see Section 3.14 on strain) by shear in the boundary layer until the whole flow solidified as heat was lost outward by conduction. Note the greatest strains (more elongate ellipses) occurred at the dyke margin boundary layers where velocity gradients were highest. The generalized shape of the whole flow boundary layer is shown by the dashed white line; it approximates to the non-Newtonian case of Fig. 4.16.

corresponding to a narrow zone in the free stream for a Newtonian fluid and a broad zone for a non-Newtonian fluid, there is no internal deformation. The principle is wonderfully illustrated by a serendipity exposure of an ancient intruded magma body, *Coward's dyke* (Fig. 4.17; see Section 5.1 on magma intrusions generally), whose internal shear deformation has been "fixed" in time due to

the flow deformation of included gas bubbles (geologists call these *vesicles*). Notice how the form of the bubbles changes from highly sheared at the margins to virtually spherical and unsheared in the middle of the former flow. The form of the profile, with a marked rigid central plug, indicates that the magma was behaving as a non-Newtonian fluid at the instant of cessation of flow.

4.5 Turbulent flow

Leonardo made many sketches of the swirling patterns of turbulent flows (Fig. 4.18a). Our own casual observations show fluid to be moving around in what initially seem confusing and possibly random patterns. However, careful analysis (Fig. 4.18b–f), including Leonardo's time-average observations, shows that the motions seem to have a well-defined coherence or structure; flow visualizations show clearly that eddies have a 3D nature with vorticity and "strands of turbulence." Turbulence is a strongly rotational phenomenon, characterized by fluctuating vorticity. Such 3D turbulent fluctuations cause local velocity gradients to be set up in the flow which work against the mean velocity gradient to remove energy from the flow. This turbulent energy is ultimately dissipated by the action of viscosity on the turbulent fluctuations.

4.5.1 More about turbulent eddies

So, we now have evidence from flow visualization that the "sinuous motion" of Reynolds (Section 4.2) is dominated by 3D eddy motions with vorticity (Figs 4.18 and 4.19). Prandtl noted these eddies and from his surface flow visualization experiments regarded them as moving fluid "lumps" (*Flussigkeitsballen*) that transferred momentum throughout the boundary layer. We use this concept to derive the basic flow law for turbulent flows in Cookies 10 and 11. We would like to know a little more about the nature of fluid eddies. For example, the following questions arise:

1 Are turbulent eddies coherent entities?
2 How do eddies relate to momentum transfer and turbulent stresses in turbulent flows?

Looking at the various flow visualizations in Fig. 4.18b–f (see also Sections 4.2 and 4.3), we observe that:

In the *xz* plane:
1 Relatively fast fluid is displaced downward from the outer to the inner flow region in curved 3D vortices. These are termed *sweep* motions.

2 At the same time, relatively slow "lumps" of strongly rotating fluid move out from the inner to the outer flow. These are termed *burst* motions.

In the *xy* plane:
3 Close to the bed there exist flow-parallel lanes of relatively slow and fast fluid that alternate across the flow. The low-speed lanes are termed *streaks*.
4 The streaks become increasingly less defined and "tangled-up" as we ascend the flow.

A 3D reconstruction of the eddies of turbulent shear flows shows the presence of large-scale coherent vortex structures within the boundary layer (Fig. 4.19). These are rather far removed from Prandtl's "lumps" of fluid and more similar to Reynolds' "sinuous" description. They have the shape of *hairpin vortices* whose "legs" are formed from the low-speed streaks (Fig. 4.20). The streak pattern is quasi-cyclic, with new streaks forming and reforming constantly across the flow as the hairpin vortices rise up, advect through the boundary layer and are destroyed in the outer flow.

4.5.2 The distribution of velocity in turbulent flows

We previously derived the distribution of velocity in laminar flow. We now need to make a similar attempt for turbulent flows, where inevitably the situation is more complicated and we must take advantage of both experimental observations and intuition. For example, at a flow boundary, eddies will be vanishingly small since here the velocity components, u and v, must be zero because of the no-slip boundary condition (Section 4.3) and for two-dimensional (2D) flow in the *xz*-plane no net vertical velocity is possible. In this region we would expect viscosity to still dominate flow resistance. Experiments confirm this, establishing a *linear* increase of velocity with height and negligible turbulent shear stresses just above the boundary. This very thin zone of flow closest to the bed (Figs 4.21 and 4.22) is known as the *viscous sublayer*.

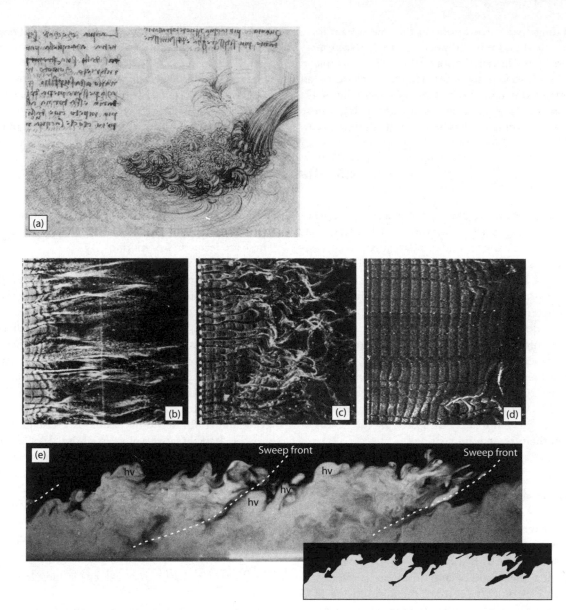

Fig. 4.18 Visualizations of turbulence structure: (a) Sketch by Leonardo to show turbulence generated at a hydraulic jump. Note impression of "coils" of turbulent eddies. (b)–(d) Views from above to show deformation of H_2-bubbles generated along a speck-insulated platinum wire on the left of each frame (see Fig. 29.2). Each instantaneous view is taken successively higher in the same flow: (b) shows sublayer streaky structure developing downflow, (c) shows streaks entangling and mixing as the vortices rise up through the boundary layer, and (d) shows a view high in the boundary layer where turbulence is restricted to a few advected "blobs." (e) Side view of smoke visualization to show the 2D structure of large-scale turbulence and the pattern of subsidiary hairpin burst vortices (hv) lifted by the upstream-inclined shear layers (dashed white lines) that define sweep inflows at the interface of major fast outer (dark) and slower (white) inner flow fluid. Smoke was released at the bed just upstream from the frame.

Making the assumption that viscous stresses dominate, the simplest option open to us is to assume that Newton's viscous stress equation, $\tau = \mu du/dz$ operates. Integration (Cookie 10) for the no-slip boundary condition, $u = 0$ at $z = 0$, yields the appropriate flow law, $u = \tau_0 z/\mu$, where τ_0 is the viscous shear stress at the boundary.

Further out in the flow, observations (Section 4.3) show a decrease in rate of change of velocity, du/dz, with height

(Fig. 4.21; see also Section 6.2). In 1904, Prandtl proposed to completely neglect the influence of viscosity away from the boundary, with the momentum transport being entirely achieved by eddies. He later made the simplest possible assumption about the decay rate, that it decreases as the inverse of distance from the bed, that is, $du/dz \propto 1/z$. This assumption leads to (Cookie 11) a logarithmic relationship for the turbulent velocity as a function of distance from the boundary that is fully supported by experimental evidence (Figs 4.21 and 4.22) (Cookie 12). The complete form of this relationship was proposed in 1925 and is commonly called the von Karman–Prandtl equation or the "*law of the wall*." The lower part of the *logarithmic layer* merges into the viscous sublayer via a *buffer zone* where the majority of turbulent stresses are both generated and dissipated in turbulent flows. The high rates of both generation and destruction of turbulent kinetic energy in this area of turbulent flow leads it to be termed as the *equilibrium layer*.

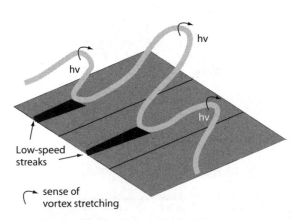

Fig. 4.19 Sketch of major hairpin vortices (hv) that dominate near-bed turbulent flows.

4.5.3 Summary of eddy motions and turbulent boundary layer structure

We may summarize the above discussion by dividing the turbulent boundary layer into two rather distinct zones:

1 An inner zone close to the bed with its upper boundary between the top of the viscous sublayer and logarithmic region of the turbulent boundary layer. The zone is distinguished by (a) being the site of most turbulence production, (b) containing low- and high-speed fluid streaks that alternate across the flow, and (c) the lift-up of low-speed streaks in areas of high local shear near the upper boundary.

2 An outer zone extending up to the flow free surface. The outer zone and (a) provides the source of the high-speed fluid of the sweep phase near its lower boundary (b) contains large vortices near the area of burst break-up that are disseminated through the outer zone and may reach the surfaces as "boils" of turbulence.

4.5.4 The simple physics of turbulence – origin of Reynolds' stresses

Reynolds' approach to the statistical study of turbulence (Section 3.11; Cookie 8) defined time-mean turbulent accelerations. Modern high-speed supercomputers tracking particle movement in turbulent flows reveal astonishing instantaneous accelerations, up to $1,500g$, due to eddy motions. The fact that time-mean turbulent fluctuations exist means that turbulence cannot be random, or else the positive and negative combinations would cancel each other out. Further, mean turbulent acceleration requires a net force to produce it; the only candidate from the equations of motion is the pressure force. The results of turbulent flow measurements suggest high contributions to local *Reynolds' stresses* from burst and sweep

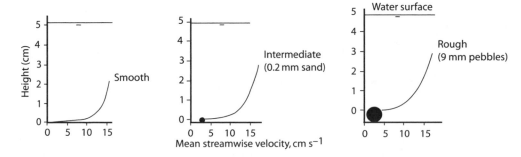

Fig. 4.20 Velocity boundary layer profiles for turbulent channel flows of water over smooth, intermediate, and rough boundaries. Reynolds number constant.

motions; more than 70 percent of the Reynolds stresses in turbulent flows are due to these events, the majority produced close to the wall in the equilibrium layer. Use of a simple quadrant diagram (Fig. 4.23) brings out the essential contrasts between burst and sweep turbulent interactions and enables us to get the signs of the stresses. Turbulent bursts are quadrant 2 events because they involve injections of slower-than-mean horizontal velocity fluid upward (i.e. instantaneous u values are negative and w velocities are positive). The instantaneous average product, $u'w'$, is thus negative and this determines that the Reynolds' stress ($-\rho u'w'$) is overall positive. It is the same for sweep motions, but here the motions are quadrant 4 events, with faster-than-average downward motion giving the negative product.

4.5.5 To illustrate the importance of turbulent stresses

Notwithstanding the viscous stress contribution in the viscous sublayer, through most of any turbulent flow field,

turbulent stresses dominate (Fig. 4.24). Just how much may be appreciated from Bradshaw's argument. For boundary layer flow, the rate of change of velocity with height h through the flow is of order $\bar{u}/h$, so that the mean viscous stress is $\mu\bar{u}/h$. If the flow has a realistic mean velocity fluctuation, u', of say 10 percent of the mean, $\bar{u}$, then $u' = 0.1\bar{u}$ and the Reynolds stress per unit volume of fluid is of order $0.01\rho\bar{u}^2$. The ratio of the magnitude of turbulent to viscous stress, that is, $0.01\rho\bar{u}^2/\mu\bar{u}/h$ is therefore $0.01\,Re$. For a geophysical flow, Re may easily be greater than 10^5 (Section 4.2) and so turbulent stresses completely dominate the boundary layer.

4.5.6 Turbulence over roughened beds

Experimental results of great practical interest have been conducted into turbulence over roughened boundaries. Increasing boundary roughness causes increasing boundary layer velocities and therefore mean boundary shear stress, as

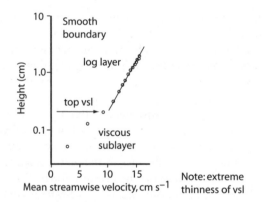

Fig. 4.21 A plot of log height versus mean velocity reveals zones where: (1) velocity varies as a function of log height, where Prandtl's "law of the wall" operates and (2) velocity increases linearly with height, defining the viscous sublayer (vsl).

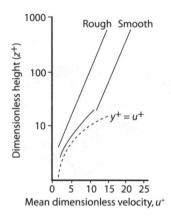

Fig. 4.22 The viscous sublayer is destroyed by the rough boundary. Here the height is nondimensionalized in $z+$ units. Mean velocity is also dimensionless, expressed in $u+$ units.

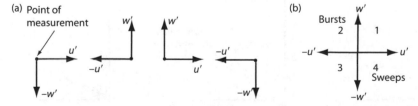

Fig. 4.23 (a) The four possible combinations of velocity fluctuations for 2D flows. (b) A quadrant of possible fluctuation possibilities. Random turbulence would generate equal likelihood of events 1–4 and no net Reynolds' acceleration. In fact, quadrants 2 and 4 dominate, giving deceleration and implying a "structure" to turbulence.

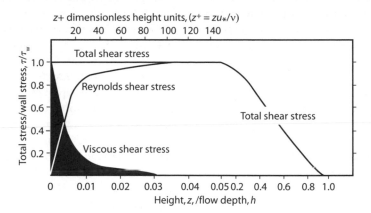

Fig. 4.24 The total shear stress in a turbulent wall flow is a combination of a viscous stress and a turbulent stress. The former rapidly tends to zero away from the viscous sublayer, while the latter dominates the flow from the buffer layer outward. The total stress decays to zero in the outer flow.

we would expect, but the turbulence intensities measured by u'_{rms} and w'_{rms} are independent of roughness conditions beyond a certain height. In the outer layer the intensity depends solely on boundary distance and shear stress and is independent of the conditions producing the stress. Closer to the bed the data separate so that, with increasing boundary roughness, the longitudinal turbulence intensity stays constant or decreases while the vertical intensity increases. We can envisage smooth-boundary viscous sublayer fluid

and the fluid trapped between roughness elements as "passive" reservoirs of low-momentum fluid that are drawn on during bursting phases. Entrainment of this fluid is extremely violent in the rough-boundary case, with vertical upwelling of fluid from between the roughness elements. Viscous sublayer streaks are poorly developed over roughened boundary flows. Faster deceleration of sweep fluid causes the decrease of longitudinal turbulent intensity and the increase of vertical turbulent intensity.

4.6   Stratified flow

Many geophysical flows occur because of instabilities due to local density contrasts (Section 4.1). A stratified flow is one which exhibits some vertical variation in its density brought about because of heat energy transfer, salinity variations, or the effects of suspended solids. If the density increases with height, for example, a desert surface heating an incoming wind, then the situation is unstable and the forward transport of fluid is accompanied by turnover motion that tries to reverse the unstable stratification. Examples of such *forced convection* are given in Section 4.20. Here we are interested in the case of a *stable stratification* in which the density decreases vertically, as might be

produced in a wind blowing over a cool surface and being heated from above (Fig. 4.25). Left to its own devices a stationary stratified fluid or a purely laminar stratified flow would simply slowly lose its density contrast by slow molecular diffusion. Once put into turbulent motion however, the stratified flow will experience burst-like coherent motions that try to lift the heavier fluid upward and sweep-like turbulent motions that will try to carry the lighter fluid downward. In both cases work must be done by the flow against resisting buoyancy forces. We might imagine in the most general way that the turbulent energy may be either capable or incapable of overturning the buoyancy.

Fig. 4.25 View of cool sea breeze system (arrow) propagating up-valley. This stratified flow is about 300 m thick advancing right to left at several meters per second velocity. Note Kelvin–Helmholtz billows at shearing interface indicative of vigorous turbulent mixing (i.e. $Ri_g < 0.25$).

In the former case the stratification tends to be destroyed by turbulent mixing, in the latter it remains there and the turbulence itself is dissipated. In both cases an inhibition against the tendency for turbulent mixing exists and a loss of turbulent energy results.

4.6.1 Criteria for shear stability of density-stratified flow

Richardson noted that stratified fluid undergoing shear has a negative upward gradient of density and that in order to

Richardson

Fig. 4.26 Any stratified turbulent flow has an accelerative (inertial) tendency to mix or destabilize any stratified fluid. As in the Reynolds number argument, call this force the *inertial term* and give its order of magnitude per unit volume as $\rho \cdot u^2 l^{-1}$.
The stratified flow also has a buoyancy force that may act to stabilize the flow or to destabilize it. A stabilizing force involves density decreasing upward. For a mean density contrast of $\Delta\rho$ across the flow the stabilizing force per unit volume acting is $-\Delta\rho g$.
The ratio of the stabilizing buoyancy force to the destabilizing inertial force is the dimensionless bulk *Richardson number, Ri*. Negative *Ri* corresponds to a destabilizing buoyancy force, positive *Ri* to a stabilizing force. To check for dimensions:

$$(\rho u^2 l^{-1})/(\Delta\rho g) = (\text{ML}^{-3}\text{L}^2\text{T}^{-2}\text{L}^{-1})/(\text{L}^3\text{M}^{-1}\text{L}^{-1}\text{T}^2) = 0$$

mix the fluid, turbulent accelerations in the boundary layer have to overcome resistance due to buoyancy (Fig. 4.26). He imagined a situation where the density contrast is very large; try as it might, a turbulent flow may never mix the denser layer upward. Conversely a low negative buoyancy is more easily overcome. Appropriate dimensions of resisting buoyancy force to applied inertial force per unit volume are $-\Delta\rho g$ for the former and $\rho u^2/l$ for the latter. Taking the ratio of these, all dimensions cancel (Fig. 4.26) and we have the simplest possible form of the bulk *Richardson Number, Ri*. The smaller the value the more likely it is that any stratified shear flow will undergo mixing and homogenization. Although this derivation has the correct basic physical principles, it somewhat ignores the physical situation envisaged, that of a shear flow with a *continuous* stable vertical variation of density (Fig. 4.27). The former is characterized by a negative velocity gradient, the latter by a negative density gradient. The two combine to give the *gradient Richardson number, Ri_g*.

4.6.2 Stratification and the phenomenon of double diffusion

The dual control of density by temperature and salinity in ocean waters leads to an interesting scenario because adjacent water masses in thermal contact lose thermal contrast much more quickly than they can lose salinity contrast. This is because the molecular diffusion (conduction) of heat is $c.10^2$ faster than the molecular diffusion of salinity. We imagine a scenario of *metastable stratification* of water layers, for example, an upper salty, warm layer has an initial density, ρ_1, less than that of a lower cool, fresh layer, ρ_2. Such a scenario is to be widely expected in the oceans as a consequence of summer evaporation and warming of surface layers, or to the inflow and outflow of contrasting water masses like that of the well-known western

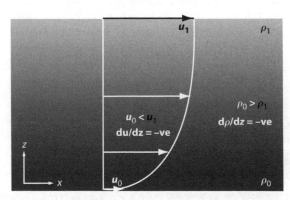

Schematic representation of a stably stratified flow undergoing shear in its boundary layer

Here the *Ri* condition is given by the ratio of density gradient times *g* over density times velocity gradient squared in symbols:

$$-(g\text{d}\rho/\text{d}z) / \rho(\text{d}u/\text{d}z)^2$$

This is the gradient Richardson Number, Ri_g. Negative Ri_g corresponds to a destabilizing buoyancy force, positive Ri_g to a stabilizing force

Fig. 4.27 We need to consider gradients of density and velocity in order to fully utilize the Richardson criterion.

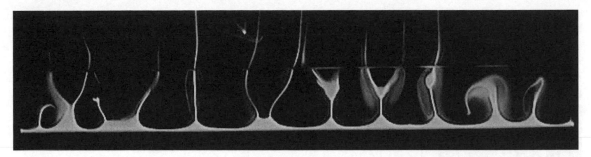

Fig. 4.28 Experimental double diffusive convection. At the beginning of the experiment the pale-colored saline water at the base of the image underlay cold freshwater stably as a continuous layer. As it was heated from below, rapid heat diffusion lowered its density below that of the freshwater. The less dense basal layer deformed into narrow ascending "fingers." In reverse (view image upside down) this is how outflowing warm, salty Mediterranean waters may react to cooling by the Atlantic. Image shows open top *Hele-Shaw cell*, length 800 mm, height 50 mm.

Mediterranean warm saline outflow into the Atlantic. The situation leads to enhanced mixing by convection, sometimes called *double-diffusive convection*, at much greater rates than mixing by molecular diffusion. In this process, more rapid heat diffusion across zones of thermal contact cause the stable stratification to break down. In our example, cooling of the saltier layer from below across the boundary of thermal contact causes the cooled saltier fluid

to fall. An intricate pattern of small-scale mixing gradually develops as moving fluid "fingers" its way downward (Fig. 4.28). Such double-diffusive instabilities can set up regular layering in the water column, with layer boundaries having high rates of change of temperature and salinity. Double-diffusion in crystallizing magma chambers is also thought to cause distinct *igneous layering* of different silicate minerals (Section 5.1).

4.7 Particle settling

It is a common occurrence for solid particles to fall through a still or moving fluid. For example, sand or silt grains settling out from the atmosphere after a dust storm, crystals settling through magma, and dead plankton settling through the ocean. In a clearly related phenomenon of motion, though of opposite sign and somewhat more complex, *gas bubbles* or immiscible liquid may rise through other liquids, expanding as they rise through coalescence and ever-decreasing hydrostatic pressure.

4.7.1 A Reynolds number for particles

It is reasonable to give a *Re* for solid, liquid, and gaseous particle motions in fluid (Figs 4.29 and 4.30). A combination of fluid and solid physical properties defines the *particle Reynolds' number*, Re_g. We use the mean particle size (diameter, d, or radius, $r = d/2$), as the length scale with which to consider flow interactions. The velocity term is the relative velocity between particle and fluid, $\pm V_p$.

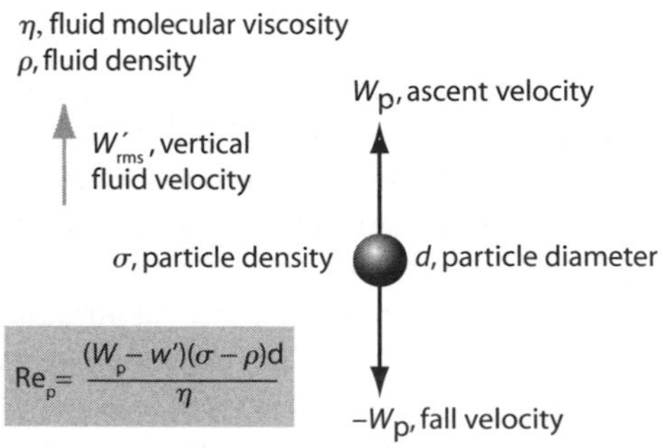

η, fluid molecular viscosity
ρ, fluid density

W'_{rms}, vertical fluid velocity

W_p, ascent velocity

σ, particle density d, particle diameter

$$Re_p = \frac{(W_p - w')(\sigma - \rho)d}{\eta}$$

$-W_p$, fall velocity

Fig. 4.29 Necessary parameters to define a particle Reynolds' number.

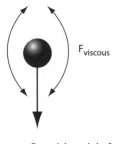

$F_{viscous}$

$-F$, particle weight force

Fig. 4.30 Necessary forces to balance to derive relation for particle fall velocity.

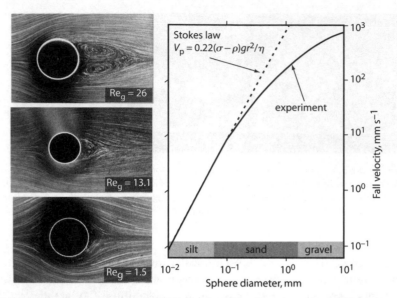

Fig. 4.31 Experimental determination of fall velocity for silica spheres in water at STP.

The mass term per unit volume is expressed as the effective density, $(\sigma - \rho)$, of the solid. Note that the sign of the overall density term defines whether ascent or descent will take place. The only appropriate viscosity term is that of the ambient fluid molecular viscosity, η.

4.7.2   Stokes law

Spheres introduced singly into a static liquid initially accelerate, the acceleration decreasing until a steady velocity, known as the *terminal*, or *fall velocity* is reached. Experimental data for smooth quartz spheres in water (Fig. 4.31), shows clearly that the fall velocity increases with increasing grain diameter but that the rate of increase gets less. Accurate prediction of the rate of fall is only possible for one highly specialized case; that of the steady descent of single, smooth, insoluble spheres through a still Newtonian fluid in infinitely wide and deep containers when the grain Reynolds number is low (<1.5) so there is no viscous flow separation (Fig. 4.31).

Concerning the general problem of predicting the rate of fall (Figs 4.30 and 4.32), we might surmise that the terminal velocity varies directly with density contrast, $(\sigma - \rho)$, size, d, and gravity, g, while varying inversely with fluid viscosity, η. We then write a functional relationship (see appendix) of general form $V_p = f(\sigma - \rho)dg/\eta$. This leaves the determination of proportionality constant(s) to experimentation. Another solution is the approach of the mathematical physicist (Stokes), to solve the problem by recourse to the fundamental equations of flow. Stokes

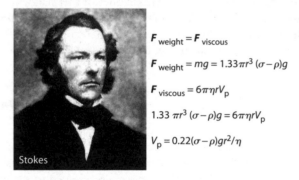

$$F_{weight} = F_{viscous}$$

$$F_{weight} = mg = 1.33\pi r^3 (\sigma - \rho)g$$

$$F_{viscous} = 6\pi\eta r V_p$$

$$1.33\,\pi r^3 (\sigma - \rho)g = 6\pi\eta r V_p$$

$$V_p = 0.22(\sigma - \rho)gr^2/\eta$$

Fig. 4.32 Stokes approach.

Box 4.1   For multiple particles, the Richardson–Zaki expression rules.

$$V_{mp} = V_p(1 - c)^n$$

where c is the volume concentration of particles, n varies between 2.32 $(Re_p > 500)$ and 4.65 $(Re_p < 0.2)$.

considered viscous fluid resistance forces only and arrived at a theoretical relation that expressed fall velocity as a function of particle and fluid properties (Fig. 4.32). During steady fall, all acceleration terms in the equations of motion vanish, leaving a balance between the solid particle's immersed weight force and the viscous drag force acting over the solid particle surface. Stokes' law very accurately predicts the fall velocity of spherical particles whose particle Reynolds number is <0.5. As viscous and then turbulent separation occurs around particles

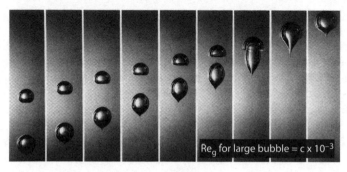

Re$_g$ for large bubble = c x 10^{-3}

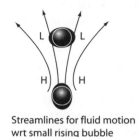

Streamlines for fluid motion
wrt small rising bubble

Fig. 4.33 Two air bubbles are rising through syrup. The larger is moving faster and catches up with the smaller. As the larger encroaches into the low velocity/high pressure field created below the smaller it is strained by extension into a prolate ellipse. The smaller is strained by its own pressure field due to motion, with a lateral flattening due to decreased spanwise pressure about its equatorial plane. Eventually the bubbles collide and coalesce, the smaller spreading over the larger.

(Fig. 4.31) the drag increases more rapidly and Stokes equation no longer applies. The periodic eddy motions shed off alternate sides of a settling particle during viscous and turbulent flow separation are known as *von Karman vortices*. In multi-particle settling the fluid streamlines of individual grains interact. Increased drag results in a decrease of fall velocity (Box 4.1).

4.7.3 Bubble motion and mutual particle interactions

The movement and deformation of growing gas bubbles in cooling, flowing magma is an important process in magmatic intrusions (see Fig. 4.17) and during volcanic

eruptions (Section 5.1). Stokes law can be used to predict the rate of ascent of the bubbles, though bubble deformation and independent controls on the rate of bubble growth during motion make the analysis a good deal more complicated than for simple solid motion. The motion of multiple bubbles involves interaction between velocity and pressure fields; this imposes strain upon the moving and interacting bubbles (Fig. 4.33). The visible straining of bubbles makes it clear that the motion of multiple solid particles, be they moving through still fluid or in a shear flow, distort the flow pressure field and exert additional pressure forces on distant flow boundaries. These affects are important in the movement and size segregation of all particles in granular and fluid-granular mixtures (Section 4.11).

4.8   Particle transport by flows

Natural fluid flows often interact with a solid boundary. If the boundary comprises sediment particles these can interact with the fluid boundary layer. The study of fluid–sediment interactions is termed *loose boundary hydraulics*.

4.8.1 Initiation and paths of grain motion

It is conventional to separate transported sediment in regular, if intermittent, contact with the bed from that traveling aloft in the flow. The former is known as *bed load*, the latter as *suspended load*. A turbulent fluid shear stress, τ_{zx}, drives the former, while a normal stress, τ_{zz}, drives the latter (Fig. 4.34; Cookie 13). For individual particles this division of sediment transport is often a statistical one, since individual particles are constantly

being exchanged between one mode of transport and the other.

Particles initially at rest under a turbulent boundary layer are subject to both shear (drag) and lift forces (Fig. 4.35). The shear force arises from gradients of viscous or turbulent shear stress in the flow boundary layer (Sections 4.3–4.5), while the lift force arises from the Bernoulli effect (Section 3.12). Both shear and lift forces act on the protruding parts of particles resting on the boundary. When the combined magnitude of these is sufficient to overcome particle inertia and friction then the particles will move. In practice (Fig. 4.36), because of variable bed particle size and particle exposure to turbulent fluctuations, the threshold condition for initiation of motion must be determined experimentally and expressed using dimensionless numbers (Fig. 4.37): threshold conditions are given as a critical value of the overall dimensionless boundary shear stress, θ. The threshold value is thus an important practical parameter in environmental engineering. A particular fluid shear velocity, u_*, above the threshold for motion may also be expressed as a ratio with respect to the critical threshold velocity, u_{*c}. This is the *transport stage*, defined as the ratio u_*/u_{*c}. Once that threshold is reached, grains may travel (Fig. 4.38) by (1) rolling or intermittent sliding (2) repeated jumps or *saltations* (3) carried aloft in *suspension*. Modes (1) and (2) comprise bedload as defined previously. Suspended motion begins when bursts of fluid turbulence are able to lift saltating grains upward from their regular ballistic trajectories, a crude statistical criterion being when the mean upward turbulent velocity fluctuation exceeds the particle fall velocity, that is, $w'/V_p > 1$.

4.8.2 Fluids as transporting machines: Bagnold's approach

It is axiomatic that sediment transport by moving fluid must be due to momentum transfer between fluid and sediment and that the resulting forces are set up by the

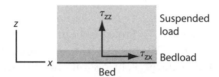

Fig. 4.34 Stresses responsible for sediment transport.

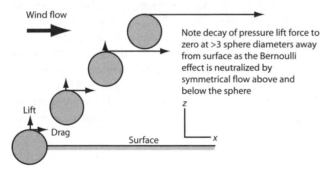

Fig. 4.35 Relative magnitude of shear force (drag) and pressure lift force acting on spheres by constant air flow at various heights above a solid surface.

Fig. 4.36 (left) W.S. Chepil made the quantitative measurements of lift and drag used as a basis for Fig. 4.35. Here he is pictured adjusting the test section of his wind tunnel in the 1950s. Much research into wind blown transport in the United States was stimulated by the Midwest "dust bowl" experiences of the 1930s.

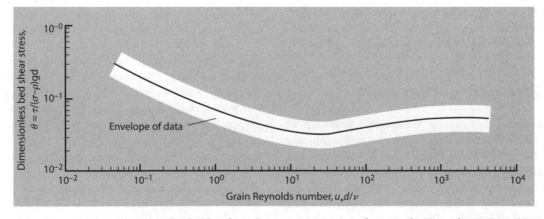

Fig. 4.37 Variation of dimensionless shear stress threshold, θ, for sediment motion in water flows as a function of grain Reynolds number. θ is known as the Shields function, after the engineer who first proposed it in the 1930s.

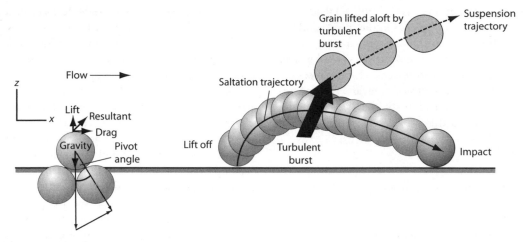

Fig. 4.38 Grain motion and pathways.

differential motion of the fluid over an initially stationary boundary. Working from dynamic principles Bagnold assumed that

1 In order to move a layer of stationary particles, the layer must be sheared over the layer below. This process involves lifting the immersed mass of the topmost layer over the underlying grains as a *dilatation* (see Section 4.11.1), hence work must be done to achieve the result.

2 The energy for the transport work must come from the kinetic energy of the shearing fluid.

3 Close to the bed, fluid momentum transferred to any moving particles will be transferred in turn to other stationary or moving particles during impact with the loose boundary; a dispersion of colliding grains will result. The efficacy of particle collisions will depend upon the immersed mass of the particles and the viscosity of the moving fluid (imagine you play pool underwater).

4 If particles are to be transported in the body of the fluid as suspended load, then some fluid mechanism must act to effect their transfer from the bed layers. This mechanism must be sought in the processes of turbulent shear, chiefly in the bursting motions considered previously (Section 4.5).

The fact that fluids may do useful work is obvious from their role in powering waterwheels, windmills, and turbines. In each case flow kinetic energy becomes machine mechanical energy. Energy losses occur, with each machine operating at a certain efficiency, that is, work rate = available power × efficiency. Applying these basic principles to nature, a flow will try to transport the sediment grains supplied to it by hillslope processes, tributaries, and bank erosion. The quantity of sediment carried will depend upon the power available and the efficiency of the energy transfer between fluid and grain.

4.8.3 Some contrasts between sediment transport in air and water flows

Although both air and water flows have high Reynolds numbers, important differences in the nature of the two transporting systems arise because of contrasts in fluid material properties. Note in particular that

1 The low density of air means that air flows set up lower shearing stresses than water flows. This means that the competence of air to transport particles is much reduced.

2 The low buoyancy of mineral particles in air means that conditions at the sediment bed during sediment transport are dominated by collision effects as particles exchange momentum with the bed. This causes a fraction of the bed particles to move forward by successive grain impacts, termed *creep*.

3 The bedload layer of saltating and rebounding grains is much thicker in air than water and its effect adds significant roughness to the atmospheric boundary layer.

4 Suspension transport of sand-sized particles by the eddies of fluid turbulence (Cookie 13) is much more

Table 4.3 Some physical contrasts between air and water flows.

Material or flow property	Air	Water
Density, ρ (kg m^{-3}) at STP	1.3	1,000
Sediment/fluid density ratio	2,039	2.65
Immersed weight of sediment per unit volume (N m^{-3})	$2.6 \cdot 10^4$	$1.7 \cdot 10^4$
Dynamic viscosity, μ (Ns m^{-2})	$1.78 \cdot 10^{-5}$	$1.00 \cdot 10^{-3}$
Stokes fall velocity, V_p (m s^{-1}) for a 1 mm particle	~8	~0.15
Bed shear stress, τ_{zx} (N m^{-2}) for a 0.26 m s^{-1} shear velocity	~0.09	~68
Critical shear velocity, u_{*c} needed to move 0.5 mm diameter sand	0.35	0.02

difficult in air than in water, because of reduced fluid shear stress and the small buoyancy force. On the other hand the widespread availability of mineral silt and mud ("dust") and the great thickness of the atmospheric boundary layer means that dust suspensions can traverse vast distances.

5 Energetic grain-to-bed collisions mean that wind-blown transport is very effective in abrading and rounding both sediment grains and the impact surfaces of bedrock and stationary pebbles.

4.8.4 Flow, transport, and bedforms in turbulent water flows

As subaqueous sediment transport occurs over an initially flat boundary, a variety of bedforms develop, each adjusted to particular conditions of particle size, flow depth and applied fluid stress. These bedforms also change the local flow field; we can conceptualize the interactions between flow, transport, and bedform by the use of a feedback scheme (Fig. 4.39).

Current ripples (Fig. 4.40c) are stable bedforms above the threshold for sediment movement on fine sand beds at relatively low flow strengths. They show a pattern of flow separation at ripple crests with flow reattachment downstream from the ripple trough. Particles are moved in bedload up to the ripple crest until they fall or diffuse from the separating flow at the crest to accumulate on the steep ripple lee. Ripple advance occurs by periodic lee slope avalanching as granular flow (see Section 4.11). Ripples form when fluid bursts and sweeps to interact with the boundary to cause small defects. These are subsequently

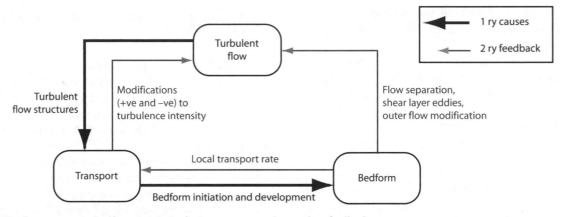

Fig. 4.39 The flow–transport–bedform "trinity" of primary causes and secondary feedback.

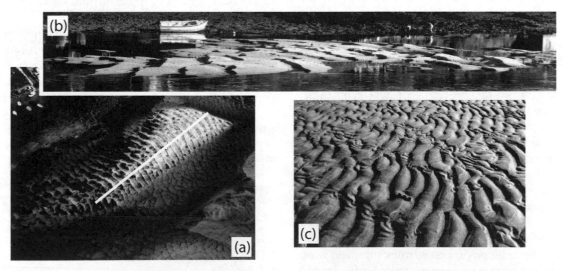

Fig. 4.40 Hierarchy of bedforms revealed on an estuarine tidal bar becoming exposed as the tidal level falls. (a) Air view of whole bar from Zeppelin. Light colored area with line (150 m) indicates crestal dunes illustrated in (b). (b) Dunes have wavelengths of 5–7 m and heights of 0.3 0.5 m. (c) Detail of current ripples superimposed on dunes, wavelengths *c.*12–15 cm.

enlarged by flow separation processes. Ripples do not form in coarse sands ($d > 0.7$ mm); instead a *lower-stage plane bed* is the stable form. The transition coincides with disruption of the viscous sublayer by grain roughness and the enhancement of vertical turbulent velocity fluctuations. The effect of enhanced mixing is to steepen the velocity gradient and decrease the pressure rise at the bed in the lee of defects so that the defects are unable to amplify to form ripples.

With increasing flow strength over sands and gravels, current ripples and lower-stage plane beds give way to *dunes*. These large bedforms (Fig. 4.40a, b) are similar to current ripples in general shape but are morphologically distinct, with dune size related to flow depth. The flow pattern over dunes is similar to that over ripples, with well-developed flow separation and reattachment. In addition, large-scale advected eddy motions rich in suspended sediment are generated along the free-shear layer of the separated flow. The positive relationship between dune height, wavelengths, and flow depth indicates that the magnitude of dunes is related to thickness of the boundary layer or flow depth.

As flow strength is increased further over fine to coarse sands, intense sediment transport occurs as small-amplitude/long wavelength bedwaves migrate over an *upper-stage plane bed*.

Antidunes are sinusoidal forms with accompanying in phase water waves (Fig. 4.41) that periodically break and

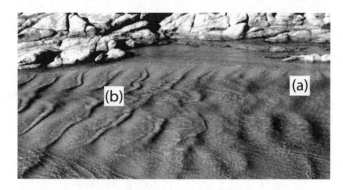

Fig. 4.41 Fast, shallow water flow (flow right to left; Froude number > 0.8) over sand to show downstream trend from (a) in-phase standing waves over antidune bedforms, to (b) downstream to upstream-breaking waves. In the next few seconds the breaking waves propagate into area (a). The standing waves subsequently reform over the whole field and thereafter the upstream-breaking cycle begins again.

move upstream, temporarily washing out the antidunes. They occur as stable forms when the flow Froude number (ratio between velocity of mean flow and of a shallow water wave, that is, $u/(gd)^{0.5}$ is >0.84, approximately indicative of rapid (supercritical) flow, and are thus common in fast, shallow flows. Antidune wavelength is related to the square of mean flow velocity.

4.9 Waves and liquids

Waves are periodic phenomena of extraordinarily diverse origins. Thus we postulate the existence of sound and electromagnetic waves, and directly observe waves of mass concentration each time we enter and leave a stationary or slowly moving traffic jam. A great range of waveforms transfer energy in both the atmosphere and oceans, with periods ranging from 10^{-2} to 10^5 s for ocean waves. They transfer energy and, sometimes, mass. The commonest visible signs of fluid wave motion are the surface waveforms of lakes and seas. Many waveforms are in lateral motion, traveling from here to there as *progressive waves*, although some are of too low frequency to observe directly, like the tide. Yet others are *standing waves*, manifest in many coastal inlets and estuaries. In the oceans, waves are usually superimposed on a flowing tidal or storm current of greater or lesser strength. Such *combined flows* carry attributes of both wave and current but the combination is more complex than just a simple addition of effects (Section 4.10). Waves also occur at density interfaces within *stratified fluids* as *internal waves*, as in the motion along the

oceanic, thermocline, oceanic, and shelf margin tides, density and turbidity currents. We must also note the astonishing *solitary waves* seen as tidal bores and reflected density currents.

4.9.1 Deep water, surface gravity waves

"Deep" in this context is a *relative* term and is formally defined as applying when water depth, h, is greater than a half wavelength, that is, $h > \lambda/2$ (Fig. 4.42). Deep water waves at the sea or lake surface are more-or-less regular periodic disturbances created by surface shear due to blowing wind. The stationary observer, fixing their gaze at a particular point such as a partially submerged marker post, will see the water surface rise and fall up the post as a wave passes by through one whole wavelength. This rise and fall signifies the conversion of wave potential to kinetic energy. The overall wave shape follows a curve-like, sinusoidal form and we use this smoothly varying property as a

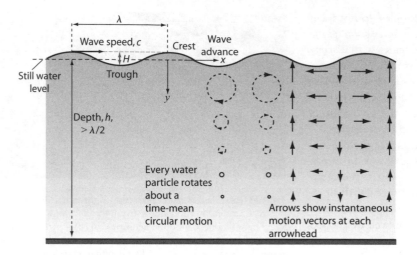

For simple harmonic motion of angular velocity, ω, the displacement of the still water level over time, t, is given by:

$$y = H \sin \omega t$$

The equations of motion for an inviscid fluid can be solved to give the following useful expression for wave speed, c:

$$c = \sqrt{g\lambda / 2\pi}$$

Since the coefficients are constant, for SI units we have:

$$c = 1.25 \sqrt{\lambda}$$

Fig. 4.42 Deep water wave parameters, circular orbitals, and instantaneous water motion vectors. Deep water waves are sometimes called *short waves* because their wavelengths are short compared to water depth.

Note: In nature, individual waves pass through wave groups traveling at speed C^{-2} with energy transmitting at this rate.

simple mathematical guide to our study of wave physics (Cookie 14). It is a common mistake to imagine deep water waves as heaps and troughs of water moving along a surface: it is just wave energy that is transferred, with no net forward water motion.

The simplest approach is to set the shape of the waveform along an *xz* graph and consider that the periodic motion of *z* will be a function of distance *x*, wave height, H, wavelength, λ, and celerity (wave speed), c. Attempts to investigate wave motion in a more rigorous manner assume that the wave surface displacement may be approximated by curves of various shapes, the simplest of which is a harmonic motion used in linear (Airy) wave theory (Cookie 14). Sinusoidal waves of small amplitude in deep water cause motions that cannot reach the bottom. Small-amplitude wave theory approach assumes the water is inviscid and irrotational. The result shows that surface gravity waves traveling over very deep water are *dispersive* in the sense that their rate of forward motion is directly dependant upon wavelength: wave height and water depth play no role in determining wave speed (Fig. 4.42). An important consequence of dispersion is that if a variation of wavelength occurs among a population of deep water waves, perhaps sourced as different wave trains, then the longer waves travel through the shorter ones, tending to amplify when in phase and canceling when out of phase. This causes production of *wave groups*, with the group speed, c_g being 50 percent less than the individual wave speeds, c (Cookie 15).

At any fixed point on or within the water column the fluid speed caused by wave motion remains constant while the direction of motion rotates with angular speed, ω; and any particle must undergo a rotation below deep water waves (Fig. 4.42). The radii of these water *orbitals* as they are called, decreases exponentially below the surface.

4.9.2 Shallow water surface gravity waves

Deep water wave theory fails when water depth falls below about 0.5λ. This can occur even in the deepest oceans for the tidal wave and for very long (10s to 100s km) wavelength *tsunamis* (see below). Shallow water waves are quite different in shape and dynamics from that predicted by the simple linear theory of sinusoidal deep water waves. As deep water waves pass into shallow water, defined as $h < \lambda/20$, they suffer attenuation through bottom friction and significant horizontal motions are induced in the developing wave boundary layer (Figs 4.43 and 4.44). The waves take on new forms, with more pointed crests and flatter troughs. After a transitional period, when wave speed becomes increasingly affected by water depth, shallow-water gravity waves move with a velocity that is proportional to the square root of the water depth, independent of wavelength or period (Cookie 16). The dispersive effect thus vanishes and wave speed equals wave group speed. The wave orbits are elliptical at all depths with increasing ellipticity toward the bottom, culminating at the bed as horizontal straight line flow representing to-and-fro motion. Steepening waves may break in very shallow water or when intense wind shear flattens wave crests (Section 6.6). In both cases air is entrained into the surface

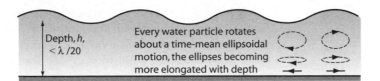

Depth, h, < λ /20

Every water particle rotates about a time-mean ellipsoidal motion, the ellipses becoming more elongated with depth

The waves move with a velocity proportional to the square root of the water depth, independent of the wavelength or period:

$$c = \sqrt{gh}$$

Fig. 4.43 Shallow water waves and their ellipsoidal orbitals. Shallow water waves are sometimes called *long waves* because their wavelengths are long compared to water depths. Note that the orbital motions flatten with depth but do not change in maximum elongation.

Note: All waves in similar water depths travel at the same speed and transmit their energy flux at this rate.

Fig. 4.44 Time-lapse photograph of shallow water wave orbitals visualized by tracer particle. This flow visualization of suspended particles was photographed under a shallow water wave traversing one wavelength, λ, left to right. Wave amplitude is 0.04λ and water depth is 0.22λ. The clockwise orbits are ellipses having increasing elongation toward the bottom. Some surface loops show slow near-surface drift to the right. This is called *Stokes drift* and is due to the upper parts of orbitals having a greater velocity than the lower parts and to bottom friction. The surface drift is accompanied by compensatory near-bed drift to the left, due to conservation of volume in the closed system of the experimental wave tank. Stokes drift without the added effects of bottom friction also occurs in short, deep water waves.

boundary layer of the water as the water collapses or spills down the wave front, thus markedly increasing the air-to-sea-to-air transfer of momentum, thermal energy, organo-chemical species, and mass. The production of foam and bubble trains is also thought to feed back to the atmospheric boundary layer itself, leading to a marked reduction of boundary layer roughness and therefore friction in hurricane force winds (Section 6.2).

4.9.3 Surface wave energy and radiation stresses

The energy in a wave is proportional to the square of its height. Most wave energy (about 95 percent) is concentrated in the half wavelength or so depth below the mean water surface. It is the rhythmic conversion of potential to kinetic energy and back again that maintains the wave motion; derivations of simple wave theory are dependent upon this approach (Cookies 14 and 16). The displacement of the wave surface from the horizontal provides potential energy that is converted into kinetic energy by the orbital motion of the water. The total wave energy per unit area is given by $E = 0.5\rho g a^2$, where a is wave amplitude (=0.5 wave height H). Note carefully the energy

dependence on the square of wave amplitude. The energy flux (or wave power) is the rate of energy transmitted in the direction of wave propagation and is given by $\omega = Ecn$, where c is the local wave velocity, and the coefficients are $n = 0.5$ in deep water and $n = 1$ in shallow water. In deep water the energy flux is related to the wave group velocity rather than to the wave velocity. Because of the forward energy flux, Ec, associated with waves approaching the shoreline, there exists also a shoreward-directed momentum flux or stress outside the zone of breaking waves. This is termed *radiation stress* and is discussed in Section 6.6.

4.9.4 Solitary waves

Especially interesting forms of *solitary waves* or *bores* may occur in shallow water due to sudden disturbances affecting the water column. These are very distinctive *waves of translation*, so termed because they transport their contained mass of water as a raised heap, as well as transporting the energy they contain (Fig. 4.45). These amazing features were first documented by J.S. Russell who came across one in 1834 on the Edinburgh–Glasgow canal in central Scotland. Here are Russell's own vivid words, written in 1844:

I happened to be engaged in observing the motion of a vessel at a high velocity, when it was suddenly stopped, and a violent and tumultuous agitation among the little undulations which the vessel had formed around it attracted my notice. The water in various masses was observed gathering in a heap of a well-defined form around the centre of the length of the vessel. This accumulated mass, raising at last to a pointed crest, began to rush forward with considerable velocity towards the prow of the boat, and then passed away before it altogether, and, retaining its form, appeared to roll forward alone along the surface of the quiescent fluid, a large, solitary, progressive wave. I immediately left the vessel, and attempted to follow this wave on foot, but finding its motion too rapid, I got instantly on horseback and overtook it in a few minutes, when I found it pursuing its solitary path with a uniform velocity along the surface of the fluid. After having followed it for more than a mile, I found it subside gradually, until at length it was lost among the windings of the channel.

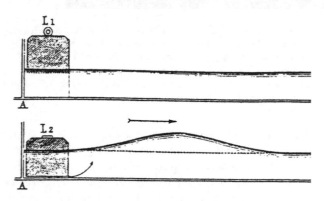

Fig. 4.45 Solitary waves: Russell's original sketch to illustrate the formation and propagation of a solitary wave. You can achieve the same effect with a simple paddle in a channel, tank, or bath. The solitary wave is raised as a "hump" of water above the general ambient level. The "hump" is thus transported as the excess mass above this level, as well as by the kinetic energy it contains by virtue of its forward velocity, c.

Briefly, a solitary wave is equivalent to the top half of a harmonic wave placed on top of undisturbed fluid, with all the water in the waveform moving with the wave; such bores, unlike surface oscillatory gravity waves, transfer water mass in the direction of their propagation. Somewhat paradoxically we can also speak of trains of solitary waves within which individuals show *dispersion* due to variations in wave amplitude. They propagate without change of shape, any higher amplitude forms overtaking lower forms with the very remarkable property, discovered in the 1980s, that, after collision, the momentarily combining waves separate again, emerging from the interaction with no apparent visible change in either form or velocity (Fig. 4.46). Such solitary waves are called *solitons*.

4.9.5 Internal fluid waves

Within the oceans there exist sharply-defined sublayers of the water column which may differ in density by only small amounts (Fig. 4.47). These density differences are commonly due to surface warming or cooling by heat energy transfer to and from the atmosphere by conduction. They may also be due to differences in salinity as evaporation occurs or as freshwater jets mix with the ambient ocean mass. The density contrast between layers is now small enough (in the range 3–20 kg m^{-3}, or 0.003–0.02) so that the less dense and hence buoyant surface layers feel the drastic effects of reduced gravity. Any imposed force causing a displacement and potential energy change across the sharp interface between the fluids below the surface is now opposed by a *reduced gravity* (Section 3.6) restoring force, $g' = (\Delta\rho/\rho)g$. The wave propagation speed, $c = \sqrt{gh}$ is now reduced in proportion to this reduced gravity, to $c = \sqrt{g'h}$, while the wave height can be very much larger. Internal waves of long period and high amplitude progressively "leak" their energy to smaller length scales in an *energy "cascade,"* causing turbulent shear that may

Solitons in shallow water
$t = 0$ $t = +1$ s

Fig. 4.46 Solitary wave A–A′ has just formed as a reflected wave from a harbor wall behind and to the left. The views show the wave moving forward ($c = 1$ m s^{-1}) through incoming shallow water waves B–B′ and C–C′ with little deformation or diminution.

ultimately cause the waves to break. This is an important mixing and dissipation mechanism for heat and energy in the oceans (Section 6.4.4).

4.9.6 Waves at shearing interfaces – Kelvin–Helmholtz instabilities

Stratified fluid layers (Section 4.4) may be forced to shear over or past one another (Fig. 4.48). Such contrasting flows commonly occur at mixing layers where water masses converge; fine examples occur in estuaries or when river tributaries join. On a larger scale they occur along the margins of ocean currents like the Gulf Stream (see Section 6.4). In such cases an initially plane *shear layer* becomes unstable if some undulation or irregularity appears along the layer, for any acceleration of flow causes a pressure drop (from Bernoulli's theorem) and an accentuation of the disturbance (Fig. 4.48). Very soon a striking, more-or-less regular, system of asymmetrical vortices appears, rotating about approximately stationary axes parallel to the plane of

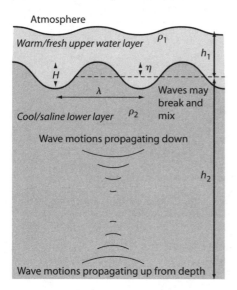

Fig. 4.47 Internal waves at a sharp density interface.

shear. These vortices are important mixing mechanisms in nature; they are called *Kelvin–Helmholtz waves*.

4.9.7 The tide: A very long period wave

The tide, a shallow-water wave of great speed $(20–200 \text{ ms}^{-1})$ and long wavelength, causes the regular rise and fall of sea level visible around coastlines. Newton was the first to explain tides from the gravitational forces acting on the ocean due to the Moon and Sun (Figs 4.49–4.52). Important effects arise when the Sun and Moon act together on the oceans to raise extremely high tides (spring tides) and act in opposition on the oceans to raise extremely low tides (neap tides) in a two-weekly rhythm. It has become conventional to describe tidal ranges according to whether they are macrotidal (range >4 m), mesotidal (range 2–4 m), or microtidal (range <2 m), but it should be borne in mind that tidal range always varies very considerably with location in any one tidal system.

An observer fixed with respect to the Earth would expect to see the equilibrium tidal wave advance progressively from east to west. In fact, the tides evolve on a rotating ocean whose water depth and shape are highly variable with latitude and longitude. The result is that discrete rotary and standing waves dominate the oceanic tides and their equivalents on the continental shelf (see Section 6.6). In detail the nature of the tidal oscillation depends critically on the natural periods of oscillation of the particular ocean basin. For example, the Atlantic has 12-h tide-forming forces while the Gulf of Mexico has 24 h. The Pacific does not oscillate so regularly and has mixed tides.

Advance of the tidal wave in estuaries that narrow upstream is accompanied by shortening parallel to the crest, crestal amplification, and steepening of the tidal wave whose ultimate form is that of a *bore*, a form of solitary wave. In a closed tidal basin a *standing wave* of characteristic resonant period, T, with a node of no displacement in the middle and antinodes of maximum displacement at the ends, has a

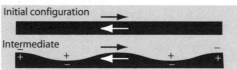

Fig. 4.48 Kelvin–Helmholtz waves formed at a sharp, shearing interface between clear up-tank moving less dense fluid and dark downtank-moving denser fluid. Note shape of waves, asymmetric upflow. Pressure deviations from Bernoulli accelerations amplify any initial disturbances into regular vortices.

wavelength, λ, twice the length, L, of the basin. The speed of the wave is thus $2L/T$ and, treating the tidal wave as a shallow-water wave, we may write *Merian's formula* as $2L/T = \sqrt{gh}$. T is now given by $2L/\sqrt{gh}$. When the period of incoming wave equals or is a certain multiple of this resonant period, then amplification occurs due to resonance, but with the effects of friction dampening the resonant amplification as distance from the shelf edge increases. The tide

occurs as a standing wave off the east coast of North America, where tidal currents are zero in the nodal center of the oscillating water near the shelf edge and maximum at the margins (antinodes) where the shelf is broadest.

4.9.8 A note on tsunami

The horrendous Indian Ocean tsunami of December 2004 focused world attention on such wave phenomenon. Tsunami is a Japanese term meaning "harbor wave." Tsunami is generated as the sea floor is suddenly deformed

Fig. 4.49 Revolution of the Earth–Moon pair.

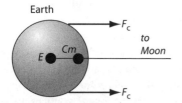

Fig. 4.50 The centripetal acceleration (see Section 3.7) causes and the centrifugal force, F_c, directed parallel to EM of the same magnitude occur *everywhere* on the surface of the Earth.

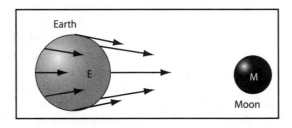

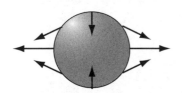

The resultant tide-producing forces

Fig. 4.51 The gravitational attraction of the Moon on the Earth varies according to the inverse of the distance squared of any point on the Earth's surface from M, the center of mass of the Moon. Hence the resultant of the centrifugal and gravitational forces is the tide-producing force.

Fig. 4.52 The magnitude of the tide producing force is only about 1 part in 10^5 of the gravitational force. We are interested only in the horizontal component of this force that acts parallel to the surface of the ocean. This component is the *tractive force* available to move the oceanic water column and it is at a maximum around small circles subtending an angle of about 54° to the center of Earth. The tractive force is at a minimum along the line EM connecting the Earth–Moon system. An equilibrium state is reached, the *equilibrium tide*, as an ellipsoid representing the tendency of the oceanic waters flowing toward and away from the line EM. Combined with the revolution of the Earth this causes any point on the surface to experience two high water and two low water events each day, the *diurnal equilibrium tide*.

by earthquake motions or landslides; the water motion generated in response to deformation of the solid boundary propagate upward and radially outward to generate very long wavelength (100s km) and long period (>60 s) surface wave trains. By very long we mean that wavelength is very much greater than the oceanic water depth and hence the waves travel at tremendous speed, governed by the shallow water wave equation $c = \sqrt{gh}$. For example, such a wave train in 3,000 m water depth gives wave speeds of order 175 m s^{-1} or 630 km h^{-1}. Tsunami wave height in deep-water is quite small, perhaps only a few decimeters. The smooth, low, fast nature of the tsunami wave means wave energy dissipation is very slow, causing very long (could be global) runout from source. As in shallow water surface gravity waves at coasts, tsunami respond to changes in water depth and so may curve on refraction in shallow water. Accurate tsunami forecasting depends on the water depth being very accurately known, for example, in the oceans a wave may travel very rapidly over shallower water on oceanic plateau. During run-up in shallow coastal waters, tsunami wave energy must be conserved during very rapid deceleration: the result is substantial vertical amplification of the wave to heights of tens of meters.

4.9.9 Flow and waves in rotating fluids

We saw in Section 3.7 what happens in terms of radial centripetal and centrifugal forces when fluid is forced to turn in a bend. In Section 3.8 we explored the consequences of free flow over rotating spheres like the Earth when variations in vorticity create the Coriolis force which acts to turn the path of any slow-moving atmospheric or oceanic current loosely bound by friction (geostrophic flows). A simple piece of kit to study the general nature of rotating flows was constructed by Taylor in the 1920s, based upon the *Couette apparatus* for determining fluid viscosity between two coaxial rotating cylinders. This consisted of two unequal-diameter coaxial cylinders, one set within the other, the outer, larger cylinder is transparent and fixed while the smaller, inner one of diameter r_i is rotated by an electric motor at various angular speeds, Ω. The annular space, diameter d, between the cylinders is filled with

Fig. 4.53 Taylor vortices produced in Couette apparatus (a) Regularly-spaced toroidal Taylor vortices and (b) Wavy Taylor vortices.

liquid of density, ρ, and molecular viscosity, μ, and a small mass of neutrally buoyant and reflective tracer particles. As the inner cylinder rotates it exerts a torque on the liquid in the annular space, causing a boundary layer to be set up so that the fluid closest to the outer wall rotates less rapidly than that adjacent to the inner wall. At very low rates of spin nothing remarkable happens but as the spin is increased a number of regularly spaced zonal (toroidal) rings, termed *Taylor cells*, form normal to the axis of the cylinders (Fig. 4.53); then, at some critical spin rate these begin to deform into wavy meridional vortices. These begin to form at a critical inner cylinder rotational Reynolds number, $Re_i = r_i \Omega d\rho/\mu$, of about 100–120, with the 3D wave like instabilities beginning at $Re_i > 130$–140. At high rates of spin the flow becomes turbulent, the 3D wavy structure is suppressed and the Taylor ring structure becomes dominant once more. Taylor cell vortex motions involve separation of the flow into pairs of counter rotating vortex cells.

4.10 Transport by waves

4.10.1 Transport under shallow water surface gravity waves

The previous sections made it clear that a sea or lake bed under shallow water surface gravity waves is subject to an

oscillatory pattern of motion (Fig. 4.54). As the velocity of this motion increases, sediment is put into similar motion. Experiments reveal that once the threshold for motion is passed then the sediment bed is molded into a pattern of

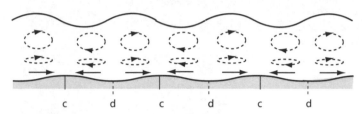

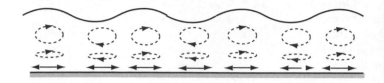

The maximum horizontal orbital velocity of a shallow water wave of surface speed

$$c = (gh)^{0.5}, \text{ is } u_{max} = H/2h\,(gh)^{0.5},$$

where H = wave height and *h* = water depth.

Fig. 4.54 The pattern of oscillatory motion under progressive surface shallow water gravity waves engenders a to-and-fro motion to any sea or lake bed. Should this bed be a loose boundary of sand, gravel, or silt then bed defects cause net sediment transport and planes of divergence (d) to convergence (c). These gradually develop into symmetrical ripple-like bedforms.

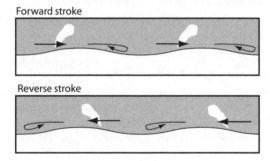

Forward stroke

Reverse stroke

Fig. 4.55 Once developed the forward and reverse portions of the to-and-fro oscillatory motion develop flow separation on the ripple lee side and a "jet" of suspended sediment upstream.

ripple forms, termed *wave-formed* or *oscillatory ripples*. The wavelength and height of these ripples, of order decimeters to centimeters respectively, reflects in a simple way the decay in the magnitude of the oscillating flow cell transmitted from wave surface to bed. The oscillatory flow induces alternate formation of closed "roller" vortices in the lee of either side of ripple crests during each forward and backward stroke of the cycle. As the oscillatory flow increases further in magnitude, the "up" part of each half-stroke sends a plume of suspended sediment into the water column (Fig. 4.55) and gradually an equilibrium suspension layer is formed that increases in thickness and concentration with increasing wave power. Experiments also reveal that wave ripples in shallow water have an inherent "wave-drift" landward (Fig. 4.56). The ripples themselves continuously adjust to changing wave period during storms (Fig. 4.57) and may reach wavelengths of up to

1 m for wave periods of >10 s. At some critical junction the increasingly 3D bed ripples are planed off and a flat sediment bed is formed under a thick layer of suspended sediment.

4.10.2 Transport under combined surface shallow water surface waves and tidal currents

The observations made on transport under progressive waves are perfectly valid for environments like lakes, but in the shallow ocean, tidal currents of varying magnitude and direction are invariably superimposed. These currents may cause net transport of suspended sediment put up into the flow by near-bed oscillatory motions. For low energy conditions over smooth flow boundaries there seems to be little overall effect of the current on near-bed values of fluid shear stress due to the waves alone. At some critical transport stage rough-bed flows show increased near-bed vertical turbulent stresses and suspended sediment concentrations: it seems that some sort of interaction is set up between the bed roughness elements, the flow, and the oscillations.

4.10.3 Transport and mixing under internal progressive gravity waves

Internal progressive gravity waves have important roles in ocean water mixing and the transport and erosion of substrates (Section 6.4.4). Vertical mixing occurs as internal

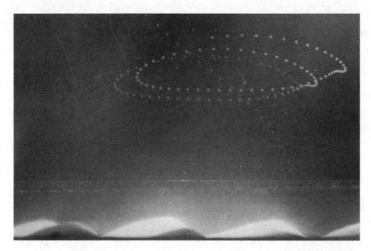

Fig. 4.56 These oscillation wave ripples formed in sand on the bed of a laboratory channel are being generated under progressive shallow water waves. Water depth is about two ripple wavelengths and the period of the surface waves is *c*.3 s. The small illuminated dots are reflected light from a small neutrally buoyant marker particle that has been photographed stroboscopically. The pattern is noteworthy for its demonstration of Stokes wave drift, whereby net forward motion occurs in shallow water waves. This engenders a net forward sediment transport vector and a forward asymmetry to the ripple forms.

Fig. 4.57 Marta paddling beside a group of spectacular steep and linear symmetrical wave formed ripples developed on sand. The ripples developed under storm wave conditions, probably with some amplification in the beach inlet.

waves break under a critical vertical gradient in imposed shear and create turbulence. Because of the Coriolis effect the efficacy of the resulting mixing process decreases equatorward. Progressive internal waves commonly develop at the shelf edge and in fjords in summer months when shelf waters are relatively undisturbed by storms and when thermal stratification is at maximum. Erosion of fine-grained substates by internal wave motion is thought to cause enhanced sediment suspension that is "captured" in the interfacial zone of influence of internal wave oscillations. Once established, these interfacial layers of enhanced concentration (termed *nepheloid layers*) may drift shoreward or oceanward. The density interfaces formed by the stratification may trap organic suspensions stirred-up from the bottom or derived from settling from the oceanic photic zone above. Combined with any tendency for summer upwelling, the sites of internal wave generation may thus focus organic productivity.

4.11   Granular gravity flow

At home we are familiar with granular flow, dawdling over the breakfast table with a jar of muesli or cereal, a pot of sugar crystals, or a salt cellar. Each of these materials is a granular aggregate, quite stable within its container walls until tilted to a certain critical angle, upon which the particles loose themselves from their neighbors and tumble down the inclined face. We sleepily observe that the grain aggregates must transport themselves with no help from the surrounding fluid, in this case, air. We deduce by observation that aggregates of particles may either be at rest in a stable fashion or else they flow downslope like a fluid. How does this behavior come about?

4.11.1   Reynolds again

As so often in this text we follow the pioneering footsteps (literally damp footsteps in this case) of Reynolds, who presented basic observations and hypotheses on the problem in 1885. Reynolds pointed out that ideal, rigid, smooth particles had long been used to explain the dynamics of matter and that more recently they formed the physical basis for the kinetic theory of gases and explanations for diffusion. He pointed out, however, that the natural behavior of masses of rigid particles, exemplified as he strode over a damp sandy beach, had a unique property not possessed by fluids or continuous solids that "consists in a definite change of bulk, consequent on a definite change of shape or distortional strain, any disturbance whatever causing a change of volume." Reynolds' walks across newly exposed but still water-saturated beach sand: "When the falling tide leaves the sand firm, as the foot falls on it the sand whitens, or appears momentarily to dry round the foot . . . the pressure of the foot causing dilatation . . . the surface of the water . . . lowered below that of the sand." Let us develop Reynolds' concept in our own way.

4.11.2   Static properties of grains

In order to simplify the initial problem, we assume, as did Reynolds, that the particles in question are perfectly round spheres. We are thus dealing with *macroscopic* particles of a size too large to exhibit mutual attraction or repulsion due to surface energies, as envisaged for atoms. While at rest a mass of such particles must support itself against gravity at the myriad of contact points between individual grains (Fig. 4.58). We can imagine two end-members for geometrical arrangement, the *ordering* or *packing*, of such spheres. The maximum possible close packing would place the spheres in cannon ball fashion, each fitting snugly within the depression formed by the array of neighbors below and above. By way of contrast, the minimum possible close packing would be a more idealized arrangement, difficult to obtain in practice, but nevertheless possible, where each sphere rests exactly above or below adjacent spheres. The reader may recognize these packing arrangements as similar to those revealed by x-ray analysis of the arrangement of atoms in certain crystalline solids, the former termed *rhombohedral* and the latter *cubic*.

Using these simple end-member models for ideal packing we can define an important static property of granular aggregates. This is solid concentration, C, or fractional *packing density*. Its inverse is $(1 - C)$, defining the *intergranular concentration, P*, termed *porosity* or *void fraction*. To calculate C we take the total volume of space occupied by the grain aggregate as a whole, as for example in some real or imaginary container of known volume, and express the fraction of its space occupied by the solid grains alone.

Fig. 4.58 (a) Mode of granular packing epitomized by this stable pyramid of cannonballs. (b) and (c) Any displacement from condition (b) to (c) must involve a dilatation of magnitude, $\Delta d = y_2 - y_1$.

The minimum possible solid concentration, $C = 0.52$, is for cubic packing ($\pi/6$) and the maximum, 0.74, is for rhombohedral packing ($1/\sqrt{2} \cdot \pi/3$). In practice, as Reynolds pointed out, natural solid concentration varies widely, but always between our upper and lower limits. Both C and P are obviously important properties of natural granular aggregates like sediment and sedimentary rock. They control the ability of such aggregates to hold fluid in their pore space, be it water in aquifers, hydrocarbons in reservoirs, or magma melt in the crust or mantle. Also, the size of the pores has an important control over rate of fluid throughflow, termed *permeability* (Sections 4.13 and 6.7).

4.11.3 Conditions for a stationary aggregate to shear and/or flow

There is only one force, gravity, that can cause the self propelled flow of grain aggregates. This may at first seem strange because it is gravity after all that is keeping the grain aggregate stable, by pulling each grain downward toward its neighbors (Fig. 4.59). But, in an experiment where we suddenly free an aggregate from a container or containing medium, the grains flow outward, shearing, colliding, and bouncing as they do so. So, it seems that the aggregate as a whole has no apparent resistance to shear! But is this not the characteristic of a fluid? How can

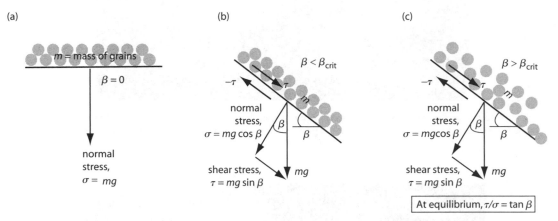

Fig. 4.59 Conditions for grain shear. (a) Grains on a horizontal surface, (b) grains on a slope just prior to granular flow, and (c) grains shearing on a slope during granular flow.

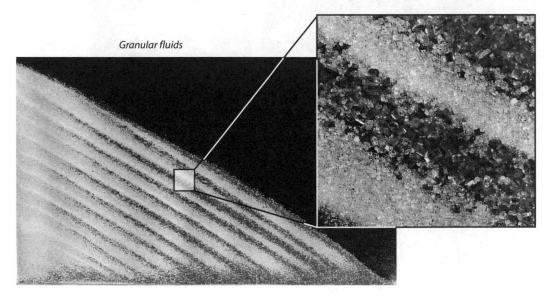

Fig. 4.60 A random initial mixture of larger sugar crystals (dark) and glass beads from a reservoir has avalanched down a 45° slope, spontaneously segregating and stratifying during transport.

a multitude of solid grains behave like a fluid? The answer is that flowing fluid behavior only occurs once a critical limit to stability has been exceeded and that it only ends once another limit is reached. The initial condition, a vertical wall of grains, was evidently in excess of the stability limit. The final conditions, defining a conical pile of grains with slopes resting at a certain average angle to the local horizontal surface, were within the limit.

In order to explain these phenomena we return to Reynolds' packing modes (Fig. 4.58). Any shear of a natural aggregate of grains ($C < 0.74$) must involve the expansion of the volume as a whole. Take the case of an array of spheres in perfect rhombohedral packing. These must be sheared and raisedup by a small average distance, Δd, over their lower neighbors before they can shear and/or slide off as a flowing mass; the grain mass suffers an

Fig. 4.61 An initial random mix of *Riojanas* beans and *Valencia* rice in a glass container is shaken at 3 Hz for 20 s. All the beans rise, magically, to the surface. Physicists use such behavior to shed light on the properties of granular fluids as analogs for the kinetic theory of gases and solids.

Fig. 4.62 Natural snow avalanches are a major hazard in mountain ski resorts. Any inclined pack of snow layers contains weak granular or refrozen horizons which are easily disturbed by ground or air vibrations. Low friction means gravity collapse can occur and the snow pack disintegrates into a granular flow whose equilibrium velocity may exceed 20 m s^{-1}.

expansion. The expansion, Reynolds' *dilatation*, of granular masses under shear, requires energy to be expended because in effect we are having to increase the solid layer's potential energy by a small amount proportional to Δd. Some force, an inertial one via Reynolds' descending foot, is required to do this. A gravity force may be more directly imagined using a variant of Leonardo's friction experiment (Section 3.9), as an initially horizontal solid body free to move rests on another fixed solid body. As the contact between the bodies is gradually steepened a critical energy threshold is exceeded, at a slope angle termed the angle of *static friction* or *initial yield*, ϕ_i. Here the block moves downslope as the roughnesses making up the contact surface dilatate. In the case of a loose aggregate, the grains flow downslope until they accumulate as a lower pile whose slope angle is now less than the initial slope threshold that caused the flow to occur in the first place. This lower slope angle, termed the angle of *residual friction* or shear, ϕ_r, is usually 5–15° less than the initial angle of yield for natural sand grains. The value $\phi_i - \phi_r$ gives the dilatational rotation required for shear and flow. Some more details on the often rather complicated controls on natural sand frictional behavior are given in Cookie 17.

4.11.4 Simple collisional dynamics of granular flows

Once in motion a granular flow comprises a multitude of grains kept in motion above a basal shear plane. An equilibrium must be set up such that the weight force of the grains is resisted by an equal and opposite force, σ, arising from the transfer of normal grain momentum onto the shear plane. This concept of *dispersive normal stress* proposed by Bagnold (Cookie 18) is analogous to the transfer of molecular momentum against a containing wall of a vessel envisaged in the kinetic theory of gases (Section 4.18). Such normal stresses have been used to explain the frequent occurrence of upward-increasing grain size, in the deposits of granular flows (see below). Marked downslope variations in sorting and grain size also develop spontaneously (Fig. 4.60): larger grains are carried further than smaller grains because they have the largest kinetic energy. This leads to lateral (downslope) segregation of grain size. More interestingly, when the larger grains have higher values of ϕ, the mixture spontaneously stratifies as the smaller grains halt first and the larger grains form an upslope-ascending grain layer above them.

The phenomenon is popularly framed in granular physics as the "Brazil nut problem," or "why do Brazil nuts rise to the top of shaken Muesli?" (Fig. 4.61).

Bagnold originally proposed that the dispersive stress is greatest close to the basal shear plane of the granular flow and that there, large particles exerted a higher stress (to the square of diameter). Hence these larger particles move upward through the flow boundary layer to equalize stress gradients. However, a second hypothesis, termed kinetic filtering, says that small grains simply filter through the voids left momentarily below larger jostling grains until they rest close to the shear plane; the larger grains must therefore simply rise as a consequence. A simple test for the rival hypotheses is to shear grains of equal size but contrasting density, since σ also depends upon grain density. It is observed that sometimes the densest grains do indeed rise to the flow surface. Further experiments with naturally varying grain density *and* size reveals variable patterns of grain segregation depending on size and density of grains and the frequency of vibration. The dispersive stress hypothesis is only partly confirmed by such

Fig. 4.63 Sand avalanches on the steep leeside slope of a desert dune. Here, repeated failure has occurred at the top of the dune face: the sand has flowed downslope as a granular fluid, "stick-slip" shearing internally to produce the observed pressure-ridges as it does so. Shear along internal failure planes causes acoustic energy signals to propagate, hence the "singing of the sands" that haunted early desert explorers.

observations: kinetic filtering is the chief mechanism for sorting and grain migration in multi sized granular flows, the commonest situation in Nature.

A further intriguing complication is demonstrated by a vibrated granular mass in a container of equal-sized grains containing one larger grain. The vibrations induce inter-granular collisions and a pattern of advection within the container, with the smaller grains continuously migrating down the walls of the container, while the larger grain, and adjacent smaller grains move up the center. Patterns also arise at the free surface of vibrating grain aggregates, the newly discovered *oscillons* creating much interest among physicists in the mid-1990s.

The wider environment of Earth's surface provides many examples of the flow of particles: witness the peri-odic downslope movement of dune sand, screen deposits, or the spectacular sudden triggering of powder snow or rock avalanches (Figs 4.62 and 4.63).

4.12 Turbidity flows

As we saw in Section 3.6, buoyancy flows in general owe their motion to forces arising from density contrasts between local and surrounding fluid. Density contrasts due to temperature and salinity gradients are common-place in the atmosphere (Section 6.1) and ocean (Section 6.4) for a variety of reasons. In turbidity flows it is sus-pended particles that cause flow density to be greater than that of the ambient fluid. In this chapter we consider sub-aqueous turbidity flows; we consider the equivalent class of volcanic density currents in the atmosphere in Section 5.1. The fluid dynamics of turbulent suspensions is a highly complicated field because the suspended particles (1) have a natural tendency to settle during flow, (2) affect the tur-bulent characteristics of the flow. The trick in understand-ing the dynamics of such flows therefore involves understanding the means by which sediment suspension is reached and then maintained during downslope flow and deposition. It is probable that natural turbidity flows span the whole spectrum of sediment concentration, but it seems that many are dominated by suspended mud- and silt-grade particles.

4.12.1 Origins of turbidity currents

The majority of turbidity currents probably originate by the flow transformation of sediment slides and slumps caused by scarp or slope collapse along continental mar-gins (Fig. 4.64). These are often, but not invariably, caused by earthquake shocks and are undoubtedly facili-tated at sea level lowstands when high deposition rates from deltas, grounding ice masses, or iceberg "graveyards" provide ample conditions for slope collapse. A role for methane gas hydrates in providing regional mass failure planes in buried sediment is suspected in some cases. Slides are thought to transform to liquefied and fluidized slumps and then to disaggregate into visco-plastic debris flows. These cannot transform further into turbulent suspensions without massive entrainment of ambient seawater, and this is not possible across the irrotational flow front of a debris flow. Instead, debris flows must transform along their upper edges by turbulent separation (Fig. 4.64).

Turbidity currents also form from direct underflow of suspension-charged river water in so-called *hyperpycnal plumes*, also better termed as *turbidity wall jets*. These have been recorded during snowmelt floods in steep-sided basins like fjords and glaciated lakes, in front of deltas, and in river tributaries whose feeder channels have extremely high loads of suspended sediments. As noted below, these freshwater underflows may undergo spectacular behavior during the dying stages of their evolution. Underflows are expected to give rise to predominantly silty or muddy turbidites.

Finally, collection of sediment by longshore drift in the nearshore heads of submarine canyons may also lead to downslope turbidity flow. The process is most efficient during and following storms and tends to lead to the trans-port and deposition of sandy sediment.

4.12.2 Experimental analogs for turbidity currents

Turbidity currents are difficult to observe in nature and to maintain in correctly-scaled laboratory experiments. We may best illustrate their general appearance by studying saline and scaled particle currents (Fig. 4.65) using lock-gate tanks or continuous underflows. In the former, as the lock-gate is removed, a surge of dense fluid moves along the horizontal floor of the tank as a density current with well-developed *head* and *tail* regions. Under these zero-slope conditions the head is usually 1.5–2 times thicker than the tail, with the ratio approaching unity as the depth of the ambient fluid approaches the depth of the density flow. Close examination of the head region shows it to be divided into an array of bulbous lobes and trumpet-shaped clefts. Ambient fluid must clearly pass into the body of the flow under the over-hanging lobes and through the clefts. A greater mixing of

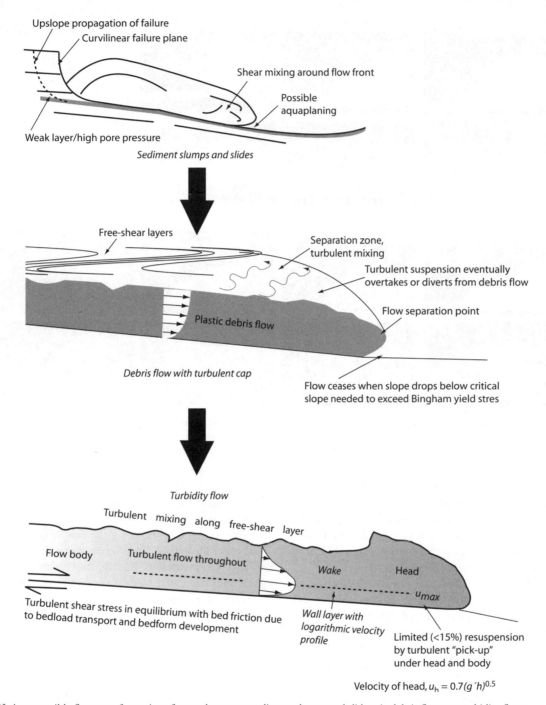

Upslope propagation of failure

Curvilinear failure plane

Shear mixing around flow front

Possible aquaplaning

Weak layer/high pore pressure

Sediment slumps and slides

Free-shear layers

Separation zone, turbulent mixing

Turbulent suspension eventually overtakes or diverts from debris flow

Flow separation point

Plastic debris flow

Debris flow with turbulent cap

Flow ceases when slope drops below critical slope needed to exceed Bingham yield stres

Turbidity flow

Turbulent mixing along free-shear layer

Flow body

Turbulent flow throughout

Wake

Head

Turbulent shear stress in equilibrium with bed friction due to bedload transport and bedform development

Wall layer with logarithmic velocity profile

u_{max}

Limited (<15%) resuspension by turbulent "pick-up" under head and body

Velocity of head, $u_h = 0.7(g'h)^{0.5}$

Fig. 4.64 Various possible flow transformations from subaqueous sediment slumps and slides via debris flows to turbidity flows.

denser and ambient fluid also takes place by entrainment behind the head. Continued forward motion of the head at constant velocity requires a transfer of denser fluid (buoyancy flux) from the tail into the head (and thus for the tail to move faster than the head) in order to compensate for boundary friction, fluid mixing, and loss of denser fluid in the head region. A steady state is brought about in flows that have a near-constant input of dense solution with time.

In Nature, steady turbidity flows might occur over a period of time as sediment-laden river water debouches into a water body and travels along the bottom as a continuous underflow. By way of contrast, surge-like

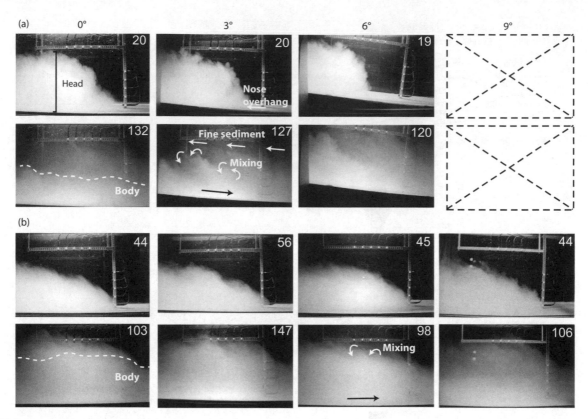

Fig. 4.65 Experimental turbidity flows (photos are *c*.0.5 m vertical extent) traveling down progressively steeper ramps (columns of slopes 0°–9°) onto flat horizontal surfaces. (a) Turbidity flow heads (first row) and bodies (second row) on increasing ramp slopes. (b) Turbidity flow heads (first row) and bodies (second row) on flat horizontal floors leading from upstream ramp slopes as indicated in (a).

turbidity flows generated from finite-volume sediment slumps or debris flows on small slopes (<1°) must decelerate because the supply of denser fluid from behind the head is finite and the buoyancy force driving the flow is insufficient to overcome frictional energy losses. The head thus shrinks until it is completely dissipated.

On slopes from at least 0.5° to 5°, head velocity of continuous underflows is independent of slope and varies according to density contrast. Head velocity is approximately 60 percent of the tail velocity in the slope range 5° to 50°, leading to the head increasing in size as it travels downslope. Entrainment of ambient fluid also causes head growth, increasingly so at higher slopes, and the momentum transferred from the current to this new fluid acts as a retarding force to counteract the buoyancy force due to any increased slope. This "steady velocity/growing head" behavior is also a characteristic of starting thermal plumes but has not been investigated from the point of view of turbidity current deposition and erosion.

Rapid dissipation of channelized turbidity flows with consequent deposition occurs as they undergo vertical expansion and lateral spreading on entering wide reservoirs

or basins. Such flows are *wall jets*, that is, aperture flows released onto the floor of large volume reservoirs (volcanic equivalents are discussed in Section 5.1). An interesting transformation takes place for the case of dissipating turbid *freshwater underflows*. These were particularly important in oceanic sedimentation during the melting phases after glaciations when vast quantities of sediment-laden freshwater were released to the oceans. In such systems, deposition progressively reduces the buoyancy force, the bulk flow density eventually decreases below that of sea water. The flow comes to a complete halt, with the now positively buoyant fluid rising upward to spread out within density interfaces in the ocean or at the surface as a plume. The process has been termed *lofting* and leads to widespread suspension and eventual deposition of suspended muds.

4.12.3 Velocity and turbulence characteristics of experimental turbidity flows

How quickly can dense underflows travel? We might guess that a solution would be to treat the surge as a moving

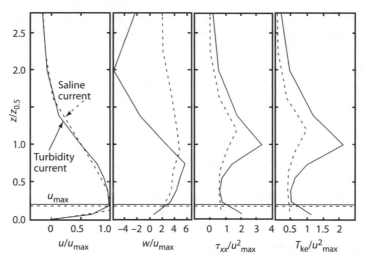

Fig. 4.66 Mean velocities (u, w), turbulent stress (τ_{xx}), and turbulent kinetic energy (T_{ke}) contrasts between experimental saline and turbidity flows of wall jet type. Dimensionless height is with reference to $z_{0.5}$, the height at which u reaches value 0.5 u_{max}.

bore or shallow-water wave. If so, the conversion of potential energy (due to mean height above the tank floor) to the kinetic energy of motion gives the velocity of motion, u, and proportional to the square root of water depth, h. We might also guess that u should depend directly upon the density difference, $\Delta\rho$, between the current and the ambient fluid, or more precisely the action of reduced gravity, g'.

Internal velocity profiles taken through experimental density and turbidity currents reveal a positive velocity gradient in the lower part of the flows; this follows the normal turbulent law-of-the-wall. Above a velocity maximum, there is a negative gradient up to the top of the flow (Fig. 4.66). The latter is due to frictional interactions and overall retardation of the flow by the ambient fluid in the form of production of large-scale eddies of Kelvin–Helmholtz type. Turbulent stresses are much increased for particulate turbidity flows over saline analogs (Fig. 4.64).

4.12.4 Reflected density and turbidity flows

Because turbidity flows derive their motive force from the action of gravity, they are easily influenced by submarine slope changes. Flows may partially run up, completely run up and overshoot, or be partially or wholly blocked, diverted, or reflected from topographic obstacles. The process of run-up and full or partial reflection ("sloshing") is particularly interesting and the effects may be seen by inserting ramps into the kind of lock exchange tanks described previously (Fig. 4.67). Run-up elevations are approximately 1.5 times flow thickness and in nature it is

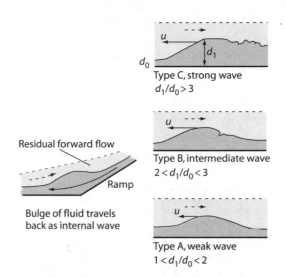

Fig. 4.67 Forward turbidity flow meets opposing topographic ramp slope and reflects back under residual forward flow as an internal solitary wave.

evident that the process can cause upslope deposition on submarine highs. Reflection may be accompanied by the transformation of the turbidity flow into a series of translating symmetrical waves, which have the properties of solitary waves or bores (Section 4.9). They travel back in the up-source direction undercutting the slowly-moving nether regions of the still-moving forward current, transporting fluid and sediment mass as they do so (Fig. 4.67). Such internal bores have little vorticity, as witnessed by their smooth forms.

4.12.5   Deposition from turbidity flows

Deposition from turbidity currents may be due to flow unsteadiness or to downstream flow nonuniformity. It is commonly thought that passage of the head is accompanied by erosion. The extent of this is a little known quantity of some importance in calculating the flux of bed material into a moving flow. Clearly any extra sediment added to the flow will cause the head of the flow to accelerate. Careful sediment budget studies of Holocene deposits in the Madeira abyssal plain reveals that 12 percent or so of turbidite volume is composed of reworked materials. This converts to about an average of 43 mm of erosion over the total areal extent of an individual deposit.

4.13   Flow through porous and granular solids

4.13.1   Flow through stationary porous solids

"Thirsty stones" are disc-shaped surfers cut from highly porous sandstone which soak up superfluous fluid spilt from drinks containers. The easy flow of fluid occurs through what we call generally a *porous medium*. Many subsurface fluid flows, of water, oil, and gas occur through such media. In general the pore fluid flow is very slow (order $<10^{-3}\,\text{m s}^{-1}$) and therefore laminar, so we can entirely neglect the kinetic contribution to flow energy. We can also neglect the effects of frictional losses at a constant rate for present purposes. From the Euler–Bernoulli energy equation (Section 3.12; Cookie 9) we are left with the total energy available to drive pore water flow as $(p/\rho) + z = \Phi$, termed the *pore water fluid potential*. From the simple working in Fig. 4.68 we see that Φ is given by the product of a constant, g, times the hydraulic head, h. Thus for practical purposes it suffices to measure the level, h, in stand pipes or wells and then to map the hydraulic water surface over the area in question (see Section 6.7.1). The pore water velocity, u, is then proportional to the *gradient*, ∇, of the head field, Φ, with flow in the direction of greatest decrease of the field. Note carefully that this gradient, called the *hydraulic gradient*, is *not* the same as the gradient of hydrostatic pore pressure (Fig. 4.68; see also Sections 3.5 and 4.1). In symbols, $\mathbf{u} = -\nabla \Phi K$, where K is the proportionality coefficient known as the *hydraulic conductivity*. The expression is universally known as *Darcy's law*, namd after its originator who was intrigued by the controls of rock hydraulic pressures on the flow of fountains in Dijon, France in the nineteenth century. There is some similarity of form of this equation

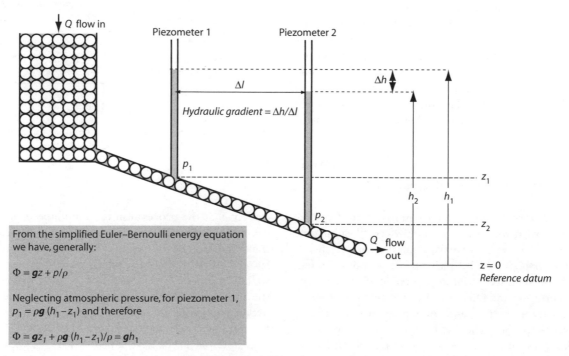

From the simplified Euler–Bernoulli energy equation we have, generally:

$$\Phi = gz + p/\rho$$

Neglecting atmospheric pressure, for piezometer 1, $p_1 = \rho g (h_1 - z_1)$ and therefore

$$\Phi = gz_1 + \rho g (h_1 - z_1)/\rho = gh_1$$

Fig. 4.68 An experiment similar to that conducted by M. Darcy in 1856 to determine the energy relations of flow through porous media.

with that for the 1D flow of heat through solid media (Section 4.18, Cookie 19). Like the heat conduction equation, it is an easy matter to solve the expression for equilibrium flow once K, a constant for a given arrangement of porosity, pore shape, pore size, fluid density, and fluid viscosity, is known. The *Kozeny–Carman equation* (not developed here) offers an approximate analytical solution to determining K as a function of these variables as long as the particle packing is random and natural homogenous isotropic porosity occurs.

4.13.2 Liquefaction and fluidization of granular aggregates

We now turn to the case when the particles in saturated granular aggregates (Section 4.11) are simply resting upon one another under gravity, with no cementing medium holding them together. It is possible to imagine the situation where throughflow might cause motion of the particles away from each other: the grain aggregate is then in a state of *liquefaction*. For example, it is a common observation after large earthquakes that sand and water mixtures are expelled (often violently) from the shallow subsurface in fountains to form sand volcanoes (Fig. 4.69). Here are the observations made by an anonymous observer of the great New Madrid earthquake of 1811–12 in the lower Mississippi Valley: "Great amounts of liquid spurted into the air, it rushed out in all quarters . . . ejected to the

height from ten to fifteen feet, and in a black shower, mixed with sand . . . The whole surface of the country remained covered with holes, which resembled so many craters of volcanoes." There is also widespread evidence for the sinking of buildings and other structures into sands during earthquakes, recalling the Biblical adage that predates modern building regulations by some millennia. Although earthquakes undoubtedly impose an effective trigger mechanism for liquefaction, they are by no means the only one. Thus eruption of sediment and water commonly occurs during river floods from the bases of artificial and natural levees.

What processes lead to production of a liquid-like state from what was previously a stable mass of water-saturated sand? There are two ways in which such a transformation can take place:

1 Upward displacement of fluid may be sufficient in itself to cause overlying grains to remain in suspension by *fluidization* or *seepage liquefaction*, conveniently defined as the process whereby a granular aggregate is converted to a fluid by flow through it. The fluidized suspension may be en masse or just restricted to selected pipe-like conduits as in the New Madrid earthquake. Fluidization in its most simple sense requires the velocity of moving pore fluid, U_f, to exceed the grain fall velocity, U_p. More generally, in order for en masse fluidization to occur, the upward-moving fluid must exert a normal stress σ_f equal to or greater than the immersed weight of any overlying sediment. This force is given by $\sigma_f = \Delta\rho g C h$, where $\Delta\rho$ is effective (immersed) grain density, g is gravity, h is overlying

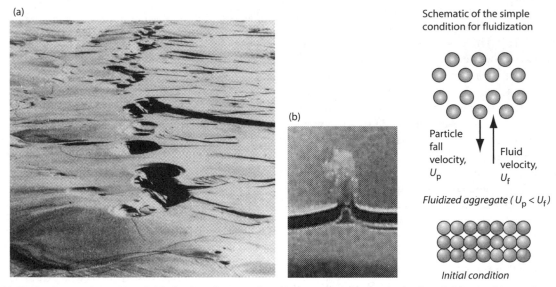

Fig. 4.69 (a) Sand volcanoes formed in a field after liquefaction induced by an earthquake, Imperial valley, California; (b) experimental fluidization and sudden conduit escape as a sand volcano eruption produced by water flow through a lower, more permeable, granular bed underlying an upper less permeable bed (dark layer).

sediment layer thickness, and *C* is fractional grain concentration. Readers may be familiar with industrial and technological uses of fluidization, notably the passage of hot gases through powdered coal beds to optimize burn efficiency. In this case gas fluidization is often accompanied by bubble formation, which is uncommon in liquid fluidization. Gas-induced fluidization is thought by some to have an important role in the maintenance of certain volcanic eruptive flows, notably *pyroclastic density currents* (Section 5.1).

2 It may result from the temporary collapse of grain-to-grain contacts in loosely packed sand. The grains shaken apart are momentarily suspended in their own porewater. The energy for granular disaggregation is provided by the high accelerations experienced during cyclic shock caused by earthquake ground motions (Section 4.17). Pressure changes due to the passage of large storm waves over a sandy or muddy (including "fluid mud") bottom may also cause liquefaction, as can the sudden arrival of a turbidity current or repeated impact of feet on saturated sands on the beach or river bed. Should the liquefied sediment be resting on a slope, however slight, then downslope flow will inevitably result, the flow transforming as it does so into a debris or turbidity flow (Section 4.12). Post-liquefaction resettlement causes a net upward displacement of pore fluid of volume proportional to the difference in pre- and post-liquefaction porosity. The spatial funneling of this flow is the cause of the violent fluidization witnessed after earthquakes (Fig. 4.69).

4.14   Fractures

In Section 3.15 the concept of brittle versus ductile behavior was explained. Brittle behavior involves the loss of cohesion in rocks when deviatoric stresses higher than the rock resistance are applied. The planes where rocks have lost their cohesion are called *fractures*. In other words the rocks break under certain conditions and fractures are the places where the rocks break. Nevertheless, the definition of fractures is a little more complicated than that. Pure brittle fractures are well-defined surfaces where the cohesion has been lost, without any distortion; if we could glue together the pieces of the broken rock, its shape would be the same as before (Fig. 4.70a). Brittle fractures are characteristic of rocks exhibiting elastic behavior and form in the upper part of the crust (about 10 km), although sudden stresses can cause brittle failure in solid viscous materials (Section 3.15). There is a broad and very intriguing transition between the brittle and ductile fields, where the definition of fractures is not so obvious; for example the material in Fig. 4.70b shows some ductile deformation accompanying the fracture since some bending occurred before the fracture was produced. In this case, joining together the fragments of the original piece of rock will not give the rock's original shape. Fractures in the brittle-ductile transition (*semi-brittle* behavior) often show intense deformation where some cohesive loss is achieved in limited, non-continuous surfaces, instead of developing a well-defined unique fracture surface. Semi-brittle fractures do not experience a total loss of coherence on well-defined surfaces. Further increase in ductility, but still in the brittle-ductile transition, produces *shear bands* where a cohort of oblique, lenticular-shaped, discrete fractures is formed (tension gashes, Fig. 4.70c). *Ductile shear zones* are characterized by intense deformation, without apparent loss of cohesion at macroscopic scale (visual or outcrop scales), concentrated in a fairly narrow band (2D) surrounded by distinctively less deformed rocks (Fig. 4.70d). Such shear zones form at deep levels in the crust.

4.14.1   Types of fractures

The first important analysis of fractures involves how the rock bodies on either side of the fracture move. According to this criterion they can be classified broadly into two

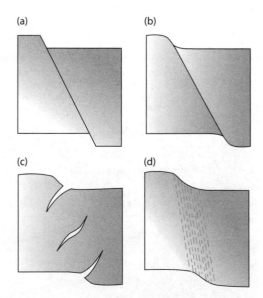

Fig. 4.70 Different kinds of shear features: (a) Pure brittle shear fracture, (b) in the brittle–ductile transition, fractures occur after some previous material distortion, (c) gash fractures in the brittle–ductile transition, and (d) ductile shear zone.

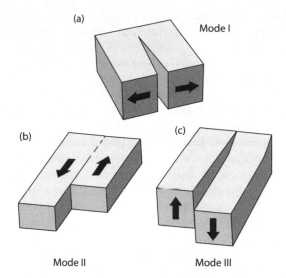

(a) Mode I

(b) Mode II

(c) Mode III

Fig. 4.71 (a) Extension fractures (Mode I) open normal to the crack, whereas, shear fractures (b) and (c) show displacements of blocks parallel to the fracture. Mode II shear fractures (b) move normal to the fracture edge and Mode III (c) move parallel to the crack edge.

types: *extension fractures* and *shear fractures*. Extension fractures or Mode I fractures (Fig. 4.71a), open perpendicularly to the fracture surface and do not experience any displacement parallel to the fracture plane. By way of contrast, shear fractures are characterized by a noticeable displacement of the blocks along or parallel to the fracture plane (Fig. 4.71b, c). This displacement can be produced perpendicular to the fracture edge (Mode II) or parallel to the fracture edge (Mode III). Faults are large shear fractures (surfaces extending from several meters to the scale of plate boundaries); the term shear fracture is mostly used for centimeter-scale fractures.

4.14.2 Extension fractures (Mode I)

Mode I fractures or extension fractures form in the principal plane of stress containing the principal axes σ_1 and σ_2, perpendicular to the direction of minimum stress σ_3. It is important to remember that in these surfaces there are no shear stress components and so there is no shear movement or displacement of blocks along the fracture surface. Extension fractures do not show offsets either side of the fracture, even at a microscopic scale. Thus extension fractures form by opening or pulling apart of blocks of rock at each side of the fracture when the tensile stress exceeds the strength of the rock and brittle deformation occurs. Such brittle deformation is characteristic of the upper part of the crust. At depth, moving into a more ductile field, extension

fractures are less common and become infilled by mineral precipitates from fluid solution, forming veins. Fractures can be observed in rock outcrops in different shapes and varieties, the commonest features are roughly planar surfaces or joints. These are cracks that extend from centimeters to hundreds of meters. They vary in shape from irregular discontinuous fractures to almost perfectly planar features. Regularly spaced, planar fracture surfaces showing roughly the same orientation form a set and are called systematic joints. On the contrary, irregular, discontinuous, arbitrarily-orientated fractures are called nonsystematic. Different sets of *joints* can be seen cutting each other in many outcrops. As they open perpendicularly to the fracture surface, the trajectories of the individual joints are not much affected by others. Nonetheless, adjustments of space once the fractures are produced may cause some minor shear displacements along the fracture surfaces. Joint surfaces can be smooth or show some interesting features such as *plumose structures*, *fringes*, or *conchoidal structures* which are very useful in discerning and describing fracture propagation. Plumose structures are linear irregularities, sometimes curved or wavy, arranged as in a feather or fan fashion, parting from a single point and ending in a narrow band or *fringe* at both sides. The fringe is formed by an array of discrete en échelon fractures (Fig. 4.72). Plumose structures form in the propagation direction of the cracks, whereas conchoidal structures are formed perpendicularly to the general direction of the plumes and can be envisaged as discontinuities in the fracture propagation. These structures can be in the form of steps (Fig. 4.72b), ribs, or ripples.

Sheet joints are subhorizontal fractures, having a tendency to parallel the topography. This arrangement gives two basic interpretations for their origin. First the parallelism can be seen as the cause of the topography, because the preexisting fractures are sites of weakness in the rock and thus control denudation patterns. A more likely explanation is that topography, being a shear free surface, affects the orientation of stresses in a principal stress plane containing two of the principal stresses. Either way, sheet joints must form when the main compressive stress is horizontal and the minimum compression is vertical which happens during crustal uplift or in general compressive tectonic settings.

Columnar joints form in lava flows and shallow intrusions as the rocks experience a volume loss by cooling in discrete domains. Commonly three conjugate sets of fractures develop, arranged in geometric patterns, mostly hexagonal as in desiccation cracks in drying wet clays. The cracks initiate at some point in the lava flow due to tensile stresses which arise because of differences in temperature and volume from the surface where the lava flow is cooler

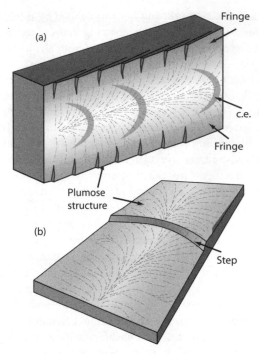

Fig. 4.72 (a) Joint surfaces show plumose and conchoidal structures (c.e.) in the main crack surface, showing the propagation of the fracture. Fringes are located at the edges of the joints and show a set of en echelon cracks. (b) Detail of a step in a conchoidal structure, cutting the plumose feature in a joint surface.

and deeper areas which are hotter and less viscous; they then propagate forming columnar features orientated roughly perpendicularly to the top and bottom boundaries of the rock body, although curved patterns of the columns are quite common (Fig. 4.73). Increments in crack propagation may result in a banding normal to the columns, displaying individual plumose structures like the ones described in joints.

Gash fractures are lenticular or sigmoid fissures (both Z and S shaped), generally filled with some mineral (most abundantly calcite or quartz) that develop in shear zones at the brittle-ductile transition (Fig. 4.74), where a combination of both ductile flow and brittle fracturing occurs. Gash veins open normal to the maximum elongation e_1 both in coaxial, homogeneous flattening (Fig. 4.75a) or noncoaxial simple shear zones (Fig. 4.75b). In noncoaxial shear zones a finite increment in stress gives way to a lenticular-shaped gash fracture set, which is orientated at an angle with respect to the shear band edges (Fig. 4.75b). The cracks are formed parallel to the $S_1 - S_3$ plane; note that gash fractures are not previous features in the shear band rotated by the shear, so carefully observe their orientation with respect to the shear sense! Remember that simple shear (see Section 3.14) is a noncoaxial strain and that any line forming an angle different to from the main shear direction rotates progressively as the main axis of the stress does. That is why a subsequent

Fig. 4.73 (a) Columnar joints form in volcanic or subvolcanic rocks by tensile stresses developed by contraction in small domains at cooling. (b) Joints form vertically in the surface and propagate down forming columnar features, as in the example of Devil's Tower in the photo (Wyoming, United States Courtersy by R. Giménez).

Fig. 4.74 Mineral-filled (calcite) gash fractures developed in Jurassic-age limestone, Delphi, Greece. Coin is 2.5 cm diameter.

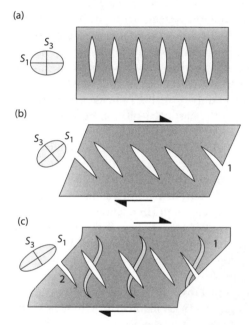

Fig. 4.75 Gashes formed by (a) homogeneous flattening and (b) simple shear, where the original fissures (1) rotate due to further shear strain acquiring (c) a sigmoidal shape whereas new formed gashes (2) are lenticular. Gashes, as all extension fractures, develop in the plane e_1–e_2, perpendicular to the direction of maximum lengthening (e_1).

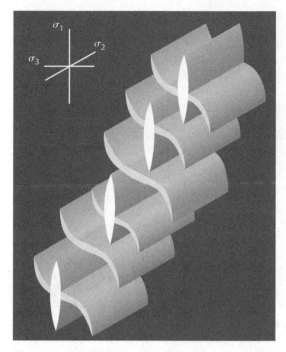

Fig. 4.76 3D geometry of gash fractures in a shear zone. Sigmoidal gashes are older than the second generation of lenticular ones. Note the orientation respect the principal stresses.

increment in strain, giving way to further stretching, will provoke a rotation in the cracks due to shearing and an extension of the cracks at the tips. These will open in the perpendicular direction to maximum elongation (the direction of the principal compressive stress axis σ_3) according to the orientation of the strain ellipse, so the cracks will show a sigmoid shape. Newly formed fissures will be lenticular, nonrotated and oriented normal to the new S_1 direction, cutting the older sigmoid gashes; so

older sets will be more rotated and the sigmoid shape more developed, due to several increments of deformation in the shear zone (Fig. 4.75c). Thus, by observing the morphology and orientation of gash fissures, the principal strain axes and therefore the direction of the principal extension, can be deduced. Also sigmoid and rotated gash veins are important in the discrimination of simple shear versus pure shear deformation.

Feather or pinnate fractures are another kind of extension fracture that form in the brittle domain associated with shear fractures and faults. Sets are arranged in an en echelon fashion forming an angle with the faults. As in all extension fractures, they are orientated parallel to the principal stress surface containing σ_1–σ_2 (Fig. 4.76) and so are good indicators of shear or sense of block motion in the fault (Fig. 4.77). The acute angle ($c.30°$–$45°$) points in the direction of block movement at either side of the fault (Fig. 4.78a). Notice that nonrotated gash veins show the same orientation (Fig. 4.78b).

4.14.3 Shear fractures (Modes II and III)

Shear fractures, like faults, are formed at an angle to the principal stress surfaces; this fact being the chief difference compared to extension fractures (Fig. 4.77). The term shear

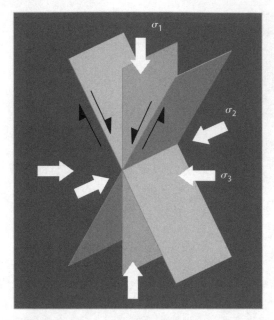

Fig. 4.77 Relations between two conjugate shear fractures and the related extension fractures formed under the same stress system. The principal compressive stresses and the shear movement are shown.

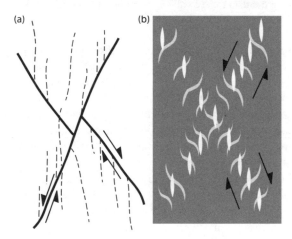

Fig. 4.78 Shear zones: (a) Conjugate brittle shear fractures and associated feather joints. (b) Conjugate shear zones at the brittle-ductile transition, showing two generations of gash veins. Note the orientation of extension and shear fractures as in Fig. 27.8.

fracture is generally used for joint-scale fractures showing discrete block displacements and which sometimes are not even discernible at outcrop scale. Some of these fractures consist of joints that experience some sliding of blocks after being formed as extension fractures, due to tectonic reactivation or simply small geometric adjustments. Shear fractures appear as smooth well-defined planes which may be arranged systematically as joints with regular spacing and with similar trends. Although shear fracture surfaces can be

absolutely featureless, *slickenlines, striations,* or *ridge and groove lineations* can be observed, reflecting the shear movement of the blocks parallel to the fracture surface. Acicular crystal fibers or slickenfibers, resulting from crystal growth in the shear direction under stress, are also common. Shear fractures can form as two conjugate sets forming an angle of approximately 30° with respect to the principal stress surface containing σ_1 and σ_2, which bisects the angle between the two fracture sets (Fig. 4.77). Shear fractures can be associated with faults (Section 4.15) as minor features, having the same orientation and origin as the faults.

4.14.4 Fracture mechanics

A great deal of knowledge concerning fracture mechanics comes from laboratory experiments. We discussed briefly in Section 3.15 the importance of experiments in the lab to observe stress–strain relations in rheology. The important field of interest in rheology is how rocks behave when subjected to stresses under certain controlled conditions of pressure, temperature, presence of fluids, etc. In this chapter we focus on the particular relation between stress and fracture formation. It is of particular interest how and why rocks break when differential stresses reach the critical value of rock resistance at the elastic or plastic limit and consequently break. Also it is important to observe under which conditions different kinds of fractures form and which are the fracture angles in relation to applied stresses. This field of knowledge, dealing with fracture or rock mechanics is very important in many applied fields, as in civil engineering, mining, and hazard controls (rock and soil stability in slopes, both natural and constructed). It has been developed over the years by both structural geologists and engineers, with mathematical models and fracture criteria of increasing accuracy.

To carry out fracture experiments, samples of different lithologies are cut using a special drilling device and smoothed at the surfaces to avoid irregularities that can cause unwanted, inhomogeneous stress concentrations. Rock samples as isotropic and homogeneous as possible are preferred. Samples are generally cylindrical, several centimeters in diameter with a length of about four or five times this amount. Both dimensions have to be measured accurately before making the experiment. Several tests can be made both by pulling apart or compressing the samples, as in the dog-bone specimens (Fig. 4.79a, b) and the Brazilian discs pressed down to achieve a perpendicular tension (Fig. 4.79c). A number of diverse apparatuses have been designed to carry out different tests which can be made in unconfined conditions or with lateral confinement

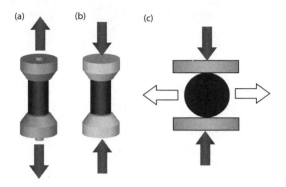

Fig. 4.79 Sample preparation for fracture tests. Dog-bone samples for (a) tensile and (b) compression tests. (c) Brazilian test in which tensile failure is achieved by pressing the sample longitudinally creating a normal tension.

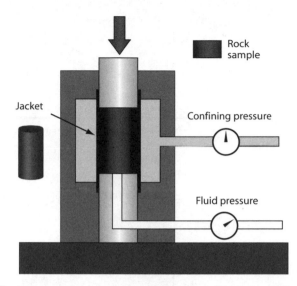

Fig. 4.80 Samples are insulated in a jacket for axial tests in which a pressurized fluid surrounding the sample is introduced. By jacketing the sample, the fluid pressure in the pores can be monitored separately from the confining pressure.

by pressurized fluid. Simpler tensile experiments involve stretching the sample from the ends under unconfined conditions (Fig. 4.79a), creating a uniaxial extension or tension in which $\sigma_1 = \sigma_2 = 0 > \sigma_3$ (σ_3 is the tensile axial stress). Similarly the sample can be compressed at the ends under unconfined conditions, defining a *uniaxial compression* test in which $\sigma_1 > \sigma_2 = \sigma_3 > 0$ (Fig. 4.79b). In rock mechanics it is of key importance to simulate the confining pressure of the surrounding rocks (lithostatic pressure) and fluids (hydrostatic pressure) in the natural environment and to understand the role of fluids contained in rock pores. To achieve this, *axial* experiments (one of the principal stresses is set and progressively increased, the other two provide the confining pressure) or true *triaxial* tests (all stresses are different, which has been achieved only recently) are defined and more sophisticated apparatus needed. Although the test rigs may vary greatly in design, a typical apparatus for an axial compression test consists of a pressure vessel in which the sample is confined and pressed down by pistons (Fig. 4.80) creating a load (σ_1). To prepare this setting the sample is protected in a weak easily deformable cylinder jacket, like rubber or copper (Fig. 4.80) which insulates it from the surrounding pressurized fluid ($\sigma_2 = \sigma_3 \neq 0$). The sample itself can contain some fluid in the interstitial pores (a hydrostatic pressure) and the jacket provides a way to control both fluid pressures separately. Temperature and both pore and confined pressures can be adjusted independently and monitored.

4.14.5 Fracture criterion and the fracture envelope

Considering the contrasting states of stress which can be simulated in diverse experiments, several kinds of fractures can be produced. Monitoring failure at different states of stress (confining pressure, main load, or axial stress) it is possible to construct *failure envelopes* and develop *fracture criteria* (Fig. 4.81). Fracture criteria are models expressed as mathematical equations based on empirical data; they can be either linear or nonlinear (parabolic, as in Fig. 4.81b). Failure envelopes represent conditions for fracturing for a particular kind of rock and can be constructed joining all points of coordinates σ_n and τ at which fractures are produced for different settings of confining pressure differential stress. The failure envelope divides the Mohr diagrams in two fields: one where the states of stress are possible or stable and a second in which the states of stress are not possible (Fig. 4.81). Failure envelopes mark the rock strength at tension or compression; over this value rock samples will break making states of stress over this value impossible. Although there are some problems inherent in the experimental technique due to the shapes of the samples, observing and measuring the angle at which fractures form is important for the construction of models. Fracture criteria have to satisfy the stress conditions to produce the fractures and predict the angle at which the fractures form with respect to the principal stress axes. In the Mohr diagram, the radius of the circle which joins the tangent to the failure envelope (in both cases linear and nonlinear) defines the angle 2θ, which can be used to obtain the orientation of the fractures. θ is the angle of the fracture surface with the normal to σ_1 (Fig. 4.81c) which is the same angle that the normal to the fracture

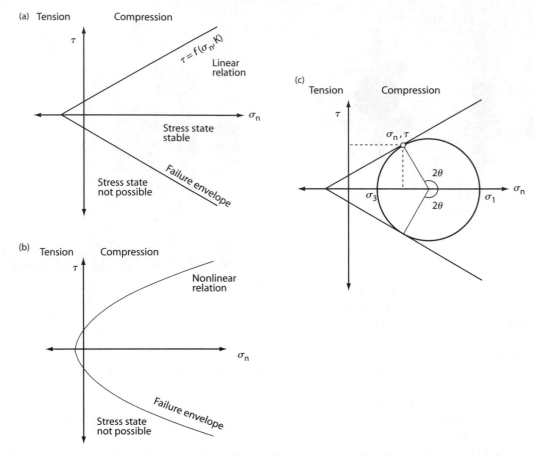

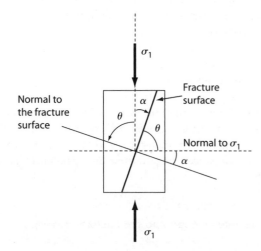

Fig. 4.81 Fracture criteria. (a) General graphic representation of a linear fracture criterion. The failure envelope at which fractures occur, separates a field of stable stresses (white) from a field of unstable stresses (shaded), where states of stress are not possible (unstable stress) because rock strength is exceeded. (b) Nonlinear parabolic fracture criterion showing the same fields as before. (c) Mohr circle showing the angles (both for positive and negative shear) at which fractures form (θ is the angle that the fracture surface forms with the normal to σ_1).

surface forms respect to σ_1 (Fig. 4.82). Fracture angle α with the principal stress σ_1 can be calculated as $\theta + \alpha = 90°$. Linear fracture criteria predict that fracture angles with respect to the main stresses remain constant for all the experiments, whereas in nonlinear criteria the angles should vary along the curve.

Several kinds of fractures can be produced in different situations (Fig. 4.83). In tests of uniaxial extension, which are carried out by stretching the sample from the long axis, without lateral confinement ($\sigma_1 = \sigma_2 = 0$; Fig. 4.83a) tensile fractures are formed. Axial compression tests are established by applying an axial load, which defines the main principal stress σ_1, in unconfined conditions ($\sigma_2 = \sigma_3 = 0$; Fig. 4.83b); this will cause vertical splitting of the sample, with fractures orientated parallel to the main applied load. Axial or triaxial settings, in which compression is carried out with some level of confining

Fig. 4.82 Geometric relation between the angle θ and the fracture angle α, which is the angle between σ_1 and the fracture surface.

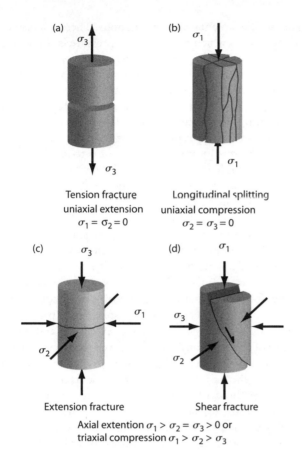

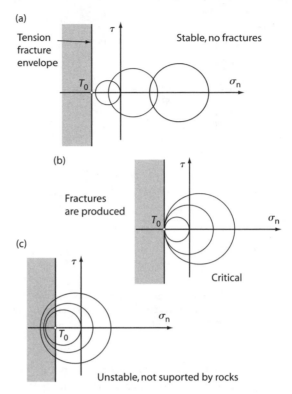

Fig. 4.83 Types of brittle fractures developed during laboratory experiments on rock samples.

Fig. 4.84 States of stress in relation with the tension fracture envelope defined by T_0, which represents the rock tensile strength. (a) Stable, (b) critical, and (c) unstable.

pressure, produce extension fractures parallel to the surface σ_1–σ_2 and perpendicular to the less compressive stress when σ_3 is the axial load (Fig. 4.83c) or shear fractures (Fig. 4.83d) which form an angle α of 30°–45° with respect to the principal compressive stress σ_1 in both axial compression and true triaxial compression experiments.

4.14.6 Tensile fractures

Rocks are weaker under tension than compression and break at lower levels of stress, typically about −5 to −20 MPa depending upon rock type. The *tensile fracture envelope* marks the conditions at which fractures are produced when the fundamental tensile strength T_0 is achieved. The envelope is a line parallel to the τ-axis that cuts the σ_n-axis in a point located in the negative or tensile field, which gives the value for the maximum normal stress that the rock can support (Fig. 4.84), which is the critical normal stress σ_c. Fractures will be produced when the Mohr circles are tangent to the tensile fracture envelope; over this value all states

of stress represented by the corresponding Mohr circles are unstable, which means that the rock will bust earlier. States of stress to the right of the envelope are stable, that is, stresses are not high enough to produce fracturing.

An experiment for tensile strength by stretching the sample can be made under unconfined conditions until a fracture is produced. This tension applied in the direction of the sample long axis will have the value σ_3 since $\sigma_1 = \sigma_2 = 0$ and tensile stresses have negative values. The Mohr circles will be bigger as the tension, and consequently the differential stress $(\sigma_1 - \sigma_3)$ increases (Fig. 4.85). When the tensile stress σ_3 reaches the critical value $\boldsymbol{\sigma_c}$ and the tensile strength T_0 is exceeded, a tensile fracture will be produced perpendicular to the stretching direction where σ_3 is located. Fracture criterion will be in the form of the equation: $\sigma_3 = \boldsymbol{\sigma_c} = T_0$ which explains the state of stress at failure and also accounts for the angle of fracturing at 0° from σ_1. Fractures will form parallel to the $\sigma_1 = \sigma_2$ plane, which is a principal stress surface, and so the value of the shear stress component is 0. Note that the fracture forms an angle $2\theta = 180°$, which means that the fracture angle α equals 0° (Fig. 4.85c). Also note that the point at which the fracture is produced is at the x-axis with coordinate values σ_3, 0. Tensile fractures are Mode I

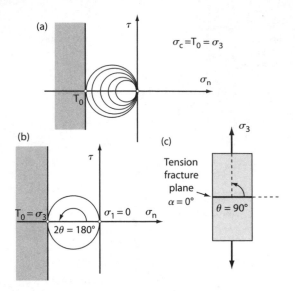

Fig. 4.85 (a) Tensile strength experiments in unconfined conditions. Stress is increased until T_0 is reached; (b) stress at failure showing the angle θ. (c) Relation between θ and the fracture angle α.

which, as explained earlier in the chapter, do not experience shear displacements or movements along the surface but open perpendicularly to the smaller compressive stress.

4.14.7 Transitional tensile fractures

It is possible to simulate transitional tensile to compressive behavior by providing a lateral confining pressure to the rock sample $\sigma_1 = \sigma_2 \neq 0$. At the beginning of the experiment the conditions for the confining pressure are set at a convenient value. The state of stress will be represented as a point in the Mohr diagram showing hydrostatic conditions. An axial extension is created (σ_3); a differential stress $\sigma_1 - \sigma_3$ will develop, representing progressively bigger Mohr circles, as σ_3 moves to the left. The sample will break when the differential stress reaches the rock tensile strength T_0. When the confining pressure is set at a moderate value (about 10 MPa for sandstones), less than about three times the magnitude of the absolute value of the tensile strength, $\sigma_1 < [3T_0]$, Mode I fractures normal to the tensile stress form (Fig. 4.86a). A different situation occurs when the confining pressure has a value between three and five times the tensile strength of the sample $[3T_0] < \sigma_1 < [5T_0]$ (Fig. 4.86b). When the confining pressure is increased in this range in successive experiments, rocks fail to define a parabolic failure envelope (Fig. 4.86b, c), which is tangent to the Mohr circle at two points for every individual experiment. These two points

define two possible fracture orientations. There is a particular situation in which the confining pressure is about five times bigger than the absolute value of the rock tensile strength; the two points of intersection between the failure envelope and the Mohr circle are located at the τ-axis; fractures displaying a combined behavior of extension and shear develop for values of $\sigma_c = 0$ and τ_c about $2T_0$ (Fig. 4.86c) where τ_c represents the cohesive strength of the rock. The relation of the critical stress to produce failure and the tensile strength is given by *Griffith's equation* (Fig. 4.86c), which also defines the parabolic curve for transitional tensile behavior.

The analysis of tensile fractures and transitional tensile fracture formation is important in the study of joints, because the most probable origin of joints having plumose and conchoidal structures is related to some tensional stress where $\sigma_c < 0$ (pure Mode I fractures). Other joints may be formed at some transitional-tensile conditions where σ_c is close to 0 and a shear displacement parallel to the fracture surface occurs (hybrid extension and shear fractures of Modes II and III). As the vertical stress is compressive due to gravity and a tensile stress is required to form joints, the best conditions for tensile fracture formation will be those of small differential stresses and high fluid pore pressures, since (Section 3.15) fluid pore pressure exerts a hydrostatic pressure that lowers the applied stress level, displacing the Mohr circles toward the left.

4.14.8 Fractures in compressive tests: the Coulomb criterion

Laboratory experiments dealing with full compressive settings to assess rock failure conditions and fracture strengths are carried out by introducing a confining pressure in rock samples of a certain lithology. This is given by $\sigma_2 = \sigma_3 \neq 0$. The sample is pressed down by a piston or a hydraulic mechanism so as to achieve an axial load, the maximum compressive stress σ_1, of higher magnitude than confining pressure. Increasing the axial load will give increasing differential stresses and bigger Mohr circles, which will be positioned at the right side of the diagram (Fig. 4.87a). Raising the load will lead eventually to failure. When the broken sample is released the fracture angle with respect to the axial load can be measured. Typical fracture angles are between 25° and 35° with respect to the axial load located along the sample long axis. Angles are very regular, with the average close to 30°, although two orientations are possible (Fig. 4.87b). Once the fracture angle (α) is measured it is easy to calculate the angle θ, which is the angle between the normal to the principal

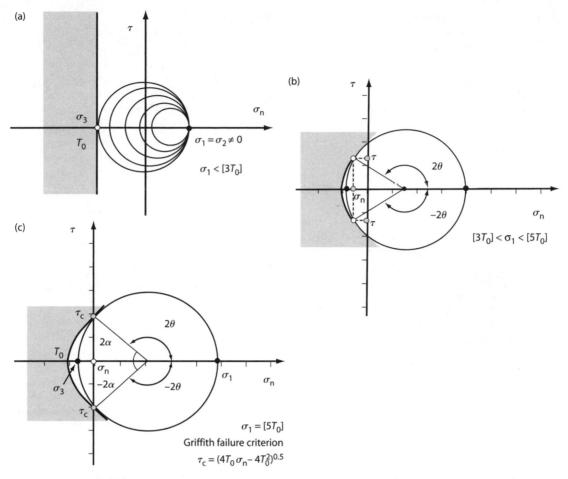

Fig. 4.86 Tensile stress experiments under confining pressure conditions. (a) For small confining pressures a tensile fracture develops at T_0. (b) Confining pressures with values between $[3T_0]$ and $[5T_0]$. (c) The maximum conditions for noncompressive fractures.

stress σ_1 and the fracture surface ($\alpha + \theta = 90°$; Fig. 4.87b). The angle 2θ (this will be 120°) can be plotted in the Mohr diagram for the two virtual fractures and the coordinates σ_n and τ at which the fractures form can be determined.

Further experiments with samples of the same lithology can be carried out by increasing the confining pressure. Generally it is been observed that the differential stress necessary to produce failure is bigger at higher values of confining pressure, so the Mohr circles represented at positions located more to the right of the diagram will be bigger. The shear fractures formed in different experiments show α angles similar to the ones previously produced, which are close to 30° ($2\theta = 120°$). This means that when the fracture points formed at every Mohr circle are joined, a straight line inclined toward the τ-axis (Fig. 4.87c) is defined. This straight line is the shear fracture envelope for compressive stress and it is known as the *Coulomb envelope*, which will represent the critical states of

stress to produce fractures for different conditions of confining pressure. For every rock there are two straight lines, one for the field of positive shear and other for negative shear. As before, this envelope separates a field where the stresses are stable or can be sustained by the rock (in white) and a field where the stresses are unstable or impossible (in gray in Fig. 4.87c).

The Coulomb envelope can be expressed by the mathematical law known as Coulomb or *Coulomb–Mohr fracture criterion*, which is defined by the equation: $\tau_c = \tau_0 + \mu \sigma_n$, where τ_c is the critical shear stress to produce fracture, σ_n is the value of the normal stress, and τ_0 and μ are two constants inherent to the rocks which varies for different lithologies and rock properties (Fig. 4.88). The value τ_0 is the rock *cohesion* or *cohesive shear strength*, measured in stress units, and its value defines the height at which the Coulomb envelope intercepts the τ-axis. Note that the cohesion represents the value of the shear strength of a rock over a surface in which no normal stress is acting,

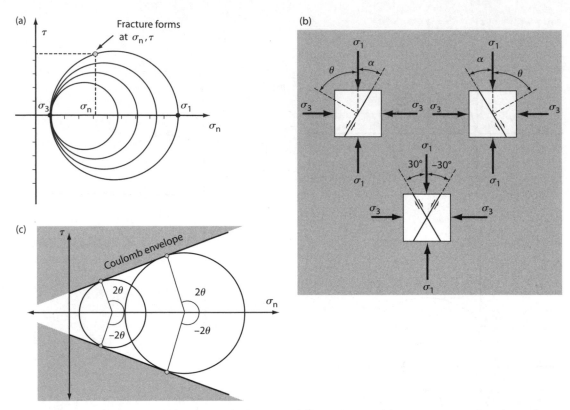

Fig. 4.87 Experiments at compression. (a) Mohr diagrams for increasing differential stress while maintaining constant the confining pressure. A shear fracture will form at a critical stress level. (b) Geometrical representation of the shear fractures and their relation to the principal compressive stresses in 2D. (c) The Coulomb fracture envelope.

so that $\sigma_n = 0$. The rate of increasing shear strength with increasing values of normal stress is represented by the constant $\mu = \tan \phi$. This constant is called the *Coulomb coefficient* or *coefficient of internal friction* and ϕ the *angle of internal friction*, which defines the slope angle of the Coulomb envelope and the minimum stress value required to overcome the rock frictional resistance and produce a fracture. Considering the Mohr circles now, fractures will be produced when the Mohr circle intercepts the two envelopes and the coordinates of the tangent points will give the values of the surface stress components at failure and also the angles of the fractures with respect to the principal compressive stress σ_1. Typical values for the angle of internal friction ϕ for most rocks are 25°–35°; the angle of internal friction is related to θ by the equation, $2\theta = 90 + \phi$ (as is obvious in the Mohr diagrams; Fig. 4.88). So, if an average of 30° is taken, the angle θ will be 60° which explains the fracture angle of 30° observed in most experiments in the Coulomb field. Note that fractures in these settings do not form at 45°, which are the planes of maximum shear stress, because the normal stress also exerts an influence. The fractures tend to form

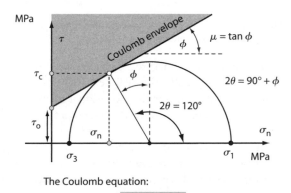

The Coulomb equation:

$$\tau_c = \tau_0 + \sigma_n \, \mu$$

Fig. 4.88 The Coulomb envelope and criterion for fracture formation in the compressive field.

at high values of shear stress and low values of normal stress.

The two possible fractures that can be produced are described as *conjugate shear planes* (Fig. 4.89). Such Coulomb fractures will form at intermediate confining pressure values and indicate a linear increase of the rock compressive strength with higher values of normal stress. It might be expected that increasing confining pressure will continue the linear behavior, but for high values of confining pressure the Coulomb tendency does not explain fracture behavior.

4.14.9 Fractures in compressive tests at high confining pressures: The von Mises criterion

Increasing confining pressure defines a transitional ductile field and eventually a ductile field in which the Coulomb

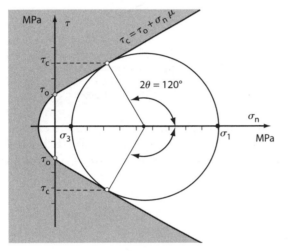

Fig. 4.89 A combined envelope for tensile, transitional, and compressive fields. The tangent points give the orientation of the conjugate shear planes at which fracture can possibly form.

criterion does not apply, as the expected increase of critical shear stress to produce fractures is not observed. Thus, the fracture envelope bends with higher confining pressures and eventually parallels to the σ_n-axis (Fig. 4.90). Consequently the Mohr circles intercept the envelope in positions which define higher fracture angles with respect to the principal stress σ_1, from the classical $\alpha = 30°$ ($2\theta = 120°$) of the Coulomb field finally to $\alpha = 45°$ ($2\theta = 90°$), which are the surfaces of maximum shear stress. Ductile shear fractures or simply ductile deformation (barreling) will form in these settings. This pair of straight lines, in which the value of the critical shear stress is constant for different values of normal stress and is independent of the confining pressure, form the *von Mises fracture envelope*. The von Mises ductile failure criterion can be expressed as, $\tau_0 = $ constant. This value of critical shear stress is the yield stress, which defines the boundary for the ductile field of deformation (see Section 3.15).

All envelopes can be put together defining the critical state of stress and the fields of stress stability and instability in all possible settings, from tensile to compressive with high confining pressure conditions. Nonetheless, as seen in rheology (Section 3.15), some other factors are important in rock behavior, such as temperature and confining pressures which both increase with depth, lowering the yield stress and allowing ductile deformation, or fluid pressure in interstitial pores (Fig. 4.90). As seen previously (Section 3.15) the effect of pore fluid pressure in fracture formation is very important as the Mohr circles are displaced toward the left by reducing the effective stress

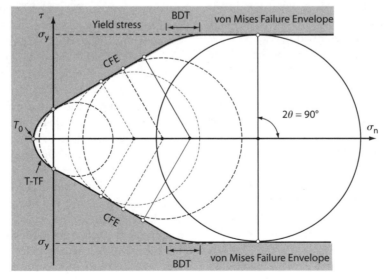

Fig. 4.90 The von Mises failure envelope for high confining pressures in relation to the other envelopes (CFE: Coulomb fracture criterion; T-TF: transitional tensile envelope; T_0: Tensile strength). The curved stretch between the Coulomb and the von Mises failure envelopes is the brittle–ductile transition (B-DT). In the von Mises field ductile failure occurs with fracture angles of 45° ($2\theta = 90°$).

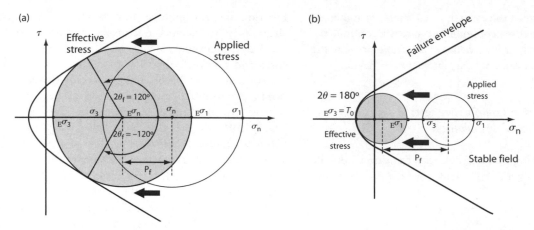

Fig. 4.91 Effect of pore fluid pressure in fracture formation. (a) With high differential stresses Coulomb fractures can be produced when the Mohr circle moves to the left by pore fluid pressure. (b) With low differential stresses, even when the applied stress may be compressive, and fully located in the field of stress stability, fluid pore pressure can reduce the effective stress displacing the circle to the tensile field and producing joints if the condition $_E\sigma_3 = T_0$ is satisfied.

to a lower level, while maintaining the differential stress(Fig. 4.91). With low differential stresses, even when the applied stress may be compressive, and fully located in the field of stress stability, fluid pore pressure can reduce the effective stress displacing the circle to the tensile field and producing joints if the condition $_E\sigma_3 = \tilde{T}_0$ is satisfied.

4.15 Faults

4.15.1 Nomenclature and orientation

Faults are fracture surfaces or zones where several adjacent fractures form a narrow band along which a significant shear displacement has taken place (Fig. 4.92a, b). Although faults are often described as signifying brittle deformation there is a transition to ductile behavior where shear zones develop instead. As described in Section 4.14, shear zones show intense deformation along a narrow band where cohesive loss takes place on limited, discontinuous surfaces (Fig. 4.92c). Faults are commonly regarded as large shear fractures, though the boundary between features properly regarded as *shear fractures* or joints is not sharply established. In any case, although millimeter-scale shear fractures are called *microfaults*, faults may range in length of order several decimeter to hundreds of kilometers: they can be localized features or of lithospheric scale defining plate boundaries (Section 5.2). Displacements are generally conspicuous (Fig. 4.93), and can vary from 10^{-3} m in hand specimens or outcrop scale to 10^5 m at regional or global scales. Faults can be recognized in several ways indicating shear displacement, either by the presence of scarps in recent faults (Fig. 4.93a and b), offsets, displacements, gaps, or overlaps of rock masses with identifiable aspects on them such as bedding, layering, etc. (Fig. 4.93c).

Fault nomenclature is often unclear, coming from widely different sources. For example, quite a lot of the terms used to describe faults comes from old mining usage, even the term *fault* itself, and the terms are not always well constrained. Fault surfaces can be inclined at different angles and their orientation is given, as any other geological surface, by the *strike* and *dip* (Fig. 4.94a). A first division is made according to the fault dip angle; *high-angle faults* are those dipping more than 45° and *low-angle faults* are those dipping less than 45°. Faults divide rocks in two offset blocks at either side of the fracture surface. If the fault is inclined, the block which is resting over the fault surface is named the hanging wall block (HWB, Fig. 4.95) and its corresponding surface the *hanging wall* (HW, Fig. 4.96); and the underlying block which supports the weight of the hanging wall is called the footwall block (FWB, Fig. 4.95); the corresponding fault surface is called the *footwall* (FW, Fig. 4.96). If homologous points previous to fracturing at each side of the fault can be recognized, the reconstruction of the relative displacement vector or slip can be reconstructed over the fault surface, both in magnitude

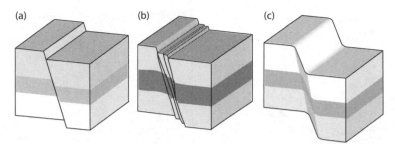

Fig. 4.92 (a) Fault, (b) fault zone, and (c) ductile shear zone. Faults are well-defined surfaces produced by brittle deformation. Weak rocks can be deformed by brittle deformation giving rise to a fault zone with multiple, closely spaced, sometimes interconnected surfaces. Shear bands develop in the ductile field.

Fig. 4.93 Faulting is marked by conspicuous shear displacements, forming distinctive features on fault surfaces like (a) bends and grooves (b) slickenlines. In (c), originally continuous bedding traces seen in vertical section show up fault displacement (all photos taken in central Greece.

and direction. The relative movement can be either parallel to the fault dip direction (*dip-slip faults*) or to the fault strike (*strike-slip faults*). Dip-slip faults show vertical displacements of blocks whereas in strike-slip faults the displacement is hori-zontal. In a composite case, the movement of blocks can be oblique; in these *oblique-slip faults* blocks move diagonally along the fault surface, allowing the separation of a dip-slip component and a strike-slip component (Fig. 4.94a). The dip–slip component

can be separated into a horizontal part which is called *heave* and a vertical part known as *throw* (Fig. 4.94b). When faults show a dip-slip movement the block which is displaced relatively downward is called *down-thrown block* (DTB, Fig. 4.94) and the one displaced relatively upward *up-thrown block* (UTB, Fig. 4.95). Blocks in strike–slip faults are generally referred to according to their orientation (for instance: north block and south block, etc.). In most cases accurate deduction of movement vectors is not

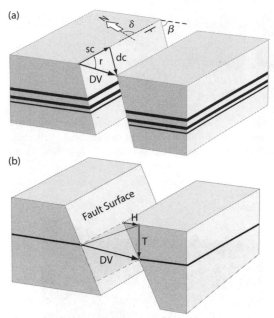

Fig. 4.94 (a) Total displacement vector (DV) in a fault (general case). If the movement is oblique, a dip component (dc) and a slip component (sc) can be defined. DV can be orientated by the rake (*r*) over the fault surface, whose orientation is given by the strike (δ) and slip (β) angles. (b) Other components can be separated from DV: the vertical offset or throw (*T*) and the horizontal offset or heave (H).

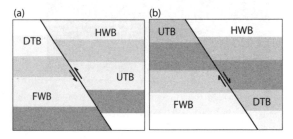

Fig. 4.95 Relative position of blocks in a fault: hanging wall block (HWB); footwall block (FWB); upthrow block (UTB); and downthrow block (DTB) in (a) a reverse fault and (b) a normal fault.

possible, and the displacement has to be guessed by the observation of offset layers. In this case the *separation* can be defined as the distance between two homologous planes or features at either side of the fault, that can be measured in some specific direction (like the strike and dip directions of the layer).

Faults initially form to a limited extent and progressively expand laterally; the offset between blocks increasing with time. The limit of the fault or fault termination, where there is no appreciable displacement of blocks is called *tip*

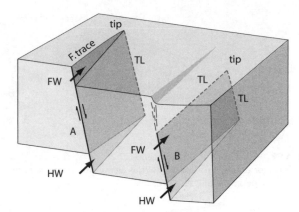

Fig. 4.96 (A) Faults have a limited extent and can cut through the surface (A) or not (B), in which case they are regarded as blind faults. Fault terminations (tip and tip lines: TL) are marked in both cases. FW marks the footwall and HW the hanging wall of the fault surfaces.

line (Fig. 4.96). In the case of faults that reach the Earth's surface, the intersection line between the fault plane and the topographic surface is called the *fault trace* and the point where the fault trace ends is called the *tip point* or *tip*. Blind faults are those which terminate before reaching the Earth's surface and although they can cause surface deformation, like monocline folds, there is no corresponding surface fault trace (Fig. 4.96) to the fault bounded at the front and upper ends by termination or tip lines.

Fault planes can have different forms. At the surface most faults appear as fairly flat surfaces (Fig. 4.97a) but at depth they can show changes in inclination. Some faults show several steps: in high angle faults, stepped segments showing a decrease in dip are called *flats* (Fig. 4.97b), whereas in low angle faults, segments showing a sudden increase in dip are called *ramps* (Fig. 4.97c). Flats and ramps give way to characteristic deformation at the topographic surface; in normal faulting, for instance, bending of rocks in the part of the *hanging wall* block located over a ramp results in a *synclinal fold*, whereas the resulting deformation over a flat is an anticlinal fold. Ramps can be also present in faults with vertical surfaces as in strike–slip faults, which are called *bends*, or orientated normal (sidewall ramp) or oblique (oblique ramps) to the fault strike. *Listric faults* are those having a cylindrical or rounded surface, showing a steady dip decrease with depth and ending in a low-angle or horizontal *detachment* (Fig. 4.97c). *Detachment faults* can be described as low-angle faults, generally joining a listric fault in the surface that separates a faulted *hanging wall* (with a set of imbricate listric or flat-surface faults) from a nondeformed footwall. Detachments form at mechanical or lithological contacts where rocks show different mechanical properties, a decrease in friction

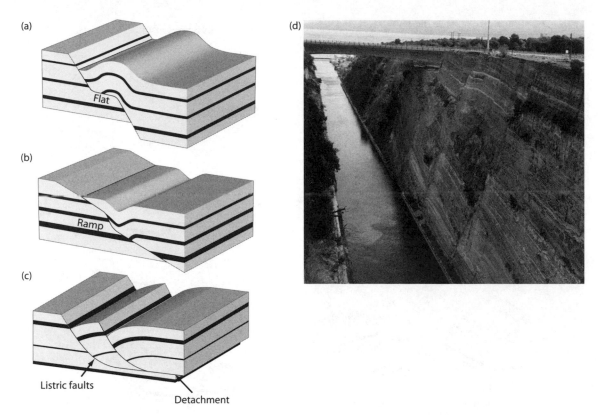

Fig. 4.97 Fault surface geometry. Faults are fairly flat at surface but at depth may show changes in the dip angle. (a) High-angle faults can have less steep reaches named flats; (b) low-angle faults can have an oversteepened reach or ramp. (c) Faults can experience a progressive decrease in dip at depth, ending in a very low angle or horizontal surface or detachment. (d) A stepped listric fault array, Corinth canal, Greece.

coefficient commonly. Secondary imbricate fault sets can be either *synthetic*, when they have the same dip sense of the main fault or *antithetic*, when they have an opposed dip direction with respect to the main fault.

4.15.2 Fault classification

Regarding the relative displacement of blocks along any fault surface, several kinds of faults can be defined (Fig. 4.98). Earlier we made a first distinction into dip-slip, strike-slip, and oblique-slip faults. Dip-slip faults, having relative block movements parallel to the dip direction, can be separated into normal faults and reverse or thrust faults according to the sense of shear (Fig. 4.98a). *Normal* faults are generally high-angle faults, with surfaces dipping close to 60° in which the hangingwall block slides down the fault surface, as the down-throw block (Fig. 4.95b). Low-angle normal faults can also form. *Reverse* and *thrust* faults are those in which the hangingwall block is forced up the fault surface, defining the up-thrown block (Fig. 4.95a). Although many authors consider both terms synonymous, a distinction between thrust and reverse faults has been made on the basis of the surface angle; the first being low-angle faults and the second high-angle faults. Strike-slip faults are those having relative movements along the strike of the fault surface (Fig. 4.98b), generally they have steep surfaces close to 90° so the terms hangingwall and footwall do not apply. There are two kinds of strike-slip faults depending on the relative shear movement; when an observer is positioned astride the fault surface, the fault is *right-handed* or *dextral* when the right block comes toward the observer and is *left-handed* or *sinistral* when the left block does (notice that it does not matter in which direction the observer is facing; Fig. 4.98). Oblique-slip faults can be defined by the dip and strike components derived from the relative movement of the blocks. Four possible combinations are represented in Fig. 4.98c as normal-sinistral, normal-dextral, reverse-dextral, and reverse-sinistral. Finally, rotational faults are those showing displacement gradients along the fault surface; they are formed when one block rotates with respect to the other along the fault surface (Fig. 4.98d).

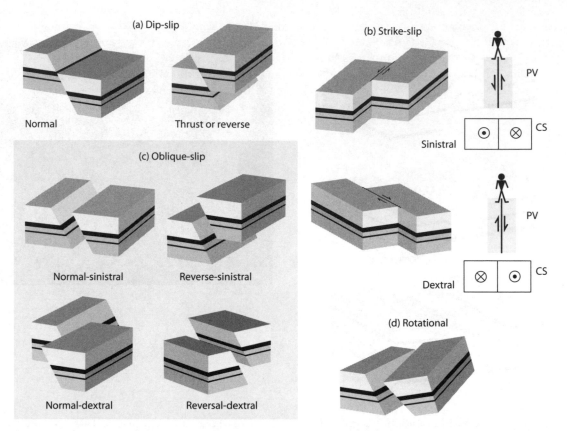

Fig. 4.98 Fault classification in relation to the relative movement of blocks along the fault surface. (a) Dip–slip faults include normal and thrust or reverse depending on the relative movement of the blocks up or down the fault surface; (b) strike–slip faults can be sinistral or dextral according to shear: in plan view (PV), if the left block of a strike-slip fault moves toward an observer straddling the fault trace (no matter which end of the fault) the fault is sinistral, whereas if the right block moves toward the observer, the fault is dextral. The notation used for shear sense in cross section, in both sinistral and dextral cases is also shown (CS). (c) Faults can show oblique-slip displacements, allowing for different combinations and, finally, (d) faults can be rotational, when the hangingwall block rotates over the footwall block.

4.15.3   Anderson's theory of faulting

In Section 4.14 we showed that for a particular stress state under certain values of confining pressure and where Coulomb's criterion applies, two conjugate fractures form at about 30° from the principal stress σ_1. Faults are shear fractures in which there is a prominent displacement of blocks along the fault surface. Consider again the nature of the stress tensor (described in Section 3.13) and remember that the principal stress surfaces containing two of the principal stresses are directions in which there are no shear stresses. Taking into consideration these facts Anderson concluded in his paper of 1905, that the Earth's surface, envisioned as the boundary layer between the atmosphere and the lithosphere, is a free surface in which no shear stresses are developed, that is, there is no possibility of sliding parallel to the surface. In this approach, atmospheric stresses are too weak to form fractures, topographic relief is negligible, and the Earth's surface is considered perfectly spherical. If the surface is a principal stress surface then the principal stress axes have to be either horizontal or vertical and two of them have to be parallel to the Earth's surface.

Anderson supposed that a hydrostatic state of stress at any point below the Earth's surface should be the common condition, such that the horizontal stresses in any direction will have the same magnitude to the vertical stress due to gravitational forces or lithostatic loading. When the horizontal stresses become different from the vertical load and a regional triaxial stress system develops, faults will form if the magnitude of the stresses is big enough. In order to have a triaxial state of stress, and considering that the vertical load remains initially constant, the horizontal stresses have to be altered in three possible ways: first, decreasing the stress magnitude by different amounts according to orientation such as the larger compressive stress σ_1 will be the vertical load and $\sigma_2 \neq \sigma_3$ horizontal stresses; second, increasing the horizontal stress levels but by different amounts so the vertical load will be the smaller stress σ_3 and $\sigma_1 \neq \sigma_2$ horizontal stresses; and third, increasing the magnitude of the stress in one direction and decreasing the stress in the other direction, so the

vertical load will be σ_2, smaller in magnitude than one of the horizontal stresses (σ_1) and larger than the other (σ_3).

Fault angles with respect to the principal stress σ_1 can be predicted from Coulomb's fracture criterion, $\tau_c = \tau_0 + \mu$ σ_n, with the coefficient of internal friction (μ) and the cohesive strength (τ_0) both depending on the nature of the rock involved. This criterion has been validated in

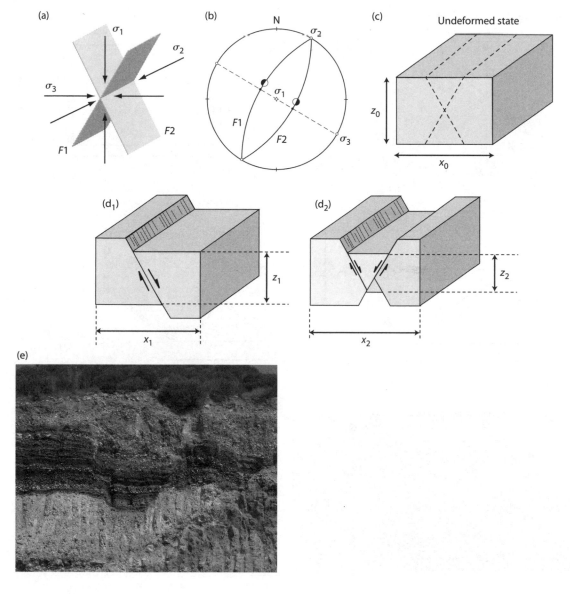

Fig. 4.99 Normal faults form to accommodate an extension in some section of the crust. (a) Anderson's model for the relation between a pair of normal conjugate faults (*F*1 and *F*2) and the orientation of the principal stress axes are shown. According to this model, normal faults form when σ_1 is vertical (this will be the orientation of the principal strain axis S_3). (b) The stereographic projection (Cookie 19) for the model in (a) is shown. (c) Considering an initial segment of the crust, normal faulting is a response of brittle deformation caused by extension, and produces a progressive horizontal lengthening and vertical shortening by the formation of new faults (d_1) and (d_2). (e) An example of normal faults cutting recent deposits (Loutraki, Greece).

numerous laboratory experiments in which the relation between the shear fractures, extension fractures, and the principal axes orientation are well established. Combining Coulomb's criterion and the nature of the Earth's surface as a principal stress surface, Anderson concluded that there are only three kinds of faults that can be produced at the Earth's surface: normal faults when σ_1 is vertical (Fig. 4.99a,b); thrust faults when σ_3 is vertical (Fig. 4.100a,b) and strike–slip faults when σ_2 is vertical (Fig. 4.101a,b). Normal faults will dip about 60° and will show pure dip–slip movements; thrust faults will be inclined 30° and will give also way to pure slip displacements, whereas strike–slip faults will have 90° dipping surfaces and blocks will move horizontally. Note the relation

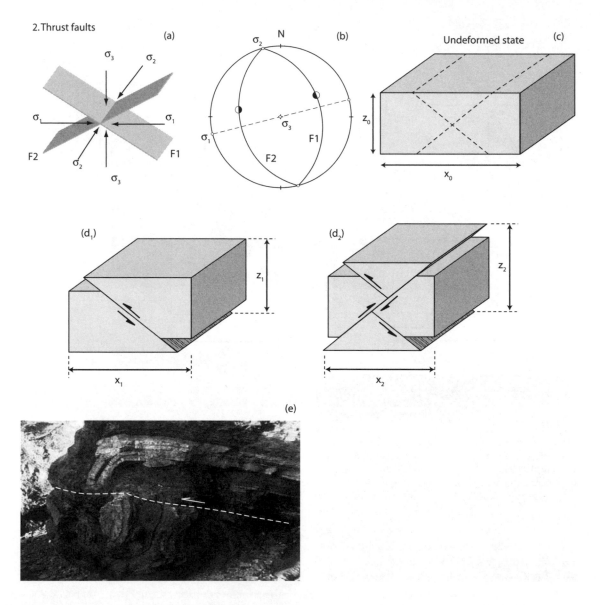

Fig. 4.100 Thrust faults form to accommodate a shortening due to compression in some sections of the crust. (a) Anderson's model for the relation between a pair of thrust conjugate faults (*F*1 and *F*2) and the orientation of the principal stress axes are shown. Thrust faults, following Anderson's model form when σ_3 is vertical (this will be the orientation of the principal strain axis S_1). (b) The stereographic projection (Cookie 19) for the model in (a) is shown. (c) Considering an initial segment to the crust, thrust faulting will form as a response of brittle deformation caused by compression, which produces a progressive horizontal shortening and vertical thickening by the formation of (d) new faults d_1 and d_2. (c) An example of reverse and thrust faults cutting recent deposits (Loutraki, Greece).

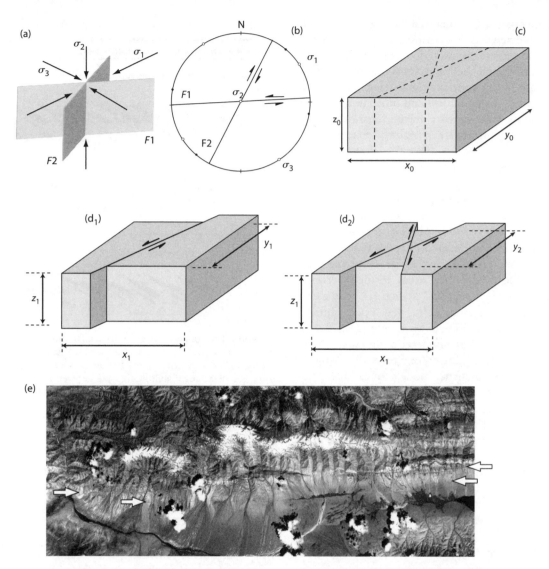

Fig. 4.101 Strike–slip faults form to accommodate deformation in situations in which an extension and compression occur in the horizontal surface in some section of the crust. (a) Anderson's model for the relation between a pair of strike-slip conjugate faults (*F1* and *F2*) and the orientation of the principal stress axes are shown. According to this model, strike-slip faults form when σ_2 is vertical (this will be orientation of the principal strain axis S_2). (b) shows the stereographic projection (Cookie 19) for the model in (a). Considering an initial segment of the crust (c), strike-slip faulting produces a progressive horizontal lengthening and shortening in directions at 90°, whereas no vertical shortening or lengthening occurs (d_1 and d_2). (e) Aerial view of strike–slip fault.

in all the models between the two conjugate faults formed and the principal stress axes. Independent of the kind of faults formed, according to Anderson's model, a pair of conjugate faults cross each other with an angle of 60°; the main principal stress σ_1 always bisects the acute angle between the faults (following Coulomb's criterion that predicts fractures produced at 30° from σ_1), σ_2 is located at the intersection of the fault planes and σ_3 is located at the bisector of the obtuse angle formed between the faults.

4.15.4 Normal faults

Normal faults form in tectonic contexts in which there is horizontal extension in the crust. As discussed previously, following Anderson's theory the larger principal stress is due to the vertical load and so the remaining axes has to be of a lesser compressive magnitude. There are a number of geologic settings in which normal faults form, both in continental and oceanic environments; the most important ones are the divergent plate margins (Section 5.2), which are subjected to extension. The main areas are continental rifting zones and extensional provinces, midoceanic ridges, back-arc spreading areas, and more local examples such as in magmatic and salt intrusions (diapirs and calderas discussed in Section 5.1), delta fronts and other areas of slope instability like cliffs which involve gravitational collapse.

Normal faults accommodate horizontal extension by the rotation of rigid blocks in brittle domains. The resulting deformation produces horizontal lengthening and vertical thinning of the crust (Fig. 4.99c,d). The combined movement of conjugate normal faults produces characteristic structures such as a succession of *horsts* and *grabens* or *half grabens*. Horsts are topographic high areas formed by the elevated footwall blocks of two or more conjugate faults; whereas grabens and half grabens are the low basin-like areas formed between horsts. Grabens are symmetrical structures with both opposite-dipping conjugate faults developed equally, whereas half graben structures are asymmetric (Fig. 5.43), being formed by a main fault and a set of minor synthetic and antithetic faults belonging to one or both conjugate sets. There are several kinematic models for normal faulting that can explain the combined movements of related faults and the observed tectonic structures formed in extensional settings. Most of the models depend on the initial fault geometry (flat, listric, or stepped). The basic movement of a pair of flat conjugate faults is depicted in Fig. 4.99. Note that progressive faulting by the addition of normal faults cannot result in unwanted gaps along the fault surfaces as will happen if both faults cut each other at the same time forming an X configuration and the central block is displaced downward. A simple model for blocks bounded by flat surfaces is the *domino model* (Fig. 4.102a,b), which involves the rigid rotation of several blocks to accommodate an extension in the same way that a tightly packed pile of books will fall to one side in the bookshelf when several bocks are removed, thereby creating horizontal space. As a result of block rotations a shear movement is formed along the initially formed fault surfaces between the individual blocks, fault surfaces suffer a progressive decrease in the dip angle, the horizontal space occupied by the inclined blocks becomes larger, and the vertical thickness decreases. A most sophisticated version of the domino model involves rotating the blocks over a listric and detachment fault (Fig. 4.102c,d). In both situations a geometric problem results in the formation of triangular gaps in the lower boundary with the detachment surface, because the blocks when rotated stand on one of their corners. Ductile flow, intrusions filling the gaps, and other defor-

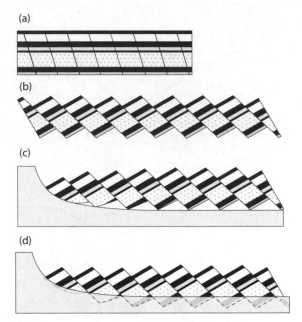

Fig. 4.102 The domino model for normal faulting. (a) Initial stage showing the position of the normal faults. (b) Rotation of blocks to accommodate the extension. (c) The domino model in relation to listric and detachment faulting showing geometric problems related to the lower block corners. (d) The same model without the bottom gaps.

mations have been invoked to solve this inconvenience. Although small-scale examples show the intact rectangular shape of the rotated blocks, seismic lines very often show the geometry represented in Fig. 4.102d, in which the blocks are flattened at the bottom to adjust to the detachment surface. This deformation can be achieved by further shearing or fracturing of the block corners.

In Section 3.14 several displacements were proposed for the deformation of blocks in listric faults. Rigid rotation or translation of the hangingwall block is not allowed as explained above, because this gives rise to gaps between the blocks. Different models (Fig. 4.103) involve distortion by internal rotation of the hangingwall block to form a rollover anticline as the blocks involved have to keep in touch along the entire fault surface (Fig. 4.103b, c). In more rigid environments, the extension can be accommodated by the formation of additional synthetic faulting in the hangingwall block, which is divided into smaller blocks that rotate in a similar way to the domino model (Fig. 4.103d). The formation of a set of imbricate synthetic listric faults can also occur; they rotate like small rock slides down the fault surface (Fig. 4.103e). An

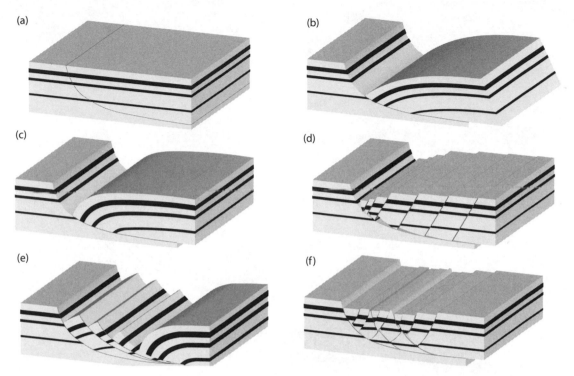

Fig. 4.103 Various kinematic models for deformations accompanying the development of normal listric faults (see text for explanations).

increase in block subsidence by sliding gives way to flattening of the block as it reaches the subsided area, whereas bedding or other initially horizontal layering becomes progressively steeper. The progressive formation of faults, younger toward the footwall is called *back faulting*. Finally, a combination of synthetic and antithetic listric faulting can be produced in the hangingwall, the adjustment of the holes between the blocks being provided by ductile deformation or minor fracturing (Fig. 4.103f).

Stepped faults showing flat and ramp geometries can develop special deformation structures and involve distinctive kinematics. The hangingwall block deforms over the steps causing synclines or anticlines if the rocks are ductile enough (Fig. 4.104). The flanks to ramp- or flat-related folds formed by bending are areas where shear deformation increases and are preferred sites for secondary faulting of the hangingwall block. Ramps change position as extension progresses by cutting sigmoidal rock slices called *horses* from the footwall block. Together all the horses form a duplex structure bounded in the upper part by a *roof fault* and at the bottom by a *floor fault*. The floor fault is active (experiencing shear displacements along the surface) as it is part of the main fault, whereas the roof fault

plays a secondary roll, being active only when the corresponding horse forms.

4.15.5 Thrust and reverse faults

Thrust and reverse faults form in tectonic settings in which a horizontal compression, defining the main principal stress (σ_1), is produced and a minor compression (σ_3) provides the vertical load. The main geotectonic settings in which thrust and reverse faults form are convergent and collision related plate boundaries. Thrusts and reverse faults in continental settings form in fold and thrust belts that can extend hundreds of kilometers. In oceanic environments they appear in accretionary wedges or subduction prisms, between the trench located at the plate boundary and a magmatic arc in both intra-oceanic and continental active margins. Thrust faulting results in crustal shortening and thickening (Fig. 4.100c, d). Thrust and fold belts are limited in front (defined by the sense of movement) by an area not affected by faulting, the *foreland*, where a subsiding basin can form by tectonic loading (Section 5.2). The area located at the back of the thrust belt is the *hinterland* (Fig. 4.105). Structures in

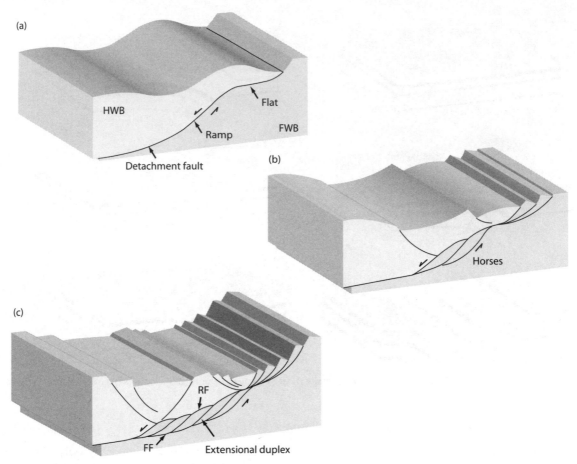

Fig. 4.104 Progressive deformation of the hangingwall block (HWB) in a normal listric fault with a ramp and detachment. (a) Bending of the hanging wall to adjust to fault surface geometry. (b) The ramp migrates as extension takes place giving way to a set of imbricated sigmoidal slices called horses (b) in the footwall block (FWB). These form together with a duplex structure at depth and a series of secondary faults in the hanging wall, defining a complex half graben with normal listric faults forming a fan (b and c). Duplex structures are bounded by two faults, the roof fault (RF) at the top and the foot fault at the bottom (FF).

thrust belts are highly asymmetrical in the direction of tectonic transport or general displacement, and generally most faults dip toward the hinterland. Locally thrust faults can form in compressive reaches of gravitational slides developed at the foot of the collapsing rock masses or other processes related to folding or igneous intrusive processes.

Reverse faults are high-angle faults, showing surfaces inclined as much as normal faults greater than or equal to 60°. They are not as common as thrusts but can be important features in many tectonic compressive settings. However they do not fit Anderson's theory of faulting in which faults formed by horizontal compression should be low-angled. Also, considering Anderson's stress conditions, reverse faults do not follow Coulomb's failure criterion either. Several explanations for the formation of high-angle reverse faults include tectonic inversion from extension to

compression regimes, and reactivation of previous generated normal faults as reverse faults. Also the curving at depth of the stress axis directions, or *stress trajectories*, can produce curved fault surfaces allowing thrust faults to evolve to reverse faults at depth and also for thrusts to evolve to high-angle faults by frontal ramping to the surface (Fig. 4.106). Diverging stress trajectories can be produced if stress gradients and differences in the state of stress exists both in the vertical and lateral directions. Thrusts generally are initiated as low-angle faults but can be subsequently deformed by compression changing the overall shape.

Compressive tectonic settings can display very complex structures with thrusts, reverse faults, and folds associated together. This style of deformation is known as *thin-skinned tectonics* because a relatively thin layer of the crust suffers intense shortening and deformation whereas the

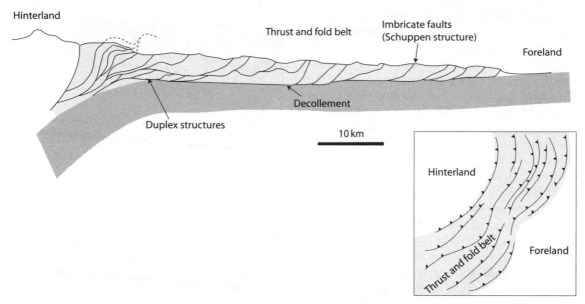

Fig. 4.105 Idealized model of a thrust and fold belt and its representation on a map. The filled triangle along the faults on the map point in the dip direction of the faults.

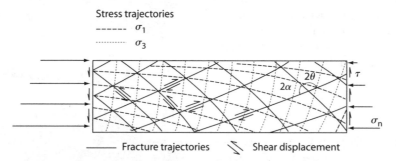

Fig. 4.106 Stress trajectories can curve at depth when there are stress gradients. Coulomb fractures will bend according to stress trajectories, which can cause the change from thrust (low-angle) to reverse faults (high-angle).

basement is mostly unaffected. This situation poses important mechanical and kinematic problems in the reconstruction of tectonic processes related to thrusting, due to the decoupling between the shortening of the basement and the cover. Common structures in thrust and fold belts are a low-angle or near horizontal basal shear plane or *decollement*, that act as detachment areas and separate a highly deformed, both folded and fractured upper part or *cover* from a relatively undeformed substratum or *basement*. The detachment is also called a *sole fault*, produced where there is a mechanical contact formed by the presence of a less frictional weak layer (typically clay, shale, or salts). Deformed rock wedges over thrust faults are often called *thrust sheets* or *nappes*. The cover is also known as an *allochthonous* terrain due to its displaced nature, forming very extensive and relatively thin triangular rock wedges that thin in the displacement or tectonic transport direction. The basement under the main decollement is often referred to as *autochthonous*, the rocks there remaining *in situ*. Erosion of part of the allochthonous terrain allows observation of the basement at the Earth's surface in so-called *tectonic windows*. Similarly, erosive remnants of an allochthonous terrain surrounded by autochthonous rocks are called *klippes*.

As in normal faults, flat and ramp geometries are common in thrust faults, lying perpendicular, parallel, or oblique to block transport direction. Commonly ramps are formed when a low-angle or horizontal fault rises to a shallower level in the crust cutting competent rocks and forming a high-angle step inclined backward with respect to the transport direction, running toward another incompetent layer where another *decollement* or flat is formed. The presence of ramps produces particular deformations in the hangingwall as described for normal faults. A very prominent structure is a syncline lying on the lower reach of the ramp surface that evolves toward an anticline located over the upper end of the ramp. As the hangingwall block climbs the footwall ramp, a syncline is formed at the toe and an anticline at the top of the ramp. Although the syncline axial surface remains in the same position, the limbs get progressively larger (Fig. 4.107). The anticlinal folds formed in the hangingwall develop ramp and flat geometries too. There are various models for fault propagation but they basically involve two kinds of thrust fault arrangement into thrust sheets. The first is formed by the faults that break the topographic surface and whose fault trace can be followed in the field. These faults can be arranged in different forms but most typical occurrences in fold and thrust belts are imbricate fans of listric faults, concave toward the hinterland, joining a basal sole fault (Fig. 4.105). These structures are known as *schuppen zones*. The second prominent structure are duplexes in which a set of horses are confined between two detachment faults, a roof fault and a foot fault. Horses forming the duplex can be inclined toward the foreland, the hinterland, or can stack vertically (Fig. 4.108).

4.15.6 Strike-slip faults

According to Anderson's theory, strike-slip faults form when the intermediate principal stress (σ_2), is vertical and due to gravitational loading, which means that in a horizontal surface of the remaining principal axis one direction experiences a compression larger than the vertical load and the other is subjected to extension or to a compressive stress less intense than the vertical load (Fig. 4.101). As a result, there is a direction of horizontal regional shortening (parallel to the direction of σ_1), normal to the direction of maximum lengthening (parallel to the direction of σ_3). There are a number of geologic settings in which strike-slip faults form, the most prominent being transform plate boundaries (Section 5.2), characterized by horizontal shearing and movement of blocks along close-to-vertical faults. These *transform faults* lie perpendicular to the spreading centers of midoceanic ridges, separating lithospheric reaches expanding at different rates. The term transform fault is used strictly for all faults affecting the whole lithosphere, which mark plate boundaries both in continental and ocean settings (Figs 4.109 and 4.110).

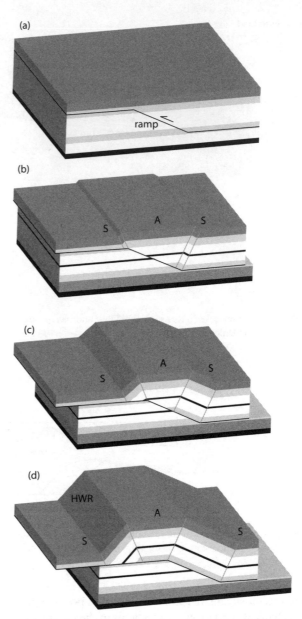

Fig. 4.107 Hangingwall deformation produced by overthrusting over a footwall block with flats and ramps. As the hangingwall block climbs the footwall ramp, a syncline (S) is formed at the toe and an anticline (A) at the top of the ramp. Although the syncline axial surface remains in the same position, the limbs get progressively larger HWR: hanging wall ramp.

Other large-scale strike-slip faults on continental settings that are not a part of plate boundaries are called *transcurrent* faults. Apart from transform plate boundaries, strike-slip faults appear in other geotectonic environments such as extensional provinces and compressive settings, like

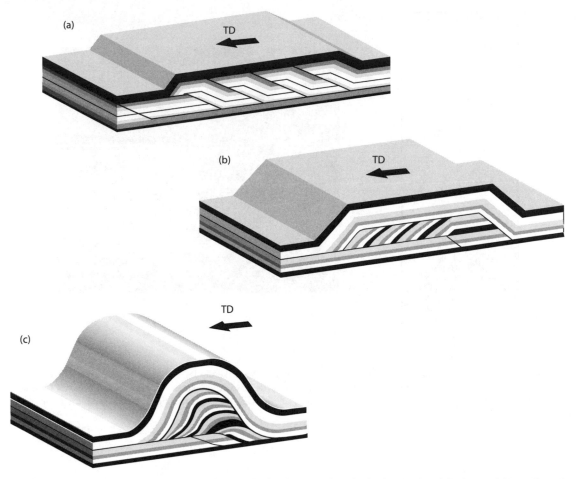

Fig. 4.108 Duplex structures in compressive settings. (a) Hinterland inclined duplex; (b) foreland inclined duplex; and (c) antiformal stack. The tectonic displacement (TD) for all three is the same, as indicated by the arrows.

mountain belts where they can be local or minor features but important in the accommodation of the overall deformation. For example, in extensional areas or compressive settings, strike-slip faults, called *transfer* faults, orientated parallel to the displacement direction, adjust the movement of half-grabens showing different polarities or separate areas experiencing different extension rates. *Tear faults* are minor strike-slip faults associated with folds, thrusts, or normal faults similar to the transfer faults, but of minor extension. Although most strike-slip faults have vertical roughly planar surfaces, forming straight traces on the surface, bends (frontal vertical ramps), and stepovers may form (Fig. 4.111). Bends and stepovers can be produced to the right or the left in both dextral and sinistral faults. These features are important because they create

special stress conditions along the faults. For example, a dextral fault having a right bend or stepover experiences extension in the bend of the offset area due to block separation during movement along the fault. Areas suffering extension along a strike-slip fault are called *transtensional* areas, the bends being extensional or releasing. Basins developed in transtensional areas are called *pull-apart basins* (Fig. 4.112). Another example illustrating a very different behavior occurs in a dextral fault with a left bend or stepover. In this case, the blocks are compressed against each other in the bended or offset area creating a *transpressive* area, and the bends or stepovers are called contractional or restraining. Transpressional and transtensional settings cause particular deformation structures called *strike-slip duplexes* or *flower structures*, defined by horsts

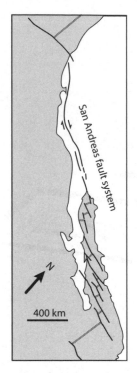

Fig. 4.109 The San Andreas fault is one of the most studied examples of an active strike slip fault system. It marks the long onshore portion of a complicated system of oceanic transform faults which displace the East Pacific Rise progressively north east in the Gulf of California and which is causing the general motion of peninsula and coastal southern California in the same direction. As indicated, the sense of motion is dextral strike slip.

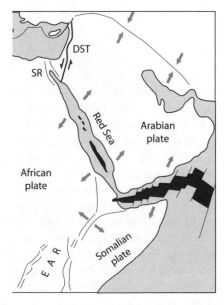

Fig. 4.110 Transform faults in the Gulf of Aden between the Arabian and Somalian plates are related to sea floor spreading. The Death Sea transcurrent (DST) fault is an example of a transform plate boundary separating the Arabian plate from African plate in a continental context. EAR – East African Rift, SR – Sinai Rift.

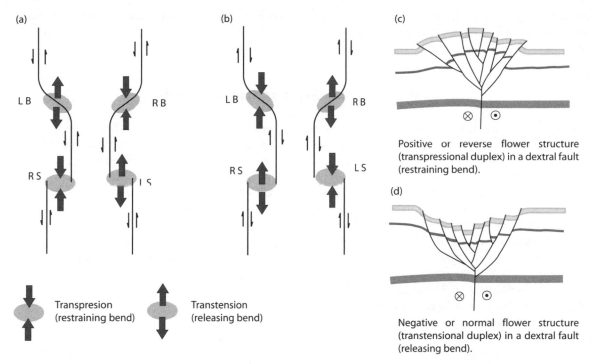

Positive or reverse flower structure (transpressional duplex) in a dextral fault (restraining bend).

Negative or normal flower structure (transtensional duplex) in a dextral fault (releasing bend).

Fig. 4.111 Bends and stepovers in (a) sinistral and (b) dextral strike-slip faults, give way to transpressional areas in restraining bends and transtensional areas in releasing bends. Strike slip duplex structures form in this area subjected to compression or tension, which are also called flower structures (RB: right bend; LB: left bend; RS: right stepover; LS: left stepover). (c) and (d) show two different strike slip duplexes in a sectional view.

Fig. 4.112 Death valley. An example of a releasing bend tectonic environment causing extension and basin formation (a) View north east towards Panamint Range. (b) Satellite image to show the central basin and bounding ranges with the Panamint range in the top left and the Armagosa Range to the right.

between strike-slip vertical faults. Transtensional areas develop horsts with a gravitational or normal component and are named normal or negative flower structures, whereas transpressional contexts give way to horsts with a negative component and the duplexes formed are called reverse or positive flower structures.

4.16   Solid bending, buckling, and folds

Folds are wave-shaped deformations produced in rocks and made visible by the deformation of planar structures such as layering in sedimentary rocks, layering and foliations in metamorphic rocks and in some igneous rocks (Fig. 4.113). Folds are some of the best described tectonic structures characteristic of ductile deformation. Individual folds can be *antiforms*, when they are convex up (A-shaped) or *synforms*, when they are concave up (Fig. 4.114). *Anticlines* and *synclines* are terms that are used to describe folds, but the meaning is quite different to antiforms and synforms. To define anticlines and synclines the age of the folded layers has to be known. Anticlines are folds that have the oldest rock layers in the fold core, concave side or inner part and the younger rocks in the outer, convex surface. Synclines are folds that have the opposite age distribution, such that the older rocks lie on the convex layer and the younger in the inner concave surface. Although in not very intensely deformed rocks, it is common to have a coincidence between anticlines and antiforms and syncline and synforms, when several folding phases occur and folds are superposed, the rocks can experience overturning, leading to a reversal in stratigraphic polarity; all four combinations are possible, with the addition of antiformal synclines and synformal anticlines.

Folds are usually arranged in fold trains in which there is a succession of antiforms and synforms. The boundary between adjacent folds is defined by the inflection points in which the bend changes polarity or sense of curvature (Fig. 4.115). As described in Section 4.15 folds can be associated with thrust faults in orogenic settings in thin-skin tectonic deformed areas, but also form in a variety of other settings in the inner areas, as the metamorphic cores, of orogenic belts. Local formation of folds

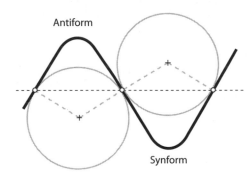

Fig. 4.114 Definition of curvature in a fold by locating a reference circle tangent to the fold sides in a line that join the middle points of the more straight parts of the fold.

Fig. 4.113 Folds are wave-shaped ductile deformations developed on layered rocks as these stratified sedimentary rocks.

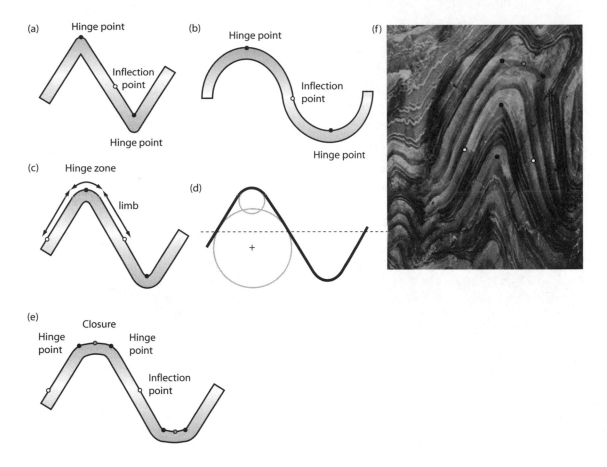

Fig. 4.115 Definition of fold geometry in two dimensions at a given transversal section of a folded surface. (a) The hinge point is the point of maximum curvature and the inflection point of minimum curvature. (b) Semicircular folds have constant curvatures and the hinge is defined in the middle point of the arch and inflection points where there is a change in the bend polarity. (c) A general case where there is a hinge zone, defined on the fold segment with a higher curvature than the reference point as shown in (d). (e) Folds with two hinge points and closure. (f) An example of folded surfaces showing different geometric elements in 2D.

can be also related to bending of a cover of ductile rocks over some rigid basement that is fractured or to the drag effect of shear movements along a fault.

4.16.1 Geometric description of folds

Folds can be described by their geometric characteristic, both in two or three dimensions. The most basic geometric elements are described in a single folded surface in two dimensions. Additional descriptions involve the 3D extension of the folded surface, and also the relation between several superimposed folded layers. Curvature of a fold may remain constant or can change. It can be defined by a

reference circle tangent to the inflection points at both sides of the fold. Tracing perpendicular lines from the inflection points will mark the center of the circle (Figs 4.114–4.115d). The *hinge point* is defined in a 2D transverse section as the location in a folded surface showing the maximum curvature. In an individual section folds can have one or several hinge points (multiple hinged folds). The point of minimum curvature between two adjacent hinge points of the same fold is called *closure* (Fig. 4.115e). In three dimensions, joining all hinge points along the surface defines the *hinge line* or *hinge* (Fig. 4.116). Low curvature areas between the hinge lines are the fold *limbs* or *flanks*. The inflection lines can be defined in three dimensions joining all inflection points

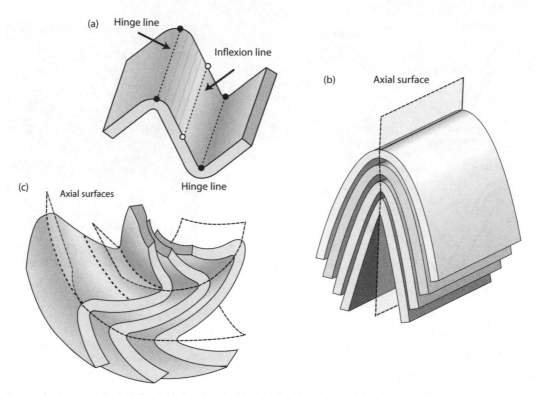

Fig. 4.116 Geometric elements of a fold in a folded surface in 3D. (a) The hinge line is defined joining all hinge points and the inflection line joining all the inflection points (b) and (c). Cylindrical folds, as in (a) and (b), have a fold axis which is any line parallel to the hinge line of constant orientation but is not located at any particular position in the fold. The fold in (c) is noncylindrical. The axial surfaces can be defined joining all hinge lines in successive superimposed folded layers; they can be flat as in (b) or curved as in (c).

along the surface. Different shapes can be expected in folds, sometimes the surface is quite rounded and defining the hinge is not straightforward, as the hinge is not located in a single point. In these situations the *hinge zone* is defined by drawing the reference circle tangent to the limbs at the inflection points; the area having less curvature than the reference circle is the hinge zone (Fig. 4.115c,d). Along a given transverse section of a folded surface, the *crest point* is the higher topographic point. Joining all crest points along all the possible transverse sections in the fold gives the crest line, the highest point in the line being the *crest line* culmination. Similarly, the lower topographic point in a section is called the *trough point* and the line along the surface joining all trough points is the *trough line*. The lowest point in this line is called the *crest line depression*. Both the crest and trough lines can be straight or curved, and may differ from the ones at adjacent folded surfaces.

Cylindrical folds have a *fold axis* that is parallel to the hinge line (Fig. 4.116). The fold axis describes the full fold surface. Folds not having an axis are noncylindrical with the exception of conical folds, whose surface is generated rotating a line but leaving one of the ends fixed in position. In conical folds the axis is like in a geometrical cone.

Considering several superimposed folded layers, other geometric elements can be defined. The *axial surface* is the surface that joins all the hinge lines in all the stacked folded layers. The shape of the axial surface can be flat or curved (Fig. 4.116). The inclination of the axial surface is called *vergence* and is a measure of fold asymmetry or shearing sense. The vergence marks the same direction as the normal to the strike of the axial surface but is defined toward the opposite sense (up dip). The intersection of the axial surface with the topography or any vertical or horizontal section is an *axial trace*. Inflection surfaces can also be defined by joining all inflection lines from several stacked folded layers.

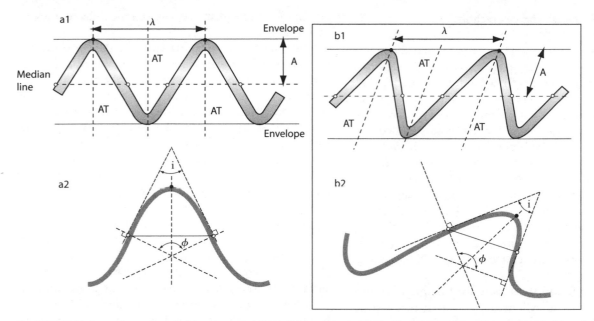

Fig. 4.117 Fold size and symmetry in (a) symmetrical folds; (b) asymmetrical folds. To define fold size the wavelength (λ) and the amplitude (A) are defined. The wavelength is measured from two consecutive antiform or synform hinges parallel to the median line. The amplitude is the distance between the median line and one of the external envelope, measured parallel to the axial trace (AT). a2 and b2 show some components to establish fold symmetry or asymmetry.

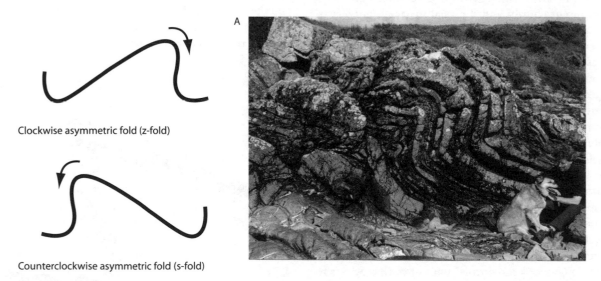

Clockwise asymmetric fold (z-fold)

Counterclockwise asymmetric fold (s-fold)

Fig. 4.118 Asymmetrical folds are defined as clockwise or z-folds and counterclockwise or s-folds. As photo shows Mike and Storm discussing an example of a z-fold (Scotland, UK).

4.16.2 Size, shape, and orientation

Folds occur over a range of sizes, from several kilometers to millimeters and are defined in two dimensions by two components, the wave length (λ) and the amplitude (A), in the same way that other wave-like forms are measured. To accurately establish both components a reference line is drawn joining all the inflection points, called the median line, and all the hinge points in both antiforms and synforms, called enveloping lines. The wave length is the distance between the hinges of two consecutive antiforms or synforms, measured in a straight line parallel to the reference lines (Fig. 4.117). The amplitude is the distance, measured parallel to the axial trace, between the median line containing the inflection points and the envelope line containing the hinge points (Fig. 4.117).

The shape of folds can be described by means of different elements. The cylindricity of a fold that can be considered an important element in fold descriptions has been illustrated previously when discussing folds containing an axis. Other elements are the fold symmetry or asymmetry, which can be given by the length and the shape of the limbs (Fig. 4.117). Symmetrical folds have equally long limbs and the axial surface is a symmetry plane that divides the fold in two halves identical in shape but mirror images. Asymmetrical folds have limbs with different lengths and the axial surface is not a symmetry plane; z-folds or clockwise folds and s-folds or counterclockwise folds can be defined (Fig. 4.118) on the basis of the limbs' rotation with respect with to a symmetric position. The asymmetry of s- and z-folds changes if we look at the folds from one side or the opposite facing along the axial surface, and so, conventionally, the sense of rotation is defined looking down the plunge of the hinge line if it is inclined. When the hinge line is horizontal some geographical reference has to be included in the description. Other elements to measure fold shapes are the tightness, the bluntness, and the aspect ratio. The tightness is defined by the interlimb angle (i, Fig. 4.117) or the fold angle (ϕ, Fig. 4.117). The limb angle is the angle that forms the tangents at each inflection point of the limbs, and the fold angle between the normal lines of both tangents to the limbs. According to these angles, folds can be classified into *acute* (when i has a value between 180° and 0° and ϕ between 0° and 180°), *isoclinal* (when $i = 0$ and $\phi = 180°$), and *obtuse* (i from 0 to −180° and ϕ between 180° and 360°). The bluntness describes the degree of roundness or curvature in the hinge zone or closure, and the aspect ratio the relation between the

amplitude and the distance between the inflection points of a fold.

Fold orientation in a 3D space is described by the orientation of the axial surface and the hinge line. Axial surface orientation is given by the strike and dip of the surface, whereas the hinge line is defined by the plunge (the vertical angle between the line with its horizontal projection) or the rake (pitch) measured over the axial surface between the hinge line, which is always located on the axial surface and a horizontal line located in the axial surface. There is also a broad nomenclature and fold classification concerning different kinds of folds according to their orientation; for example, *upright folds* are those having vertical axial surfaces; in these the hinge line can be horizontal, inclined, or vertical. Folds having horizontal axial surfaces are called *recumbent folds* (the hinge line is always horizontal) and finally, folds having inclined axial surfaces are called *steeply, moderately,* or *gently inclined folds* depending on the inclination. Inclined folds can have horizontal or inclined hinge lines. When the dip of the axial surface and the hinge line are equal in angle and orientation, the folds are called *reclined*.

Other classifications are based on geometric properties of the folded surfaces. One of the most commonly used is the Ramsay classification of folds (Figs 4.119 and 4.120), which is based on the definition of three geometrical elements: the dip isogons, the orthogonal thickness, and the axial trace thickness. To trace the dip isogons, first the axial trace and a normal line to it are plotted. The normal is the reference line to define different angles (α, Fig. 4.120).

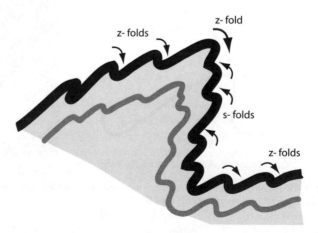

Fig. 4.119 Parasitic folds can be superimposed on larger symmetrical or asymmetrical folds. Note the change from z- to s-folds at both sides of the larger fold. The photo shows an example of some folded layer displaying parasitic folds.

Fold class	Dip isogon geometry	Orthogonal thickness	Axial trace thickness
1A	convergent	increases	increases
1B	convergent	constant	increases
1C	convergent	decreases	increases
2	parallel	decreases	constant
3	convergent	decreases	decreases

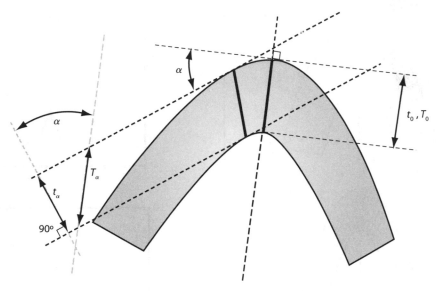

Fig. 4.120 Definition sketch for the geometric elements described for Ramsay's fold classification. The table shows all fold classes included and their principal characteristics.

4.16.3 Kinematic models

The basic deformation model for the formation of a fold is the *flexural folding* of a rock layer, which produces class 1B or parallel folds, which are those that preserve homogeneous thickness along the layer. There are two mechanisms give rise to flexural folding; bending and buckling. Bending is formed when pairs of forces or torques equal in magnitude and opposed are applied normal or at high angles to points of a layer, producing the rotation that causes the bend of wave instability to form (Fig. 4.122). Typical examples include the formation of folds in sedimentary layers located over faulted rigid basements, motion of the blocks on each side of fault. Buckling consists of the application of forces equal and opposed at the ends of a layer. Forces are applied parallel to the layer extension, producing a compression, which forms a bend in the layer. Buckling is one of the chief folding mechanism in fold and thrust belts in orogenic settings (Fig. 4.123).

Flexural folding has two principal modes: orthogonal flexure (Fig. 4.124) and flexural shear (Fig. 4.125). Orthogonal flexure is a kinematic model in which the outer convex surface of the layer experiences an increase in length whereas the inner concave surface is shortened. The stretched and shortened parts of the fold are separated by a neutral surface that maintains the original length. This folding model is called orthogonal flexure because lines initially perpendicular to the layer surfaces remain perpendicular in the deformed state. Flexural shear or flexural flow is achieved by simple shearing parallel to the surface of discrete segments of the folded layer. Individual surfaces slide like a deck of cards when folded without experiencing shortening or lengthening. Folds can further evolve after being formed by flexural folding by homogeneous flattening, which can produce thinning or thickening of parts of the fold, giving folds of classes 1A or 1C. Folds can be further deformed or exaggerated by flattening without changing their basic geometry (Fig. 4.126).

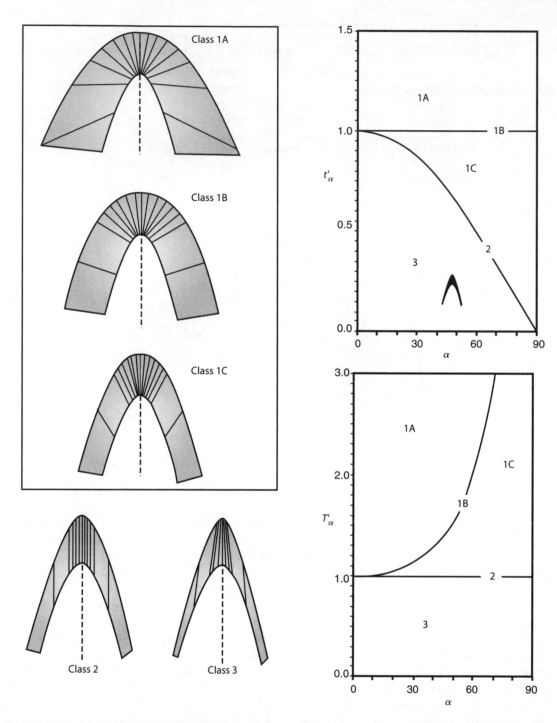

Fig, 4.121 Classes of folds described in Ramsay's classification in relation to changes in dip isogons, orthogonal thickness, and axial trace thickness.

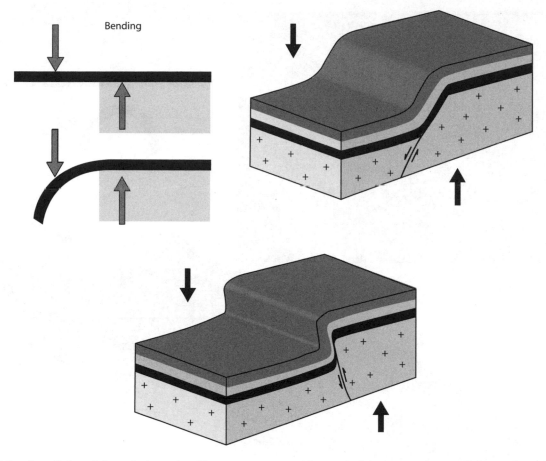

Fig. 4.122 Bending of a layer is formed when pairs of forces or torques equal in magnitude and opposed are applied normal or at high angles to points of the layer, producing the rotation that causes the bend of wave instability. Typical examples include the formation of ductile deformations by folding of a sediment cover over a faulted rigid basement, as a result of block displacements.

4.17 Seismic waves

In addition to molecular-scale motions characteristic of different thermal states, Earth materials are in constant 3D motion, termed background *seismic "noise."* This is usually of a few seconds period and of such tiny amplitude (order of 10^{-5} mm) that we are usually completely unaware of its existence. Nowadays in addition to natural causes (like thermal stresses, tides, breaking waves, and winds), many familiar human-induced ground vibrations contribute to seismic noise, like the passage of vehicles. Such seismic noise triggers periodic instabilities in moving and still fluids, preventing, for example, the accurate modern-day determination of the transition to turbulence in Reynolds' old laboratories adjacent to Manchester's busy Oxford Road. Yet periodic ground motions of the most violent kind are more familiar to many who live within areas prone to earthquakes (Fig. 4.127). These

ground motions are due to the direct deformation of the rocks surrounding a fault that has broken surface or which is located close to the surface. At the surface around the *epicentral* region of an earthquake, the direct ground motions that originate close to the deep source, the *focus* or *hypocenter*, cause seismic waves to be generated with periods of 0.5–20 s. These are only revealed by sensitive instruments called *seismographs* (Fig. 4.128) that are designed to transmit, amplify, and record the passing wave motions sufficiently so that they can be analyzed (although Theseus was reputed to possess the ability to sense incoming seismic waves).

In general the periodic higher frequency components of Earth's seismic motion are due to processes of rock rupture; testament to the ability of tectonic forces at work in outer Earth being able to strain rocks beyond their

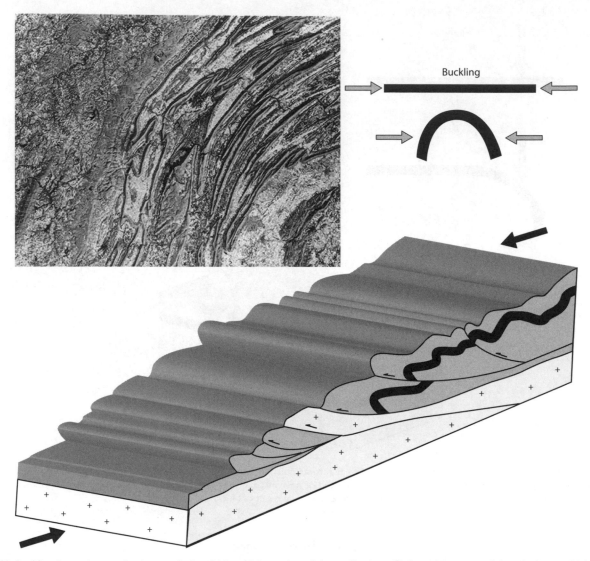

Fig. 4.123 Buckling is another mechanism producing folds, which consists of the application of balanced forces parallel to the layer, which consequently form a bend produced by the compression. Buckling is the chief folding mechanism in fold and thrust belts in orogenic settings. The photo shows an satellite view of the Appalachian belt (USA), where several kilometer-scale folds can be distinguished.

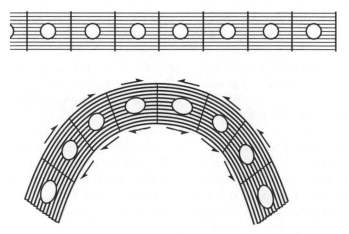

Fig. 4.124 Orthogonal flexure.

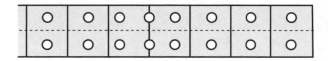

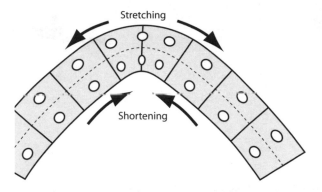

Fig. 4.125 Flexural shear folding.

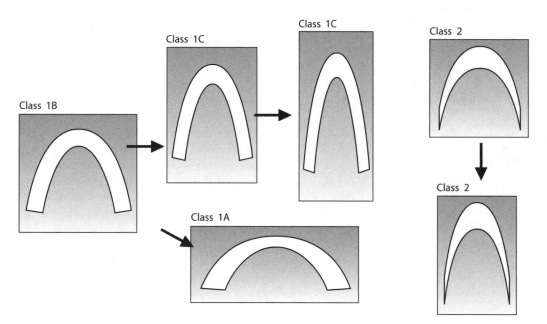

Fig. 4.126 Homogenous flattening of previously folded layers.

ability to resist. This commonplace exceedance of the elastic limit (Section 3.15) is a feature of the brittle behavior of rock as it stretches, expands, compresses, and twists in response to tectonic and thermodynamic stresses; faults, folds, and metamorphism (changes of rock state) are the geological end-result of the cumulative effects of all this motion (see Sections 4.14–4.16). Rocks may also be broken by explosive fragmentation accompanying volcanic eruptions (Section 5.1). All such rupture processes involve the release of seismic elastic energy, estimated as an average total of $7.5 \cdot 10^{17}$ J annually. Much of the energy is transmitted with the seismic waves that propagate outward radially from earthquake epicentral zones. A very strong (and very rarely occurring) earthquake triggered by

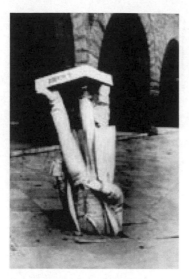

Fig. 4.127 Testament to the passage of strong surface earthquake waves – the fallen statue of Agassiz by Stanford University arches after the 1906 San Francisco earthquake.

Fig. 4.129 An historic seismic record: facsimile of the 1889 Tokyo earthquake recorded by Paschwitz at Potsdam, Germany.

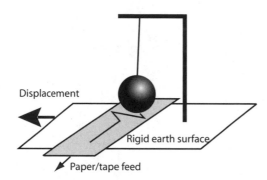

Fig. 4.128 Diagrammatic representation of a simple pendulum seismometer whose support moves with any ground motion; the inertial reaction of the suspended mass is recorded as a time series on a moving paper or tape feed.

rupture along a major fault may liberate up to 10^{19} J of energy.

4.17.1 Early clues concerning the nature of seismic waves

Ingenious inertial seismometers from second century AD China were constructed on the basis that first motion seismic waves were directional. Nowadays direct visualization of ground motions induced by passing waves close to epicentral locations can be recorded by the inertial reactions of supermarket trolleys as recorded by security cameras. The near-simultaneous arrival of the first motions of strong seismic waves at two observatories in north and east Germany in 1889 (Fig. 4.129) first established that wave-like disturbances had spread globally (126° of longitude and 17° of latitude), seemingly through *and* over the earth as a packet of energy during 2 h of recording from a teleseismic (remote) source, in this case the great Tokyo earthquake. Systematic observations of such earthquake records led Oldham in 1897 to recognize that the wave-like tremors consisted of first and second "preliminary tremors," followed by much larger amplitude waves. The inference was that the first arrivals were the faster, having traveled through Earth as *body waves*. It was therefore logically deduced that the longer the time of transit from the earthquake source the longer was the time difference separating these early arrivals from the large amplitude disturbances that traveled as *surface waves*. By plotting records from numerous earthquakes of first and second tremor arrival times as a function of angular distance (1° of arc = 111 km) subtended between source, Earth's center, and recording station, a systematic separation of the two early arrival wave packets was seen by Oldham in his famous paper of 1906 (Fig. 4.130). Not only that, but the separation sensibly increased with distance traveled, though not as a linear trend for it seemed that the first arrival tremors were transmitting relatively much more slowly with distance traveled beyond about 130°. More startlingly, at about 120° of spread the second tremor arrivals had stopped completely, to reappear and be delayed by up to 10 min at about 150° arc length. Oldham deduced that the Earth must have a massive dense core with a well-defined boundary against an outer shell. Approximate wave speeds appropriate to travel in the Earth's mantle computed from Oldham's data are tabulated in Fig. 4.130.

4.17.2 Modern parlance for seismic waves and their magnitude

Seismic tremors propagate elastic energy from a natural rupture source (fault) or human-made explosion as oscillatory 3D motion. The wave-like tremors spread radially as *body* and *surface waves*, losing energy as they spread geometrically outward. The lower amplitude body waves spread faster along curved paths deeper within Earth (Fig. 4.131) because of elastic compression. Body waves are themselves separable (Fig. 4.132) into higher speed compression-rarefraction waves (termed primary, pressure, push–pull, or usually just *P-waves*) that physically resemble sound waves and slower, transversely oscillating waves (termed secondary, shake, shear, or just *S-waves*). Unlike *P*-waves, the latter cannot travel through fluid, but they may travel through partially molten solid, slowing down (attenuating) as they do so. Like deep-water waves

(Section 4.9), surface seismic waves are *dispersive* in that their speed of travel (*celerity*) depends upon wavelength. They are divided into two groups, those with a horizontal to- and fro-oscillation normal to the direction of travel (*Love waves*) and those with an orbital motion in the direction of travel (*Rayleigh waves*), the latter resembling the orbitals observed in ocean water surface layers induced by waves (Section 1.35). The horizontally vibrating *S*-waves also share with light waves the property of *polarization*, the separation of oscillations into vertical (*SV*-waves) and horizontal (*SH*-waves) planes in this case, generated as the waves strike internal discontinuity surfaces within Earth. Complete seismogram wave arrivals are illustrated in Fig. 4.133. Like light and water waves, seismic waves also *reflect* and *refract*, observing *Snell's laws* as they do so (Figs 4.134–4.136).

4.17.3 Speeds and types of seismic wave interactions with internal discontinuities

In order to determine depth to possible internal discontinuities in Earth's interior the velocity of seismic waves must first be determined. For penetration of acoustic waves generated by various artificial energy sources through water, including the well-known sonar, this is no problem. Experiments in water bodies of known depth yield the figure for acoustic velocity in seawater of about $1.5 \ \mathrm{km \, s^{-1}}$, the acoustic wave energy traveling outward from source as straight rays. However, for the largely unknown rocks of Earth's interior the pioneers of seismology had to make use of what is termed an "inverse problem" approach, that is, they had to deduce the velocities from a knowledge of the variation of travel time, T, with distance, λ, since $T = f(\lambda)$. We have seen already that such travel time plots can be generated and that for *P*- and

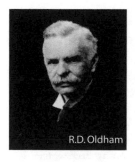

$\Delta°$	P	S
30	9.2	5.5
50	9.3	5.3
70	10.4	6.5
90	11.1	6.9
1° arc = 111 km		km s⁻¹

R.D. Oldham

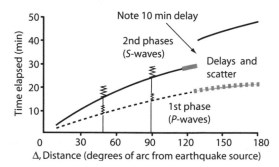

Fig. 4.130 Summary of data Oldham used to "x-ray" Earth. Note the delays and scatter of *P*-waves recorded at large arc distances and stepped slowdown of *S*-waves beyond about 130° arc: both indicative to Oldham of a central dense core, later shown to be liquid. The distant *S*-waves are now known to be reflected mantle *S*-waves, the direct waves not traveling through the liquid core. Note > velocities with Δ.

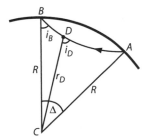

Travel time, T, varies with the angle Δ, so that the ray parameter, $p = \mathrm{d}T/\mathrm{d}\Delta$.

Fig. 4.131 For any curved ray like *AB*, speed, V, varies along the path. For any point, D, of a ray Snell's Law defines a constant ray parameters, p, such that $r_\mathrm{D}/V_\mathrm{D} = R \sin i_\mathrm{B}/V_\mathrm{B} = p$.

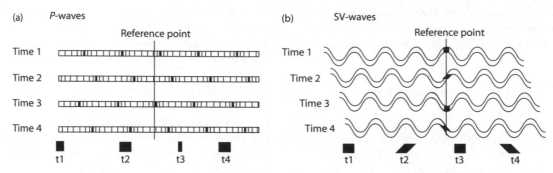

Fig. 4.132 (a) P-waves; rarefraction-compression elastic deformation shown by the evolving size of the black rectangular element for successive times at the reference point; (b) SV-waves: vertical shear elastic deformation.

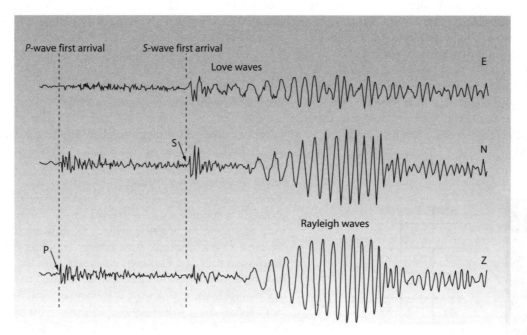

Fig. 4.133 These are seismograms from a moderate earthquake (magnitude 5.1) in the Norwegian Sea recorded in Germany. Each seismic trace shows the component waveforms recorded by seismometers of different orientation. The top two record horizontal ground motions in E–W and N–S directions, respectively, while the lower records the vertical "up–down" ground motion. In these cases the E–W seismometer has nicely picked out the horizontal E–W motion of the Love wave package. The N–S seismometer has a very clear S-wave first arrival. The vertical ground motions induced by the steeply inclined *P*-wave signal are well captured by the third seismometer, as is the vertical Rayleigh wave train. The records illustrate the long duration of the damaging surface wave signals compared to *P*- and *S*-waves.

S-waves different relationships must apply. The inverse approach determines the gradient of travel time with distance at the point of interest and relates this to the ratio of radial distance and velocity (Fig. 4.137).

The pioneers of seismology made the simplifying assumption that seismic energy was transported as linear rays moving with constant average velocity characteristic of a particular rock medium, that is, they envisaged a complete analogy between seismic rays and light rays. Subsequent research proved that seismic velocities gener-

ally increased with depth (Box 4.2) and that the rays were curved, concave side up (Fig. 4.138). Not only that, but seismic wave energy is reflected and refracted in complex ways across internal discontinuities (Fig. 4.136), a phenomenon that leads not only to elucidation of the internal planetary structure of Earth using accurate travel time data, but also to the location of subsurface geological structures containing vast economic reserves of oil and gas.

The speed of travel of seismic waves is controlled by elastic properties, in particular the *bulk modulus, K,* and

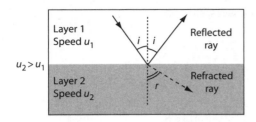

Snell's laws: 1. reflection angle, i, equals incident angle, i
2. refraction angle, r, and incident angle, i are related by: $\sin i/u_1 = \sin r/u_2$

Fig. 4.134 Seismic reflection from an interface (like the *MOHO*) and refraction across it.

Mohorovicic

There is a critical angle, i_c that enables refraction at r = 90° and hence a wave path along the interface to observer at x_2

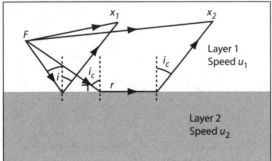

Fig. 4.135 Critical seismic refraction across an interface like the MOHO.

the *shear modulus*, G, and the density of the substances that they pass through. We have seen (Section 3.15) that the bulk modulus relates the change in hydrostatic pressure, P, in a block of isotropic material to the change in volume, V, that is, $K = dP/dV$. Both solids and fluids are compressible and hence both can sustain P- and S-waves. The shear modulus is the ratio between the shear stress, τ, and the shear strain, γ, in a cube of isotropic material subjected to simple shear, that is, $G = \tau/\gamma$. G is thus a measure of the resistance to deformation by shear stress, in a way equivalent to the viscosity in fluids. Since fluids like air and water cannot support shear motions by finite strain, G is zero and hence S-waves cannot travel through them. The expressions for wave speeds (Box 4.2) are rather surprising in that they depend inversely upon density and since we think that this property of rocks generally increases with depth it might imply that seismic wave speeds decrease with depth in the earth. However the values of K and G both depend to a large degree upon density and increase more rapidly with depth than density does (Box 4.2; compare with estimates from Oldham's original data given in Fig. 4.130), giving the required

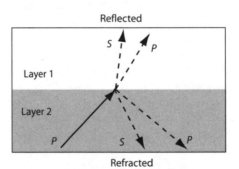

Any obliquely incident body wave may generate both refracted and reflected P **and** S waves from a layer density discontinuity.

Fig. 4.136 Incident body waves generate a variety of reflected and refracted phases.

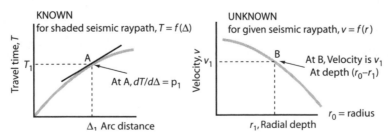

Fig. 4.137 The seismic "inverse problem."

Box 4.2 Variation of density, elastic modullii, and body wave speeds with depth

depth km	ρ kg m^{-3}	p kbar	K kbar	G kbar	$u_{P\text{-}wave}$ km s^{-1}	$u_{S\text{-}wave}$ km s^{-1}
20	2900	6	1315	441	8.11	4.49
80	3370	25	1303	674	8.08	4.67
150	3370	48	1287	665	8.03	4.44
500	3850	171	2181	1051	9.65	5.22
1000	4580	386	3519	1874	11.46	6.40
2000	5120	869	5132	2462	12.82	6.93
3000	10007	1472	6581	0	8.25	0
4200	11510	2631	10814	0	9.69	0
5000	12090	3204	12740	0	10.27	0
6200	13080	3631	14236	1756	11.26	3.66

$$u_{P\text{-}wave} = (K + 1.33G/\rho)^{0.5}$$

$$u_{S\text{-}wave} = (G/\rho)^{0.5}$$

increase of wave speed with depth required by seismological observations. It is simpler in a way to plot seismic wave velocity as a function of density alone; the relationship is linear, of the form $u = a\rho + b$, for both experimental S- and P-waves passing through crustal rocks of density $\rho < 3{,}500$ kg m^{-3}: it is sometimes known as *Birch's law*.

The wholly unique nature of seismic waves lies in the products of their interactions with internal discontinuities. An obliquely incident ray, like the P-wave illustrated in Fig. 4.135, produces not only reflected and refracted P-waves but also a reflected and refracted S-wave. The latter somewhat surprising metamorphosis is not so startling when one realizes that because the obliquely incident P-ray is traveling into a rock discontinuity, differential shear takes place along the plane and a shear wave is thus generated. For a normally incident P-wave of course, no such shear can take place and a simple P-wave reflection takes place. When the obliquely incident ray is a vertically polarized S-wave (SV-wave) then both reflected and refracted P- and SV-waves result. However, if the incident ray is horizontally polarized (SH-wave) then no compressions or differential shear can be generated across the parallel discontinuity and only reflected and refracted SH-waves are generated.

Clearly, given the complexity of wave types and the various possible transit paths through the Earth layers it would be sensible to have a common notation to describe wave attributes. The chosen code (Fig. 4.138) is based around the P- and S-wave classification. Simplest of all are P- and S-waves that leave the earthquake focus and travel entirely within the mantle to any remote recording station.

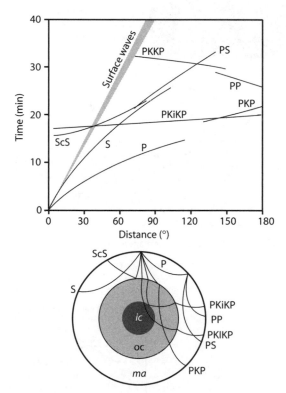

Fig. 4.138 The nomenclature of certain seismic waves and their travel times for distance from an earthquake focus (ma: mantle, oc: outer core, ic: inner core).

A *P*-wave that reflects from the outer core boundary is termed a *PcP* phase. Should the *P*-wave penetrate the liquid outer core (Section 4.17.4) then it is designated a *K* phase; if it transits the inner core then it is given an *I* code. The overall code for a *P*-wave that has passed from the mantle to traverse just the outer core is thus *PKP*. The passage into inner core means it has two *K* transits and two mantle transits and overall it is termed *PKIKP*. Should the outer core wave reflect from the inner core back through the outer core and the mantle, then it is given the notation *PKiKP*. Reflected waves from the surface have the repeated suffix, like *PP* or *SS* for two loops and *PPP* or *SSS* for three loops and so on. Should conversion have taken place upon reflection of refraction then the original code is followed by the transformed code, like *PS, SP,* and the like.

4.17.4 Internal structure of Earth from seismic waves

We demonstrated in Section 1.4 that values computed for Earth mass require a very dense interior and intimated in Section 1.5 and earlier in this chapter that the planetary interior was made up of well-defined concentric shells or layers. It was seismology that revealed the existence of a layered Earth. We briefly highlight the major developments in historical order.

We have already seen how Oldham's travel time data (Fig. 4.130) led him to recognize in 1906 that a central core existed with a sharp and distinct physical character that caused it to slow down *P*- and *S*-waves entering from arc distances of greater than 105°. We know now that the core-blocking shadow between 105° and 142° slows all refracted *P*-waves and excludes all direct *S*-waves because

of the largely fluid nature of the outer core. The interface is the site of generation of the Earth's magnetic field and is the ultimate site of submerged lithospheric plate, the so-called slab graveyard of the *D* seismic layer.

In 1909 Mohorovicic used *P*-wave travel time data from Balkan earthquakes and the concept of a critical reflection angle (see Fig. 4.135) to determine that a fundamental and sharp change in velocity delimited a step-change in rock density at about 50 km depth. This is now known as the famous *Moho* interface between mantle and crust, detectable by either refracted or reflected seismic waves (Fig. 4.134), the wave speeds increasing by 25 percent across it. This variation of transmissibility matches that predicted between dense ($\rho_m = 3{,}300\ \mathrm{kgm}^{-3}$) silicate upper mantle rich in minerals like olivine and spinel and less dense ($\rho_c = 2{,}800\ \mathrm{kgm}^{-3}$) silicate lower crustal rocks of general granitic composition rich in feldspar and quartz. The Moho depth is now known to be typically about 30 km, though thickened crust occurs in many mountain belts to the maximum of 50 km as determined by Mohorovicic in the Balkans. To their great delight, geologists can directly recognize the Moho (confirming it as a very sharp interface) within mountain ranges where gigantic faults have thrust it up toward the surface during past tectonic plate collisions. Fragments of mantle may also turn up in volcanic vents.

In 1936 Lehmann recognized *P*-wave arrivals (termed by her as P_1' and P_2') at large arc separations (Fig. 4.139) that were refracted on arriving at the mantle/core boundary and also subsequently upon leaving it. More significantly there were clear *P*-wave records within Oldham's shadow zone of reflections (105–142° arc separation) that could only have come as reflected phases from the outer surface of a solid inner core. Such P_3'-waves (now called PKKP)

Assume *P* rays are straight and with constant mantle (10 km s⁻¹) and core (8 km s⁻¹) velocities.
Rays 1 and 2 are entirely within the mantle (these we now call *P* waves).
Rays 3 and 7 are diffracted at the core:mantle boundary and focused toward the antipodes (these we call PKP waves).
Rays 4 and 6 are nearly normally incident on the core:mantle boundary (these we now call PKIKP waves).
The zone between ray 2 and 3 is the conventional "shadow zone", between about 105° and 142° arc distance.
Within the "shadow zone" arrivals like 5 (termed *P'* originally by Lehmann) but now known as PKiKP were interpreted to have reflected from a solid inner core of radius about 0.2 whole Earth radii.

Lehmann

The critical evidence came from seismic records (seismograms) like this one at Sverdlovsk Observatory located 135° arc distance from the focus of the New Zealand earthquake, that is, within the general *P*- and *S*-wave shadow zone. Lehmann, s *P'* waves are near first arrivals at these localities (arrowed).

Fig. 4.139 The discovery of the solid inner core – Lehmann's 1936 discovery, logic, and analysis.

arrived at these shadow-zone seismological stations well before any surface waves.

In the 1960s the last major interface within Earth to yield its secret to seismologists was a subtle but profound change in upper mantle mechanical strength, now known as the Low Velocity Zone (LVZ). It defines the uncoupling interface between strong lithospheric plate and weak asthenospheric mantle; it is thus the fundamental dynamic interface that enables plate tectonics to operate. The rigid lithospheric plates (Section 5.2) slide around at velocities of a few centimeters per year on the lubricating LVZ layer because it contains a tiny but significant proportion of molten rock. The LVZ was recognized because the partial melt slightly slows down (by $c.1$ percent) the passage of both P- and S-waves across it (see data in Box 4.2).

We must also mention the discovery, by a combination of seismology and experimental rock physics, of two discontinuities in the mantle (see Fig. 2.7) that owe their origins to mineral *phase changes*. These are changes to the arrangement of the atomic lattice (not chemical change), involving a reorganization involving closer packing due to the increasing pressure. For a very crude analogy think about changing cubic to rhombic packing, as defined in Section 4.11. The first of these phase changes occurs at about 410 km depth in the mantle, where at $c.14$ GPa pressure and temperature at $c.1,700$K, the common mantle Fe–Mg silicate mineral olivine (Sections 1.2 and 5.1) changes to a more densely packed structure of the chemically equivalent mineral *spinel*; the density increase is 6–7 percent. This causes both P- and S-wave velocities to increase across the discontinuity. A larger and more impor-

tant density and velocity change occurs at 660 km depth at $c.23$ GPa pressure and temperature $c.1,900$K, when the spinel structure in turn transforms to the denser *perovskite* phase. This discontinuity is taken as the boundary between the upper and lower mantle. As we shall see in Section 5.2, the discontinuity was once considered inviolate to the downward passage of lithospheric slab; nowadays seismology tells us that slabs may pass clean through to the core–mantle boundary.

4.17.5 Earthquake seismology

The second major achievement of seismologists after elucidation of the internal structure of a layered Earth has been the theory of plate tectonics, though other disciplines, notably geomagnetism, contributed vital clues as to the kinematics and physical processes involved. One seismological clue came from the accurate determination of the magnitude and geography of earthquakes as shown in the map of Fig. 4.140 for over 30,000 earthquakes. *Intensity scales* for earthquakes have been widely used and these relate to the visible damage and environmental effects felt by humans during the earthquake. The *Mercalli scale* is one such intensity indicator. Using instrumental records the magnitude of any earthquake (Box 4.3) must reflect the amplitude, A, of the seismic waves produced by it. Richter originally proposed the logarithmic scale of earthquake magnitude, M_L, that nowadays bears his name: "on the Richter scale." It must be one of the most reported phrases in the human language! The earth-

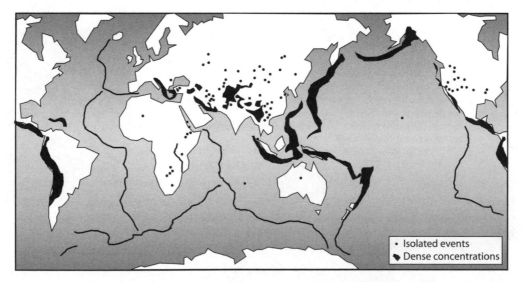

Fig. 4.140 World seismology: Concentration of major earthquake epicenters (over 4.5 magnitude) for 14 years.

quake locations in Fig. 4.140 are for those with "Richter" magnitudes greater than 4.5. The Richter magnitude scale is logarithmic to take account of the large variation in wave amplitudes from the smallest to largest earthquakes: the logarithm means that a $M_L = 4$ earthquake is 10 times greater than a magnitude $M_L = 3$ event and 100 times less than a $M_L = 6$ event. All such *magnitude scales* (Box 4.3) are arbitrary in some way for they depend upon the type of wave selected to represent the magnitude; either body and surface waves may be used for the original Richter scale but the technique is only suitable and accurate for local (within 500 km of epicenter) events. The surface-wave magnitude is M_s and is measured from the maximum horizontal magnitude of the Rayleigh surface wave signal in the range of periods 17–23 s. It is useful for relatively shallow foci earthquakes (order 50 km deep or less). It needs corrections for distance traveled from focus to recorder,

for focal depth, and sometimes for local conditions. Since deep focus earthquakes do not generate particularly impressive surface waves, the body-wave magnitude, m_b, is also in use. It is more correctly referred to as the Gutenberg–Richter scale of body-wave magnitude and makes use of *P*-, *PP*, and *S*-wave signals of 12 s period.

The magnitude of any earthquake is perhaps best understood physically by considering the magnitude of the fault rupture that produces it. The extent of the slipping motion that occurs is given by the area of the fault plane involved in the motion, A, and the magnitude of the motion, the measured rupture displacement, h. A may be routinely determined from the depth of the mainshock focus and from the pattern of aftershocks. When the two observable parameters are multiplied by the shear modulus (see above and Section 3.13), K, the total magnitude of the earthquake may be considered as a moment of force, rather like the force acting at the end of a lever arm. Thus we have the *seismic moment* as $M_0 = KAh$ with units of Newton meter that we can relate directly and proportionally to the energy released by an earthquake (Fig. 4.141). It may not be possible to measure the rupture displacement part of the seismic moment expression for many fault ruptures, for example, those in remote locations or underwater, but it has the tremendous advantage that it can also be calculated by an integration of the whole seismogram of an earthquake. It is thus widely used as the basis for the *moment magnitude* scale, M_w, of earthquakes. Nowadays in research it is common to quote both M_s and M_w for a particular earthquake event; the two values are not usually identical.

> **Box** 4.3 Earthquake magnitude expressions have a general form $M = \log_{10}(A/T) + q(\Delta, h) + a$, where A is maximum wave amplitude in 10^{-6} m, T is wave period in seconds, q is a correction factor to describe wave decay with distance (Δ) and focal depth, h and a is some constant. Thus $M_s = \log_{10}(A/T) + 1.66\log_{10}\Delta + 3.3$.
> Moment magnitude is $M_w = 0.66\log_{10}M_o - 10.7$.

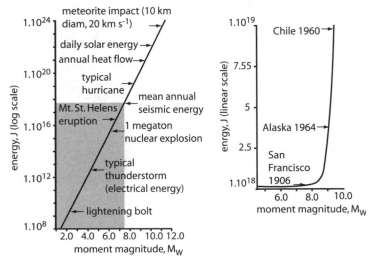

Fig. 4.141 Measures of energy generated by earthquake compared to other energetic natural phenomena.

4.17.6 Earthquakes and strain

In the introduction to this chapter we noted that earthquakes were marked by release of seismic energy along shear fracture planes. This energy is released partly as heat and partly as the elastic energy associated with rock compression and extension. In the elastic rebound theory of faults and earthquakes the strain associated with tectonic plate motion gradually accumulates in specific zones. The strain is measurable using various surveying techniques, from classic theodolite field surveys to satellite-based geodesy. In fact, the earliest discovery of what we may call *preseismic strain* was made during investigations into the causes of the San Francisco earthquake of 1906, when comparisons were made of surveys documenting *c*.3 m of preearthquake deformation across the San Andreas strike-slip fault. We have already featured the results of modern satellite-based GPS studies in deciphering ongoing regional plate deformation in the Aegean area of the Mediterranean (Section 2.4). All such geodetical studies depend upon the elastic model of steady accumulating seismic strain and displacement. But then suddenly the rupture point (Section 3.15) is exceeded and the strained rock fractures in proportionate or equivalent magnitude to the preseismic strain. This *coseismic deformation* represents the major part of the energy flux and is dissipated in one or more rupture events (order 10^{-2}–10^{1} m

slip). The remainder dissipates over weeks or months by aftershocks as smaller and smaller roughness elements on the fault plane shear past each other until all the strain energy is released. If the fault responsible breaks the Earth's surface then the coseismic deformation is that measured along the exposed fault scarp whose length may reach tens to several hundreds of kilometers.

Different types of faults give rise to characteristic first motions of *P*-waves and it is this feature that nowadays enables the type of faulting responsible for an earthquake to be analyzed remotely from seismograms, a technique known as *fault-plane solution*. Previously it was left to field surveys to determine this, often a lengthy or sometimes impossible task. The first arrivals in question are those up or down peaks measured initially as the first *P*-wave curves on the seismogram record (Fig. 4.142). It is the regional differences in the nature of these records caused by the systematic variation of compression and tension over the volume of rock affected by the deformation that enables the type of faulting to be determined. This is best illustrated by a strike-slip fault where compression and tension cause alternate zones of up (positive) or down (negative) wave motion respectively as a first arrival wave at different places with respect to the orientation of the fault plane responsible (Fig. 4.142). When plotted on a conventional lower hemisphere stereonet (Cookie 19), with shading illustrating compression, the patterns involved are diagnostic of strike-slip faulting.

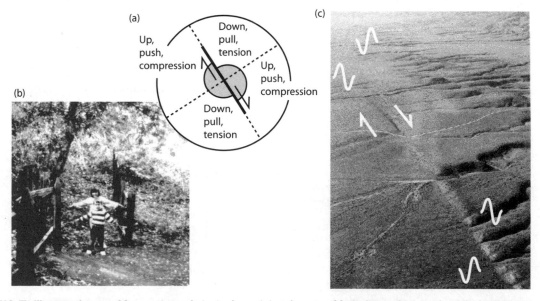

Fig. 4.142 To illustrate the use of first motion polarity in determining the type of fault slip, in this case the right-lateral San Andreas strike-slip fault; (b) 1906 San Francisco quake ground displacement; (c) San Andreas dextral strike-slip fault and schematic first *P*-wave arrival traces.

4.18 Molecules in motion: kinetic theory, heat conduction, and diffusion

We have so far discussed flow in terms of bulk movement and mixing but there are also a broad class of systems in which transport of some property is achieved by differential motion of the constituent molecules that make up a stationary system rather than by bulk movement of the whole mass. Such systems are not quite in equilibrium, in the sense that properties like temperature, density, and concentration vary in space. For example, a recently erupted lava flow cools from its surfaces in contact with the very much cooler atmosphere and ground. A second example might be a layer of seawater having a slightly higher salinity that lies below a more dilute layer. The arrangement is dynamically stable in the sense that the lower layer has a negative buoyancy with respect to the upper, yet over time the two layers tend to homogenize across their interface in an attempt to equalize the salinity gradient at the interface. In both examples there is a long-term tendency to equalize properties. In the first it is the oscillation of molecules along a gradient of temperature and in the second the motion of molecules down a concentration gradient. But how fast and why do these processes occur?

4.18.1 Gases – dilute aggregates of molecules in motion

The gaseous atmosphere is in constant motion due to its reaction to forces brought about by changes in environmental temperature and pressure. Volcanic gases also move in response to changes brought about by the ascent of molten magma through the mantle and crust. When we study the dynamics of such systems we must not only pay attention to such bulk motions but also to those of constituent molecules that control the pressure and temperature variations in the gas. Compared to any speed with which bulk processes occur, the internal motions of stationary gases involve much higher speeds. The view of a gas as a relatively dilute substance in which its constituent molecules move about with comparative freedom (Section 2.1) is reinforced by the following logic:

1 A mole of a gas molecule is the amount of mass, in grams, equal to its atomic weight. Nitrogen thus has a mole of mass 28 g, oxygen of 32 g, and so on. Any quantity of gas can thus be expressed by the number, n, of moles it contains.

2 A major discovery at the time when molecular theory was still regarded as controversial, was that there are always exactly the same number of molecules, 6×10^{23}, in one mole of any gas. This astonishing property has come to be known as Avagadro's constant, N_a, in honor of its discoverer. It implied to early workers in molecular dynamics that molecules of different gases must have masses that vary directly according to atomic weight, for example, oxygen molecules have greater mass than nitrogen molecules.

3 Following on from Avagadro's development, it became obvious that Boyle's law (Section 3.4) relating the pressure, temperature, volume, and mass of gases implies that for any given temperature and pressure, one mole of any gas must occupy a constant volume. This is 22.4 L ($22.4 \cdot 10^{-3}$ m^3) at 0°C and 1 bar.

4 It follows that each molecule of gas within a mole volume can occupy a volume of space of some $4 \cdot 10^{-26}$ m^3.

5 Typical molecules have a radius of some 10^{-10} m and may be imagined as occurring within a solid volume of some $4 \cdot 10^{-30}$ m^3.

From these simple considerations it seems that a gas molecule only takes up some 10^{-4} of the volume available to it, reinforcing our previous intuition that gases are dilute. The phenomenon of molecular diffusion in gases, say of smell or temperature change, occurs extremely rapidly in comparison to liquids because of the extreme velocity of the molecules involved. Also, since gaseous temperature can clearly vary with time, it must be the collisions between faster (hotter) and slower (cooler) molecules that bring about thermal equilibrium. And since heat is a form of energy it follows that the motion of molecules must represent the measure of a substance's intrinsic or internal energy, E (Section 3.4). Let us examine these ideas a little more closely.

4.18.2 Kinetic theory – internal energy, temperature, and pressure due to moving molecules

It is essential here to remember the distinction between velocity, u, and speed, u. If we isolate a mass of gas in a container then it is clear that by definition there can be no net molecular motion, as the motions are random and will cancel out when averaged over time (Fig. 4.143). Neither can there be net mean momentum. In other words gas molecules have zero mean velocity, $u = 0$. However, the randomly moving *individual* molecules have a mean speed, u, and must possess intrinsic momentum and therefore also mean kinetic energy, E. In a closed volume of any gas the idea is that molecules must be constantly bombarding the walls of the container – the resulting transfer

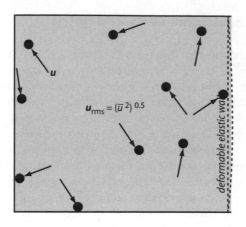

$u_{rms} = (\bar{u}^2)^{0.5}$

deformable elastic wall

In this thought experiment the container has its right hand wall as an elastic membrane. Individual gas molecules are shown approximately to scale so that the average separation distance between neighbors is about 20 times molecular radius. The individual molecules all have their own instantaneous velocity, u, but since the directions are random the sum of all the velocities, Σu, and therefore the average velocity must be zero. This is true whether we compute the average velocity of an individual molecule over a long time period or the instantaneous average velocity of a large number of individual molecules.

The arrows denote instantaneous velocities. Nevertheless the gas molecules have a mean speed, u, that is not zero. This is because although the directions cancel out the magnitudes of the molecular velocities, that is, their speeds, do not. In such cases we compute the mean velocity by finding the value of the mean square of all the velocities and taking the square root, the result being termed the root-mean-square velocity, or u_{rms} in the present notation.

This is NOT the same as the mean speed, a feature you can easily test by calculating the mean and rms values of , say, 1, 2, and 3.

The internal energy, E, of any gas is the sum of all the molecular kinetic energies. In symbols, for a gas with N molecules:

$$E = N(0.5\, mu^2_{rms})$$

Or we may alternatively view the molecular velocity as a direct function of the thermal energy:

$$u^2_{rms} = 2E/mN$$

Fig. 4.143 Molecular collisions and the internal thermal energy of a gas. One molecule is shown striking the elastic wall, which responds by displacing outward, signifying the existence of a gaseous pressure force and hence molecular kinetic energy transfer.

or flux of individual molecular momentum is the origin of gaseous pressure, temperature, and mean kinetic energy (Fig. 4.144). These properties arise from the mean speed of the constituent molecules: every gas possesses its own internal energy, E, given by the product of the number of molecules present times their mean kinetic energy. In a major development in molecular theory, Maxwell calculated the mean velocity of gaseous molecules by relating it to a kinetic version of the ideal gas laws, together with a statistical view of the distribution of gas molecular speed. The resulting *kinetic theory of gases* depends upon the simple idea that randomly moving molecules have a probability of collision, not only with the walls of any container, but also with other moving molecules. Each molecule thus has a statistical path length along which it moves with its characteristic speed free from collision with other molecules: this is the concept of *mean free path*. Since gases are dilute the time spent in collisions between gas molecules is infrequent compared to the time spent traveling between collisions. Thus the typical mean free path for air is of order 300 atomic diameters and a typical molecule may experience billions of collisions per second. Similar ideas

have informed understanding of the behavior, flow, and deformation of loose granular solids, from Reynolds' concept of dilatancy to the motion of avalanches (Section 4.11).

4.18.3   Heat flow by conduction in solids

In solid heat conduction, it is the molecular vibration frequency in space and time that varies (Fig. 4.145). Heat energy *diffuses* as it is transmitted from molecule to molecule, as if the molecules were vibrating on interconnected springs; we thus "feel" heat energy transfer by touch as it transmits through a substance. In fact, all atoms in any state whatsoever vibrate at a characteristic frequency about their mean positions, this defines their *mean thermal energy*. Vibration frequency increases with increasing temperature until, as the melting point is approached, the atoms vibrate a large proportion of their interatomic separation distances. Conductive heat energy is always transferred from areas of higher temperature to areas of lower temperature, that is, down a temperature gradient, dT/dx, so as to equalize the overall net mean temperature.

2D elastic collision between a molecule and wall

Signs and coordinates

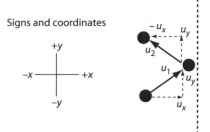

Before collision
$u_1 = u_x + u_y$

After collision
$u_2 = -u_x + u_y$

Momentum change is thus
$\Delta P = mu_2 - mu_1 = m(-u_x + u_y) - m(u_x + u_y)$
or
$\Delta P = -mu_x - mu_x = -2mu_x$

And Momentum transfer is
$\Delta P = -(-2mu_x) = 2mu_x$

The overall pressure, force per unit area, acting on any surface is given by the contribution of all molecules colliding with the wall in unit time. This number will be half of the total molecules, N, in any volume, V (the other half traveling away from the wall over the same time interval). The pressure is $0.5(N/V)(2mu_x)$. An N is given by $u_x \, dt$ and $p = mu_x^2 N/V$. Finally, since $u_x^2 = 1/3u_{rms}^2$ and $u_{rms}^2 = 2E/mN$, we have the important result that:

$$pV = 2/3(E).$$

Fig. 4.144 Origin of molecular pressure and its relation to internal thermal energy: link between mechanics and thermodynamics.

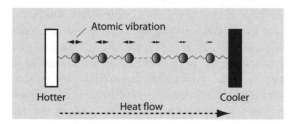

Fig. 4.145 Conductive heat flow in solids is movement of heat energy in the form of atomic vibrations from hot areas to cool areas so as to reduce temperature.

A steady-state condition of heat flow occurs when the quantity of heat arriving and leaving is equal. Many natural systems are not in steady state, for example, the cooling of molten magma that has risen up into or onto the crust (Fig. 4.146; Section 5.1) and in such cases the physics is a little more complicated (Cookie 20).

The rate of movement of heat by conduction across unit area, Q, is controlled by a bulk thermal property of the substance in question, the thermal conductivity, k, so that overall, for steady-state conditions when all temperatures are constant with time, $Q = -k\mathrm{d}T/\mathrm{d}x$ (Fig. 4.147). Conductivity relates to spatial rate of transfer, the efficiency of a substance to transfer its internal heat energy from one point to another. Heat transfer may also be expressed via a quantity known as the *thermal diffusivity*, κ (kappa; dimensions L^2T^{-1}), defined as $k/\rho c$, where ρ is the density and c is the specific heat (Section 2.2). It indicates the time rate of heat energy dissemination, being the ratio

Fig. 4.146 Bodies of molten magma intruded into the crust like the dyke shown here (see Section 5.1) or extruded as lava flows cool by conduction of heat energy outward into adjacent cooler rocks (or the atmosphere in the case of lava). The rate of cooling and the gradual decay of temperature with time may be calculated from variants of Fourier's law of heat conduction (see Cookie 20).

between conductivity (rate of spatial passage of heat energy) and thermal energy storage (product of specific heat capacity per unit mass and density, that is, specific heat per unit volume). Thermal diffusivity gives an idea of how long a material takes to respond to imposed temperature changes, for example, air has a rapid response and mantle rock a slow one. This leads to a useful concept concerning the *characteristic time* it takes for a system that has been heated up to return to thermal equilibrium. Any system has a characteristic length, l, across which the heat

energy must be transferred. This might be the thickness of a lava flow or dyke, the whole Earth's crust, an ocean current, or air mass. The conductive time constant, τ, is then given by l^2/κ.

4.18.4 Molecular diffusion of heat and concentration in fluids

In fluids it is the net transport of individual molecules down the gradient of temperature or concentration that is responsible for the transfer; the process is known as *molecular diffusion*. As before, the process acts from areas of high to low temperature or concentration so as to reduce gradients and equalize the overall value (Fig. 4.148). For temperature the rate of transfer depends upon the thermal conductivity, as for solids, but the process now occurs by collisions between molecules in net motion, the exact rate depending upon the molecular speed of a particular liquid or gas at particular temperatures. For the case of concentration the overall rate depends on both the concentration gradient and upon molecular collision frequency and is expressed as a *diffusion coefficient*. The rate of molecular diffusion in gases is rapid, reflecting the high mean molecular speeds in these substances, of the order several hundred meters per second. The rapidity of the process is best illustrated by the passage of smell in the atmosphere. By way of contrast the rate of molecular diffusion in liquids is extremely slow.

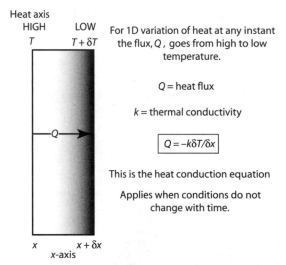

Heat axis

HIGH   LOW

T $T + \delta T$

For 1D variation of heat at any instant the flux, Q, goes from high to low temperature.

Q = heat flux

k = thermal conductivity

$$Q = -k\delta T/\delta x$$

This is the heat conduction equation

Applies when conditions do not change with time.

x $x + \delta x$

x-axis

Fig. 4.147 ID heat conduction.

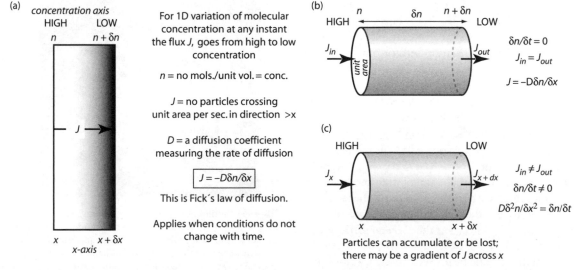

(a)

concentration axis

HIGH   LOW

n $n + \delta n$

For 1D variation of molecular concentration at any instant the flux J, goes from high to low concentration

n = no mols./unit vol. = conc.

J = no particles crossing unit area per sec. in direction >x

D = a diffusion coefficient measuring the rate of diffusion

$$J = -D\delta n/\delta x$$

This is Fick's law of diffusion.

Applies when conditions do not change with time.

x $x + \delta x$

x-axis

(b)

n δn $n + \delta n$

HIGH   LOW

J_{in} J_{out}

unit area

$\delta n/\delta t = 0$

$J_{in} = J_{out}$

$$J = -D\delta n/\delta x$$

(c)

HIGH   LOW

J_x J_{x+dx}

x $x + \delta x$

$J_{in} \neq J_{out}$

$\delta n/\delta t \neq 0$

$$D\delta^2 n/\delta x^2 = \delta n/\delta t$$

Particles can accumulate or be lost; there may be a gradient of J across x

Fig. 4.148 Molecular diffusion occurs in liquids and gases as translation of molecules from high concentration/temperature areas to low concentration/temperature areas so as to eliminate gradients. The rate of diffusion is rapid for gases and slow for liquids (a) Fick's law of 1D diffusion, (b) Derivation: Steady state diffusion (time independent), and (c) time variant diffusion (time/space dependent).

4.18.5 Fourier's famous law of heat conduction

Illustrated (Fig. 4.148) are the two cases of heat conduction and molecular diffusion for (1) steady state, with no variation in time and (2) the more complex case where conduction or molecular diffusion depends upon time. In the latter case, some mathematical development leads to a relationship in which the temperature of a cooling body varies as the square root of time elapsed (see Cookie 20).

4.19   Heat transport by radiation

4.19.1 Solar radiation: Ultimate fuel for the climate machine

Solar energy is transmitted throughout the Solar System as electromagnetic waves of a range of wavelengths, from x-rays to radio waves, all traveling at the speed of light. The Sun's maximum energy comes in at a short wavelength of about 0.5 μm in the visible range. Much shorter wavelengths in the ultraviolet range are absorbed by ozone and oxygen in the atmosphere. The magnitude of incoming radiation is represented by the *solar constant*, defined as the average quantity of solar energy received from normal-incidence rays just outside the atmosphere. It currently has a value of about 1,366 W m^{-2}, a value which has fluctuated by about ±0.2 percent over the past 25 years. As discussed below it is possible that over longer periods the irradiance might vary by up to three times historical variation.

Although the outer reaches of the atmosphere receive equal amounts of solar radiative energy, specific portions of the atmosphere and Earth's surface receive variable energy levels (Fig. 4.149). One reason is that solar radiative energy is progressively dissipated by scattering and absorption en route from the top of the atmosphere downward. Since light has to travel further to reach all surface latitudes north and south of a line of normal incidence, it is naturally weaker in proportion to the distance traveled. The fraction of monochromatic energy transmitted is given by the *Lambert–Bouguer absorption law* stated opposite (Box 4.4). Further latitude dependence of incoming solar energy received by Earth's surface arises from the simple fact that oblique incident light must warm a larger surface area that can be warmed by normally incident light. In addition to mean absorption of energy by atmospheric gases, radiative energy is also reflected, scattered, and absorbed by wind-blown and volcanic dust and natural and pollutant aerosol particles in the atmosphere. The amount of dust varies over time (by up to 20 percent or more), exerting a strong control on the magnitude of incoming solar radiation. Because of scattering, absorption, and reflection, it is usual to distinguish the *direct radiation* received by any surface perpendicular to the Sun from the *diffuse radiation* received from the remainder of the atmospheric hemisphere surrounding it. Continuous cloud cover reduces direct radiation to zero, but some radiation is still received as a diffuse component.

4.19.2 Sunspot cycles: Variations in solar irradiance and global temperature fluctuations

The extraordinary dark patches on the face of the otherwise bright sun are visible when a telescopic image is projected onto a screen and viewed. The dark blemishes

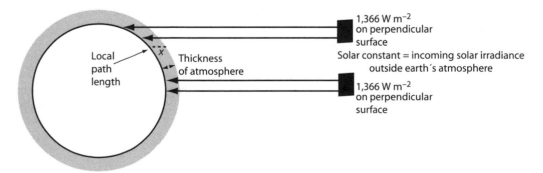

Fig. 4.149 Higher latitude radiation travels further through the atmosphere and is thus attenuated and scattered more. The more attenuated higher latitude radiance must also act upon a larger earth surface area.

Box 4.4

Box 4.4(a) Lambert–Bouguer absorption law.
$\delta = \exp(-\beta x)$
The fraction of energy, δ, transmitted through the atmosphere depends on the path length, x, and an absorption coefficient, β, whose value at sea level is about 0.1 km^{-1}.
In 10 km of travel, only $1/e$ (37%) of energy remains.

Box 4.4(b) *Other relevant aspects regarding Solar radiation*
1 The solar radiation "constant" has probably decreased over geological time since Earth nucleated as a planet. This has severe implications for estimates of geological palaeo climates.
2 Sunspots cause variations in the incoming solar energy.
3 The number of sunspots seem to vary over about an 11-year cycle. There is increasing evidence that a longer term variability has severe effects on the global climate system for example, the 80-year long Maunder Minimum in sunspots coincides with the "Little Ice Age" of northern Europe.

---

are not fixed and though cooler than surrounding areas the sun's irradiance is increased due to unusually high bursts of electromagnetic activity from them, with solar flaring generating intense geomagnetic storms. The dark patches were well known to ancient Chinese, Korean, and Japanese astronomers and to European telescopic observers from the late-Medieval epoch onward: nowadays they are termed *sunspots*. We owe this long historical record to the dread with which the ancient civilizations regarded sunspots, as omens of doom. Systematic visual observations over a *c*.2 ky time period reveals distinct waxing and waning of the area covered by sunspots. An approximately 11-year waxing and waning sunspot cycle is well established, with a longer multidecadal *Gleissberg cycle* of about 90 years also evident. Because the electromagnetic effects of sunspot activity reach all the way into Earth's ionosphere, where they interfere with (reduce) the "normal" incoming flux of cosmic rays, longer-term proxies gained from measuring the abundance of cosmogenically produced nucleides (like ^{14}C preserved in tree-rings) accurately push back the radiation record to 11 ka. What emerges is a fascinating record of solar misbehavior, culminating in the record-breaking solar activity of the last 50 or so years, which is the strongest on record, ever. This increased irradiance is thought to contribute about one-third to the recent global warming trend. But this estimate is model driven: what if the models are wrong? A chilling thought is the fact that the global "Little Ice Age" of 1645–1715 correlates *exactly* with the sunspot minimum named the *Maunder minimum*.

4.19.3 Reflection and absorption of radiated energy

The Sun's radiation falls upon a bewildering array of natural surfaces; each has a different behavior with respect to incident radiation. Thus solids like ice, rocks, and sand are opaque and the short wavelength solar radiation is either reflected or absorbed. Water, on the other hand, is translucent to solar radiation in its surface waters, although when the angle of incidence is large in the late afternoon or early morning, or over a season, the amount of reflected radiation increases. It is the radiation that penetrates into the shallow depths of the oceans that is responsible for the energy made available to primary producers like algae. It is useful to have a measure of the reflectivity of natural surfaces to incoming shortwave solar radiation. This is the *albedo*, the ratio of the reflected to incident shortwave radiation. Snow and icefields have very high albedos, reflecting up to 80 percent of incident rays, while the equatorial forests have low albedos due to a multiplicity of internal reflections and absorptions from leaf surfaces, water vapor, and the low albedo of water. The high albedo of snow is thought to play a very important feedback role in the expansion of snowfields during periods of global climate deterioration.

4.19.4 Earth's reradiation and the "greenhouse" concept

Incoming shortwave solar radiation in the visible wavelength range has little direct effect upon Earth's atmosphere, but heats up the surface in proportion to the magnitude of the incoming energy flux, the surface albedo, and the thermal properties of the surface materials. It is the reradiated infrared radiation (Fig. 4.150) that is responsible for the elevation of atmospheric temperatures above those appropriate to a gray body of zero absolute temperature. It was the *savant*, Fourier, who first postulated this loss of what he called at the time, *chaleur obscure*, in 1827. We now know that the reradiated infrared energy

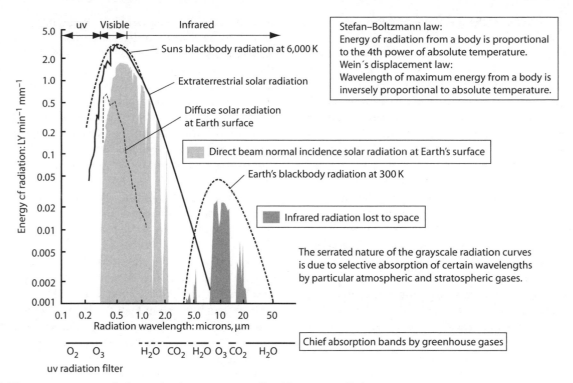

Fig. 4.150 The great energy transfer from solar short wave to reradiated long wave radiation.

flux is of the same order as that received from the Sun at the Earth's surface. Some of this energy is lost into space for ever but a significant proportion is absorbed and trapped by the gases of the atmosphere and emitted back to Earth as counter radiation where together with absorbed shortwave radiation it does work on the atmosphere by heating and cooling it. During this process water vapor may condense to water, or vice versa, and the effects of differential heating give rise to density differences, which drive the general atmospheric circulation. The insulating nature of Earth's atmosphere, like that of the glass in a greenhouse, is nowadays referred to as the "greenhouse" effect. The general concept was originally demonstrated by the geologist de Saussure who exposed a black insulated box with a glass lid to sunlight, then comparing the elevated internal temperature of the closed box with that of the box when open. Thus it is the absorption spectra of our atmospheric gases that ultimately drives the atmospheric circulation (Fig. 4.150). Water vapor is the most important of these gases, strongly absorbing at 5.5–8 and greater than 20 μm wavelengths. Carbon dioxide is another strong absorber, but this time in the narrow 14–16 μm range. The 10 percent or so of infrared radiation from the ground surface that escapes directly to space is mainly in the 3–5 and 8–13 μm wavelength ranges.

4.20 Heat transport by convection

Convection is the chief heat transfer process above, on and within Earth. We see its effects most obviously in the atmosphere, for example, in the majestic cumulonimbus clouds of a developing thundercloud or more indirectly in the phenomena of land and sea breezes. It is fairly obvious in these cases that convection is occurring, but what about within Earth? It is now widely thought that Earth's silicate mantle also convects, witnesses the slow upwelling of mantle plumes and motion of lithospheric plates. But exactly how do these motions relate to convection? We shall return to the question below and in later chapters (Sections 5.1 and 5.2).

4.20.1 Convection as energy transfer by bulk motion

We have seen previously that the heat transfer processes of radiation and conduction cause the temperature and internal energy of materials to change. Convection depends upon these transfer processes causing an energy change that is sufficient to set material in motion, whereby the moving substance transfers its excess energy to its new surroundings, again by radiation and conduction. We stress that the convection process is an indirect means of heat transfer; convection is not a fundamental mechanism of heat flow, but is the result of activity of conduction or radiation. When convection results from an energy transfer sufficient to cause motion, as for example in a stationary fluid heated/cooled from below or heated/cooled at the side, we call this *free (or natural) convection*. Alternatively, it may be that a turbulent fluid is already in motion due to external forcing independent of the local thermal conditions. Here fluid eddies will transport any excess heat energy supplied along with their own turbulent momentum. Convective heat transfer, such as that accompanying eddies forming in the turbulent boundary layer of an already moving fluid over a hotter surface is termed *forced convection* (or sometimes as *advection*).

4.20.2   Free, or natural, convection: Basics

The fundamental point about convection is that it is a buoyant phenomenon due to changed density as a direct consequence of temperature variations. We have seen previously (Section 2.1) that values for fluid density are highly sensitive to temperature. Thus if we consider an interface between fluids or between solid and fluid across which there is a temperature difference, ΔT, caused by conduction or radiation, then it is obvious that the heat transfer will cause gradients in both density and viscosity across the interface. These gradients have rather different consequences.

1 The gradient in density gives a mean density contrast, $\Delta\rho$, and a gravitational body force, $\Delta\rho g$ per unit volume, that plays a major role in free convection. The density contrast should also apply to the acceleration-related term in the equation of motion (Box 4.5) but since this complicates matters considerably, any effect on inertia

is conventionally considered as negligible by a dodge known as the *Boussinesq approximation*. This assumes that all accelerations in a thermal flow are small compared to the magnitude of g.

2 The gradient in viscosity on the other hand will cause a change in the viscous shear resistance once convective motion starts. The extreme complexity of free convection studies arises from considering both gradients of density and viscosity at the same time; the Boussinesq approximation assumes that only density changes are considered.

The magnitude of density change is given by $\alpha\rho_o\Delta T$, where α is the coefficient of thermal expansion and ρ_o is the original or a reference density. The term $g\alpha\rho_o\Delta T$ then signifies the buoyancy force (Section 3.6) available during convection and is an additional force to those already familiar to us from the dynamical equations of motion developed previously (Section 3.12). When the fluid is warmer than its surroundings the buoyancy force is overall positive: this causes the fluid to try to move upward. When the net buoyancy force is negative the fluid tries to sink downward.

In detail it is extremely difficult to determine the velocity or the velocity distribution of a freely convecting flow. This is because of a feedback loop: the velocity is determined by the gradient of temperature but this gradient depends on the heat moved (advected) across the velocity gradient! So we must turn to experiment and the use of scaling laws and dimensionless numbers such as the Prandtl and Peclet numbers discussed below.

4.20.3   The nature of free convection

A simple example is convection in a fluid that results from motion adjacent to a heated or cooled vertical wall. In the former case, illustrated for heating in Fig. 4.151, the thermal contrast is maintained as constant and the heat is transferred across by conduction. As the fluid warms up immediately adjacent to the wall it expands, decreases in

---

Box 4.5 Equation of motion for a convecting Boussinesq fluid.

$A_{\text{CCELERATION}} = P_{\text{RESSURE FORCE}} + V_{\text{ISCOUS FORCE}} + B_{\text{UOYANCY FORCE}}$

Time : Temperature balance equation for a convecting Boussinesq fluid

$\Delta T = C_{\text{ONDUCTION IN}} + I_{\text{NTERNAL HEAT GENERATION}} - H_{\text{EAT ADVECTION OUT}}.$

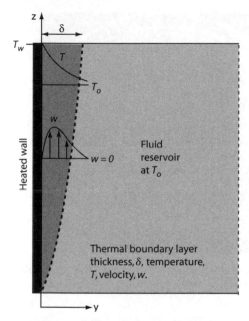

Fig. 4.151 Development of a free convective thermal boundary layer in a wide fluid reservoir adjacent to a vertical heated wall.

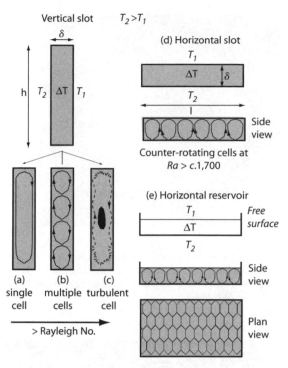

Fig. 4.152 Convection in vertical slots and in horizontal slots and reservoirs.

density, and when the buoyancy force exceeds the resisting force due to viscosity it moves upward along the wall at constant velocity, with the overall negative buoyancy force in balance with pressure and viscous forces. At this time, the background heat being continuously transferred across the wall by conduction, a portion is now transporting upward by convection within a thin *thermal boundary layer*. The general form of the boundary layer and of the temperature and velocity gradients across it are illustrated in Fig. 4.151. This situation encourages us to think about the possible controls upon convection and upon the nature of the associated boundary layers, for it must be the balance between a fluid's viscosity and thermal diffusivity that controls the degree and rate of conduction versus convection of heat energy and therefore the rate of transfer of temperature and velocity. We might imagine that when the viscosity: diffusivity ratio is high then the velocity boundary layer is thick compared to the temperature one, vice versa for a low ratio. In detail the prediction of boundary layer properties depends critically upon whether the flows are laminar or turbulent, hence the consideration of a thermal equivalent to *Re*.

The foregoing analysis has been rather dry and a little abstract and does scant justice to the interesting patterns and scales of free convection. That the process is hardly predictable and achievable by molecular scale motions is illustrated by the great variety of natural thunderclouds or by laboratory flow visualizations. Once heated or cooled

by conduction the moving fluid takes on extraordinary forms. We illustrate convective flows within vertical or horizontal wall-bounded slots and in open containers (Fig. 4.152). Here the convection takes the form of single (Fig. 4.152a) or multiple (Fig. 4.152b) vertical cells, turbulent vertical cells (Fig. 4.152c), nested counter rotating cells seen as polygons in plan view (Figs 4.152d, e and 4.153) or multiple parallel convective cells or rolls that adjust to both the shape of the containing walls and the presence of a free surface (Figs 4.154 and 4.155). The polygonal convective cells may form under the influence of variations in surface tension caused by warming and cooling and are termed Bérnard convection cells. Perhaps the commonest form of convection in nature involves the heating of a fluid by a point, line, or wall source to produce laminar or turbulent *thermal plumes* (Figs 4.156 and 4.159). Such plumes play an important role in the vertical transport of heat in the Earth's mantle, oceans, and atmosphere.

4.20.4 Forced convection through a boundary layer

In forced convection, the motive force for fluid movement comes from some external source; the fluid is forced to transfer heat as it flows over a surface kept at a higher

Fig. 4.153 View from above of Bénard convection cells in a thin layer of oil heated uniformly below: the convection is driven by inhomogeneities in surface tension rather than buoyancy. The hexagonal cells with flow out from the centers are visualized by light reflected from Al-flakes.

Fig. 4.154 Circular buoyancy-driven convection cells in silicone oil heated uniformly from below in the absence of surface tension.

Fig. 4.155 Rayleigh–Bénard convection cells in a rectangular box filled with silicone oil being heated uniformly from below. The convection is due to buoyancy in this case.

Fig. 4.156 Isotherms in a plume sourced from a heated wire and shown by an interferogram. Plume grows outward as the ⅖ power of height.

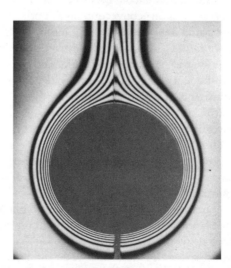

Fig. 4.157 Isotherms of a laminar plume formed by convection around a heated cylinder in air.

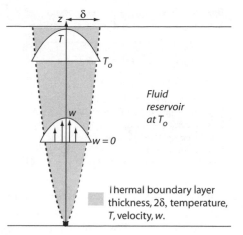

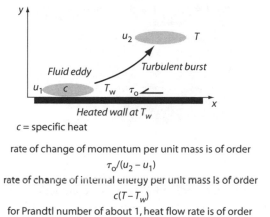

c = specific heat

rate of change of momentum per unit mass is of order

$$\tau_0/(u_2 - u_1)$$

rate of change of internal energy per unit mass is of order

$$c(T - T_w)$$

for Prandtl number of about 1, heat flow rate is of order

$$c(T - T_w)\tau_0/(u_2 - u_1)$$

Fig. 4.158 Development of a thermal plume generated from a heated point source, T_p.

Fig. 4.160 Visualization of Reynolds' analogy between thermal and momentum flux.

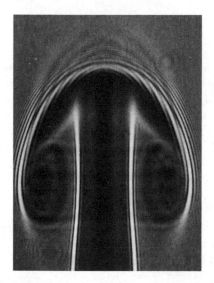

Fig. 4.159 The starting head vortex and the feeding axial column of a laminar plume.

temperature than the fluid itself (Fig. 4.160). The process is highly important in many engineering situations when relatively cool fluids are forced through or over hotter pipes, ducts, and plates. In natural situations we might envision heat transfer into a cool wind forced by regional pressure gradients to flow over a hot desert surface. In such convection the buoyancy force is small compared to that due to fluid inertia and thus the flow of heat has negligible effect on the flow field or the turbulence. Heat supplied by conduction to the boundary of flowing fluid must pass through the boundary layer. The major barrier to passage will be resistance to convective motion established by the viscous shear layer. Laminar flows at low Re, where there is no motion normal to the boundary surface, must transfer the excess heat entirely by conduction. They consequently have very much lower heat transfer coefficients than high Re turbulent flows, which have very thin viscous sublayers. In such turbulent flows, once through the thin sublayer barrier, heat is rapidly disseminated as convective turbulence by upward-directed fluid bursts (Section 4.5) shed off from the wall layer of turbulence (Fig. 4.160).

4.20.5 Generalities for thermal flows

Reynolds himself established the relationship between heat flow and fluid shear stress. Known now as "*Reynolds' analogy*" this involves a comparison of the roles of kinematic viscosity and thermal diffusivity when these two properties of fluids have approximately similar values (Box 4.6). Reynolds could proceed with his analogy because, as we mentioned in Section 3.9, Maxwell had previously viewed molecular viscosity as a diffusional momentum transport coefficient, analogous to the transport of conductive heat by diffusion. What is more natural than to express the ratio of kinematic viscosity, ν, to thermal diffusivity, D_{td}, as a characteristic property of any fluid: ν/D_{td}, is termed the *Prandtl number*, *Pr* (Fig. 4.160), whose value is usually quoted for thermal flows of particular fluids. To compare the behavior of different fluid flows, not just the fluids themselves, we make a more direct analogy with Re (remember this expression is uL/ν). The required thermal equivalent to Re, uL/D_{td}, is termed the *Peclet number, Pe*,

Box 4.6 Some Prandtl numbers for common fluids.

Fluid	Prandtl no
Air	0.71
Steam	0.93
Water	7.0
Crude oil	1000

Box 4.7 Rayleigh number: Ratio of buyoancy to viscous and thermal diffusivity

$$Ra = g\alpha(\Delta T)d^3 / \nu\kappa$$

α = expansion coefficient,
ΔT = temperature difference across fluid,
d = distance across fluid,
ν = kinematic viscosity,
κ = thermal diffusivity.

giving the ratio of advection to conduction of heat. At small values of Pe the flow has a negligible effect on the temperature distribution, which can be analyzed as if the fluid were stationary. Finally, there is a criterion, the *Rayleigh number*, that establishes whether convection is possible at all (Box 4.7). This is useful for remotely determining whether convection can occur in Earth's mantle, for example (Section 5.2). For convection in a horizontal slot Ra must exceed about 2,000, a value thought to be far exceeded in the mantle.

Further reading

Fishbane *et al.* (cited for Part 3) is again useful for basic physics. Basic concepts in fluid mechanics have never been better explained than by A. H. Shapiro in *Shape and Flow* (Doubleday, New York, 1961). Introductory fluid dynamics presented in a careful, rigorous way, but without undue mathematical demands, features in B. S. Massey's *Mechanics of Fluids* (Van Nostrand Reinhold, 1979) and M. W. Denny's *Air and Water* (Princeton, 1993). Beautiful and inspirational photos of fluid flow visualization may be found in M. Van Dyke's *An Album of Fluid Motion* (Parabolic Press, 1982) and M. Samimy *et al.*'s *A Gallery of Fluid Motion* (Cambridge, 2003). The topic of gravity currents in all their various forms is dealt with in J. Simpson's elegant and clearly written (with many superb photographs) *Gravity Currents* (Cambridge, 1997). Folds and faults are related to stress and strain as in G. H. Davies' and S. J. Reynolds's *Structural Geology of Rocks and Regions* (Wiley, 1996), R. J. Twiss and E. M. Moores' *Structural Geology* (Freeman, 1992), and J. G. Ramsay's and M. I. Huber's *The Techniques of Modern Structural Geology*, vol. 2 (Academic Press, 1993). Seismology is clearly introduced and explained in B. A. Bolt's Inside the Earth (Freeman, 1982) and the concepts beautifully illustrated in his more popular *Earthquakes and Geological Discovery* (Scientific American Library, 1993).

5 Inner Earth processes and systems

5.1 Melting, magmas, and volcanoes

The ancient Greeks supposed that a river of melt, shifting according to Poseidon's whims, ran under the Earth's surface, periodically rising to cause volcanic eruptions and violent earthquakes. We have seen evidence (Section 4.17) that most of the mantle and crust of the outer Earth is solid, exhibiting elastic or plastic behavior and transmitting P and S waves. Yet the Low Velocity Zone marking the top of the asthenosphere has a tiny amount of melt, sufficient to slow seismic waves somewhat and to enable plate motion over it (see Section 5.2). On the other hand, more than 1,500 Holocene-active volcanoes (Fig. 5.1) give

first hand evidence for localized accumulations of abundant magma not far below the surface. *Magma* is a high temperature, multiphase mixture of crystals, liquid, and vapor (gas or supercritical fluid). It is impossible to measure its temperature or other physical properties directly, for once it has flowed out of a volcanic vent as lava it will have cooled somewhat, begun to crystallize, and would have lost dissolved gas phases. We have to make recourse to experiments that show at atmospheric pressure, typical basalt magma is at about 1,280°C with a viscosity of around 15 Pa s.

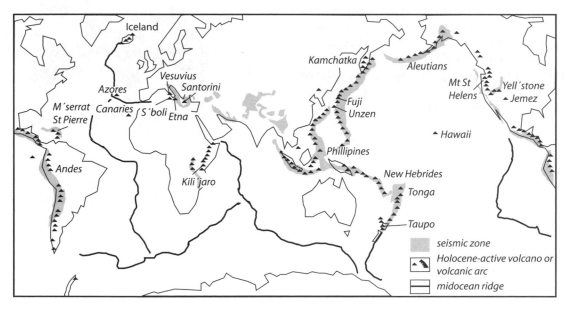

Fig. 5.1 Map showing summary world seismic belts (14 year record of $M > 4.5$) and the location of selected Holocene-active volcanoes and the major volcanic arcs.

5.1.1 Difficult initial questions and early clues

We need to ask a number of exploratory questions about magma genesis. Why, where, and how does melting of Earth's crust and mantle occur? Does magma exist as continuous or discontinuous pockets? Why and how does magma rise to the surface?

We know heat escapes from the Earth at a mean flux of some 65 mW m^{-2} (Chapter 8). But this global mean value allows for local areas of much higher flux. The geographical distribution of active volcanoes and geothermal areas shows that the local production of enhanced heat energy and subsurface melting is far from accidental or random: it usually occurs associated with areas of plate creation along the midocean ridges (Iceland) or destruction along the subduction zone trenches (Section 5.2; Fig. 5.1). Therefore we conclude that melting is also associated with these large-scale processes. Exceptions, as always, disprove this rule and so we also need to look with particular interest at those prominent volcanic edifices that occur far from plate boundaries, like the Canary Islands and Hawaii. Why does melting occur there?

We can gather clues as to the nature of magma from observing different styles of volcanic activity. Quiescent volcanoes often gently discharge gases like steam, CO_2, and SO_2 from craters or subsidiary vents called *fumaroles*. So, we infer that magma must also contain such gas phases, presumably in dissolved form under pressure, and that the gases can discharge passively. Volcanic eruptions of lava (Fig. 5.2) are themselves often passive; thus a Hawaiian volcano emits molten lava easily as rapidly moving flows. On the other hand, eruption may be far from passive; Vesuvian or Surtseyan explosions (Fig. 5.3) blast material vertically into the stratosphere as massive plumes or laterally as horizontal jets hugging the ground. Strombolian eruptions (Fig. 5.4) shower molten material periodically skywards for a few hundred meters in a fire fountain. Why this diversity of volcanic behavior into flow, blast, and fountain? A first clue came from observations made by geologists of the types of rock produced by these various styles of eruption. There is a wide range of possible chemical composition of magma, with more than a dozen main chemical elements and a score or more of minor (trace) elements involved, for our purposes we need simply to divide magmas and igneous rocks into three types (Fig. 5.5), according to their silica content – *acid, intermediate*, and *basic*. Acidic volcanic rocks rich in silica (>63 percent SiO_2), called *rhyolites*, are comparatively rare as volcanic flows. Rocks with intermediate amounts of silica (52–63 percent SiO_2), called *dacites* or *andesites*, often with minerals containing tiny amounts of water in their atomic

Fig. 5.2 Thermal imaging view of three cinder cones and associated breaching lava flow A. Note the lava levees bordering the upper channel conduit and flow wrinkles on the lobate lava fan margin. A younger flow (black) has breached the end of the levee system at B. C–E are older flows. Kamchatka, Russia.

Fig. 5.3 Explosive eruption column (2 km high) and accompanying base surge blast, Capelinhos volcano, Azores, October 1957. The central part of the Surtseyan eruption column is an internal core-jet rich in dark-colored volcanic debris. The base surge is steam-dominated.

lattices, tend to occur as the products of violent blasts. Rocks solidified from melts that passively flow as lavas tend to have the lowest amount of silica (<52 percent SiO_2); these are the ubiquitous *basalts*. Basalt flows are also the products of submarine volcanoes at midocean ridges.

Although hidden from our direct view by thousands of meters of ocean, these contribute by far the most voluminous proportion of volcanic products to the surface each year. The overall proportion of acid : intermediate : basic volcanics erupted each year is about 12 : 26 : 62 percent.

Despite the obvious surface manifestations of volcanic activity, the majority of melt (around 90 percent)

generated in the mantle and crust remains below surface forming slow-cooled *plutonic igneous rock* in the form of masses called *plutons*. Some is squirted from consolidating plutons into vertical or subvertical cracks as *dykes*, or nearer the surface as horizontal *sills*, both of which may feed surface volcanoes. Plutons, dykes, and sills are very common in the upper crust, as seen in deeply eroded mountainous terranes like the Andes or Rockies. We would like to know why such large volumes of former melt remain below the surface.

5.1.2 Melting processes

We have seen in our consideration of the states of matter (Section 3.4) that thermal systems transfer energy by changing the temperature or phase of an adjacent system or by doing mechanical work on their local environment. For melting to occur, a solid phase may be converted to a liquid by (1) application of temperature or pressure, (2) temperature retention with only minor heat loss due to work done by internal energy on expansion during adiabatic ascent, and (3) reduction in local melting point by addition of aqueous or volatile fluxes. We further amplify these reasons below.

Concerning heat energy, a certain amount, the latent heat of fusion, L_f (Section 3.4), is needed to melt crystalline rock. This amount can be measured in a calorimeter apparatus by comparing the heat released on melting silicate crystals or rock with amorphous silicate glass of

Fig. 5.4 Typical nighttime view of Stromboli fire fountain erupting from vent three, May 1979. Note parabolic ballistic trajectories of volcanic ejecta. Two Figures silhouetted for scale.

(a)

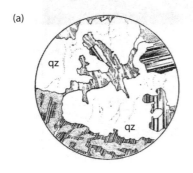

(b)

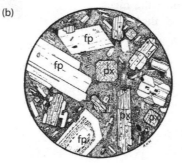

(c)

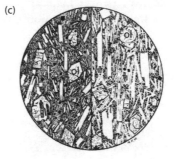

Granite with coarse equant crystals of clear quartz (qz) and shaded alkali feldspars (the laminae in the latter are twin planes or compositional layers)

Andesite lava showing well-developed phenocrysts of feldspar (fp) and pyroxenes (px) set in a very finely crystalline to glassy groundmass

Two half-views of olivine basalts, with well-developed phenocrysts of olivine (ol) and lath-like feldspars set in finely crystalline to glassy groundmass

Fig. 5.5 Sketches of microscopic fabric (fields of view about 5 mm diameter) and mineral phases of common igneous rocks that have crystallized from cooling melts.

identical composition. A selection of values for L_f is shown in Box 5.1. Because, melting of a given volume of solid cannot be achieved instantaneously, even if a homogenous mineral or elemental solid is involved, we need concepts to express the onset of melting and its completion: these are *solidus* and *liquidus* respectively. We generally draw the solidus and liquidus as lines on temperature : pressure graphs or on phase diagrams. The solidus line thus indicates the temperature at which a rock begins to melt (or conversely becomes completely solid on cooling) and the liquidus line is the temperature at which melting is complete (or conversely at which solidification begins on cooling). As an example, we can follow the solidus of basalt on the *P–T* diagram of Fig. 5.6.

Since most rocks are chemically different and may be comprised of various mineral species or minerals free to vary in composition, the onset of melting or the process of crystallization on cooling is complex. Major progress in understanding the processes of melting and crystallization of natural silicates were made by N.L. Bowen in experiments conducted in the early twentieth century (Figs 5.7 and 5.8). To illustrate this, consider one of Bowen's earliest triumphs, an explanation of the variation in behavior of the simplest possible rock made up of only *olivine*, an iron–magnesium silicate, whose composition is free to vary between 100 percent iron silicate (representing a mineral phase called *fayalite*) and 100 percent magnesium silicate (the mineral *forsterite*). The olivine system is obviously of major importance because it makes up a major mineral phase of the Earth's ocean crust. Minerals like olivine that are able to vary in their solid composition between two end-members like this are quite common in nature (the common feldspar minerals are another) and are said to exhibit *solid solution*. A solid solution is like any alloy, bronze, solder, or pewter for example, where the metal ions can mix freely in most proportions since they are of

similar size and charge. However, since the Mg^{2+} ion in forsterite is somewhat smaller than the Fe^{2+} ion in fayalite, it is held more tightly by atomic bond energy into the silicate crystal lattice and therefore melts at a higher temperature; olivines composed of pure Mg^{2+} and Fe^{2+} thus melt at about 700°C apart. Now, take a 50 : 50 combination of Fe^{2+} and Mg^{2+} silicate in an olivine solid volume and heat it up at atmospheric pressure to 1400°C (Figs 5.7 and 5.8). The composition of the initial melt, or partial melt, produced from such an olivine will tend to be

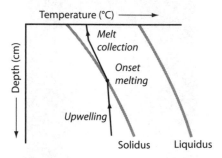

Fig. 5.6 To show solidus, liquidus, and an adiabatic melting curve as mantle rock is elevated by convection, partially melts and rises to surface.

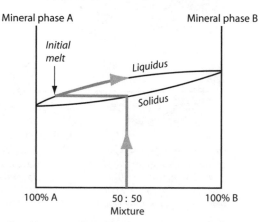

Fig. 5.7 Melting relations in a binary silicate solid solution series.

Box 5.1 Latent heat of melting (cal g^{-1}) for some important silicate minerals.

Mg-olivine	208
Fe-olivine	108
Clinopyroxene	146
Orthopyroxene	85
Garnet	82
Ca-Feldspar	67
Na-Feldspar	52
K-Feldspar	53

richer in Fe^{2+} than Mg^{2+}. As melting proceeds, the whole melt progressively enriches in Mg^{2+} until it matches the initial 50 : 50 mixture and melting of the initial solid volume has become total at the liquidus. Experiments over a range of initial compositions enable us to define a *phase diagram* showing the range of solidus and liquidus appropriate to a whole solid solution series. Similar principles govern the behavior of binary or ternary mixtures of mineral phases.

Thus far, we have considered melting temperatures as if they were unaffected by pressure. In fact, for mantle rock there is a strong change of dry solidus temperature with pressure. dT/dP is positive for the dry solidus of most key silicate minerals of the Earth's mantle (e.g. Fig. 5.6) and for the *garnet peridotite* composition (this is equivalent to an *ultramafic rock* with $c.90$ percent of Fe- and Mg-bearing minerals) that best seems to satisfy constraints for mean mantle composition.

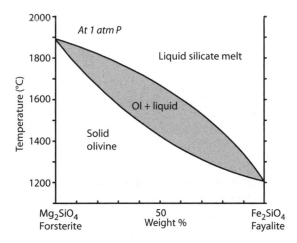

Fig. 5.8 Phase relations in the olivine solid solution series at 1 atm pressure.

5.1.3 Water, melting, and the terrestrial water cycle

Water exerts a profound influence on both the melting point (Fig. 5.9c) and strength of crustal and mantle rocks. The presence of H_2O in silicate melts is thought to cause depolymerization by breaking the Si–O–Si bonds, leading to the marked decreases in viscosity and melting temperature observed experimentally. For example, in order to give a 20 percent melt fraction, the temperature of anhydrous granite at 10 kbar pressure has to be about 900°C; the addition of 4 percent by weight of water decreases the required temperature to about 600°C. For basalt, the effect is even more startling for the positive gradient of the dry solidus noted above is reversed and at Moho depths of 35 km the saturated wet solidus temperature is reduced from $c.1150$°C to 650°C.

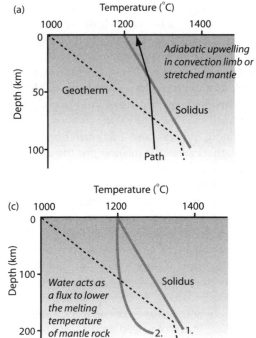

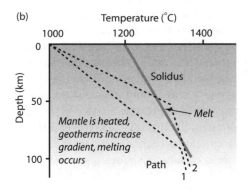

(a) The situation in the rising limb of a major convection cell under a midocean ridge or in stretched lithosphere.

(b) Mantle heating above a plume head causes geotherms to intersect solidus.

(c) The asthenosphere above a subduction zone may melt if there is sufficient flux of water from mineral dehydration reactions, especially the breakdown of serpentinite minerals.

Fig. 5.9 Various scenarios for the production of melt from mantle rocks.

The amount of ambient water present in the mantle as a whole is thought to be $c.0.03$ weight percent, so the average basaltic melt produced at the midocean ridges is largely anhydrous. Most interstitial water taken with ocean crust and sediments into *subduction "factories"* is rather efficiently processed back into the atmosphere and terrestrial environment via arc volcanoes and subsurface magma bodies. It has been calculated that of about 10^{12} kg of water taken into the subduction zones of the world every year, >92 percent is recycled in arc volcanism. This is just as well, because without recycling, the water-rich oceanic crust would effectively drain the oceans in only 10^9 years. How exactly the majority of this water is recycled by *Cybertectonica* (Section 1.6.7) shall be briefly explored below.

5.1.4 Why and where does melting occur in the Earth's crust and mantle?

The mobility of the Earth's convecting mantle (Sections 4.20 and 5.2) means that there are ample opportunities for large-scale circulation to cause hotter material to rise up from below. The process may be part of the large-scale flow to the midocean ridges (Section 5.2; Fig. 5.10), with melt volumes produced at rates of $c.25$ km³ a⁻¹. Or it may be on a more regional scale, around thermal plumes (Fig. 5.9b) whose head area may be up to of order 10^4 km². In either case, the melting associated with the slow upward motion by plastic flow, of order 10^{-2} mm a⁻¹, is coincidental. It occurs because of what has been termed *decompression melting* under conditions approximating the adiabatic thermal transformation discussed in Section 3.4. Remember that a volume undergoing adiabatic transformation is treated as being thermally isolated from its surrounding environment. In the adiabatic rise and thus decompression of deep mantle rock, despite some energy loss due to work done in expansion, the rising and expanding hot rock loses so little heat that it eventually intersects the mantle solidus (Fig. 5.9a) thereby causing melting. In this case, the adiabatic transformation is possible because of the very low thermal diffusivity (Section 4.18) of mantle material. The work done in expansion, as the pressure decreases upward, requires a certain amount of internal heat energy to be expended but this has very little effect on the temperature of a rock volume. The temperature path illustrated in Fig. 5.9a slopes gently negative to illustrate the point, with the actual solid adiabatic gradient, dT/dz, given by the expression $g\alpha T/c_p$, where T is the initial solid temperature, α = volume coefficient of thermal expansion and c_p is the isobaric specific heat

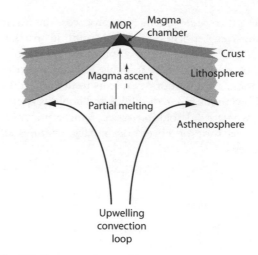

Fig. 5.10 Decompression melting under a midocean ridge magma chamber. Volcanism at the midocean ridges is by far the most voluminous on Earth.

capacity. Computed values of dT/dz for mantle peridotite are about $0.4°C$ km⁻¹. In the rising decompressing mantle, numerical calculations indicate that substantial partial melt fractions (25 percent) can be produced over 20 km or so near the surface. The partial melt fraction produced at the solidus is of anhydrous basalt composition and its intrusion and extrusion at the Earth's midocean ridges leads to the formation of new lithospheric plate (Section 5.2).

The second cause of melting (Figs 5.9c and 5.11) explains the large-scale distribution of volcanoes and melt zones associated with volcanic arcs, such as those around the Pacific "ring-of-fire" (Fig. 5.1) and returns to the subtheme of water in melts outlined above. Melting in arcs is associated with the sinking of lithosphere plates back into the mantle via *subduction zones* (Section 5.2); but it is not the sliding process and frictional heat generation (see below) that causes the melting, for the descending plate is actually quite cool, and remains so for considerable depths. Rather, it is the transformation of the oceanic mantle of the descending slab that causes melting in the overriding plate. The transformation involves a mineral group called *serpentinite*, which forms in the suboceanic mantle as olivine is altered by deep penetration of water along fracture zones and by subsea convection. Serpentinite contains up to 12 percent by weight of water in its mineral lattice. As the descending slab heats up, but still well below the limits of the mantle solidus, it loses its structural water at 400–800°C under pressures of 3–6 GPa. The water percolates upward, perhaps aided by pressure changes in fractures opened during deep

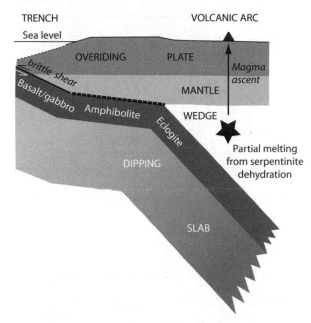

TRENCH

VOLCANIC ARC

Sea level

OVERIDING PLATE

brittle shear

Basalt/gabbro

Amphibolite

MANTLE

Magma ascent

Eclogite

WEDGE

DIPPING

Partial melting from serpentinite dehydration

SLAB

Fig. 5.11 Volcanic arc magmatism results from the fluxing effects of water released into the overiding plate as serpentinite dehydrates in a descending lithospheric slab.

dissipation of heat upward along a fault zone may decrease the efficiency and occurrence of the mechanism.

A final mechanism is thought to be responsible for widespread deep melting and continental crustal fusion in mountain belts caused by massive *overthrusting* of one crustal terrane upon another on deep thrust faults. This process acts quickly, at horizontal velocities appropriate to colliding plates (order of 10^{-1} m a^{-1}), and places crustal rocks rich in radioactive elements under other crustal rocks whose ambient temperature is that of their truncated subsurface geotherms. As always, any crustal melting that might result will be aided by the presence of water in the system and also by the rapidity of the faulting movements in relation to the thermal diffusivities of the rocks involved. The process is thought to have caused the fusion of continental crust under mountain belts like the Himalayas and the production of viscous acidic magmas that slowly crystallize to granitic rocks rich in potassium-bearing radioactive minerals like the mica and muscovite.

The approximate annual amounts of melt produced and attributed to the first two mechanisms above are indicated in Box 5.2. Fluxes from third and fourth are unknown since the melt remains subsurface.

earthquakes (called "seismic pumping") and mixes with the plastic mantle olivine of the continental lithosphere of the overriding plate. This causes the melting point of the mantle to fall and its mechanical strength to drop drastically. The resulting partial melting and melt migration eventually leads to generation of water-rich intermediate magmas characteristic of volcanic arcs.

The third melting mechanism notes the local coincidence of certain magmatic bodies, chiefly ancient magmatic plutons exposed by deep erosion, with strike-slip faults and appeals to the transformation of mechanical work to heat energy during deep faulting to cause melting. The magnitude of thermal energy produced is given by the mechanically equivalent acceleration times the velocity of the fault surface motion. This shear heating during earthquakes is of order τu, in Watts, where τ is the frictional shear stress on the fault surface and u is the mean velocity of its motion. As long as the heat energy is retained locally due to low thermal diffusivities of the rocks involved, then the temperature can build up with the possible occurrence of local melting. Temperature build up is aided by a thermal feedback process such that any increase in local strain rate caused by lowering of viscosity at the heightened temperature releases even more heat and this continues until melting occurs after a few million years. However, the presence of circulating fluids and their role in the

5.1.5 Melt material properties

Adjacent quadrivalent silicon cations, Si^{4+}, in silicate melts enter into shared coordination with four surrounding oxygen ions to form silica–oxygen tetrahedra. Adjacent tetrahedra share O ions and also join to aluminum ions in linked rings. The linked groups are said to be in a state of *polymerization* and are a feature of silicate melts. It is the continuous, polymer-like, linkage of oxygen ions (up to 15 or so tetrahedral lengths may be involved) that seems to control important physical properties; the greater the silica content and degree of group polymerization, the greater the viscosity and higher the solidus temperature. Alkali and alkali earth cations like Ca, Na, and K, together with nonbridging O anions and OH$^-$ reduce the degree of

Box 5.2 Global melt fluxes.

Total volume of oceanic plate added as melt at MORs:
c.25 km^3 a^{-1}

Total volume of oceanic plume-related intraplate volcanic melt:
c.1–2 km^3 a^{-1}

Total of volcanic arc melt: c.2.9–8.6 km^3 a^{-1}

Total of continental intra-plate melt: c.1.0–1.6 km^3 a^{-1}

polymerization and thus cause a reduction in viscosity and melting temperature. Also, water has a corrosive effect on Si–O bonds and has a key role in lowering rock melting points (and thus a major role in determining the rheology of the lithosphere). The logical extreme of these trends is pure Si–O melt with a continuously polymerized structure, which when crystallized gives rise to the continuous and rather open (i.e. not populated by heavy metallic cations) framework atomic structure of the mineral quartz (pure silica dioxide). This accounts for quartz's great durability, chemical stability, hardness, low density, low conductivity, low thermal expansion coefficient, and high melting point.

Viscosity and *density* are the two material properties of melts and magmas that largely control mobility, eruptive behavior, and other processes like crystal settling. Following our previous general discussion of viscosity (Section 3.9) it will come as no surprise to learn of the strong temperature control upon silicate melt viscosity, illustrated for basaltic melts in Fig. 5.12. To this, we must add the effect of Si content and pressure (Fig. 5.13); note the approximately three order of magnitude increase in viscosity for more silica-rich melts (andesite) over those of basaltic composition. Density increases with decreasing silica content and is strongly dependent upon pressure (Fig. 5.14); note in particular the rapid increase of density at about 15 kbar indicative of a fundamental structural change in the atomic ordering of silicate melts at these confining pressures. This is indicative of the presence of *eclogite* melt, a phase change to denser atomic ordering from normal basaltic melt. Even solid basalt undergoes this phase change (no overall chemical change is involved) to denser eclogite as ocean crust is taken deep into the mantle during subduction.

5.1.6 Flow behavior and rheology of silicate melt

Even the lowest viscosity basalt melt flows at low Reynolds number in a laminar fashion (Table 4.1). This considerably simplifies calculations concerning mean velocity profiles and internal stresses for such flows, for we can solve the equations of motion simply and with minimum approximations (Cookie 10). However, as complications arise such flows are considered in more detail:

1 Most melt flowing within conduits, certainly in the upper crust, is at temperatures much higher than that of the ambient rocks through which it moves. Therefore gradients of temperature in space and time in flow boundary layers will also cause gradients in viscosity.

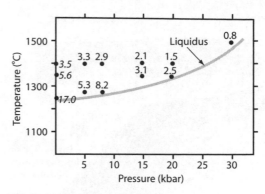

Fig. 5.12 Variation of basalt melt dynamic viscosities (Pa s) with *T* and *P*.

Note: Experimental data; lava results are much greater due to cooling and crystallization

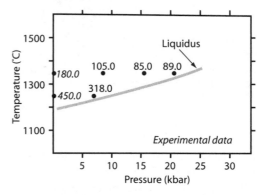

Fig. 5.13 Variation of andesite melt dynamic viscosities (Pa s) with *T* and *P*. Viscosity of basalt melt is of order 2 magnitudes less than for andesite because of the greater SiO_2 content of the latter. For a given *T*, viscosity of both melts generally decreases slowly with >*P*. For a given *P*, viscosity of both melts decreases with >*T*.

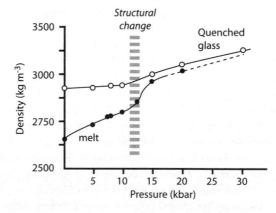

Fig. 5.14 Variation of basalt melt and quenched glass densities with *P*. Density of basalt melt is up to 15percent greater than that of the quenched glass at *P* < 15 kbar. Density of basalt melt increases with >*P*, the rate of change increasing rapidly at about 12 kbar due to structural changes in the melt. Density of basalt glass slowly increases at *P* > 12 kbar.

Rhyolite dome

Fig. 5.15 Viscosity of acidic magma is several orders of magnitude greater than that of basalt or andesite, with the result that acidic lavas are much rarer, the melts tending to intrude and extrude as lava domes, like this Alaskan example.

2 Rate of flow may control the rheological properties in a mechanism known as *thixotropy*; there is evidence that at low strain rates ($<10^{-5}$ s^{-1}) flowing melt is Newtonian in behavior (Section 3.15) while at higher rates non-Newtonian flow occurs due to the straining fluid affecting degree and orientation of silica tetrahedral chains and polymerization.

3 As the solidus is approached, especially during magma melt extrusion as lava, *Bingham behavior* (Section 3.15) occurs due to the onset of transition to crystalline solid structure. The properties of melts with yield stresses are considerably different, leading to morphological surface features like levees (Fig. 5.2). Acidic melt may be so viscous that it extrudes locally as an expanding dome (Fig. 5.15).

4 Flowing melt may contain variable proportions of suspended crystals that have precipitated from the cooling melt elsewhere. As we have seen previously (Section 3.9), suspended solids cause appreciably enhanced viscosities during shear flow (the Einstein–Roscoe–Bagnold effect). In particular, the shearing of solid suspensions give rise to a variable shear resistance depending upon the concentration of solids and shear rate. There is no evidence that the presence of solids *per se* can cause Bingham behavior; the undoubted presence of yield stresses in erupting lava flows must be due to other structural mechanisms affecting silicate melt.

5 Exsolution of volatile gases and water vapor, either as continuous phases or as bubbles, will cause the viscosity of the melt fluid to increase rapidly.

5.1.7 Melt segregation, gathering, migration, and transport

A key stage in melting is when a partial melt becomes sufficiently voluminous within the solid framework of melting rock to be able to flow away under any existing net force due to tectonics or the vertical gradient of gravitational stress. The process of *in situ* melt volume increase within a source region is called *melt segregation*. The initial melt in any crystalline substance occurs as thin films around the crystal boundaries of minerals (Fig. 5.16). When these boundary layers have dilated sufficiently, melt may overcome viscous resistance and thereafter flow. A pleasurable analog is when a sucked lollipop becomes warm enough to reach a critical stage between solid and liquid, the interstitial liquid melt can then be sucked off. A more prosaic example is the analogous situation of fluid flow through the connected pores of an aquifer rock.

Once melt has segregated in sufficient quantities it will gather and migrate in response to local pressure gradients, just like any other fluid. However, during the natural melting process, the melt itself produces a stress field independently of the state of ambient stress, for there is a substantial volume increase on melting, *c.*16 percent at 40 kbar, for common source minerals. The resulting pore fluid stresses (pore pressures; Section 3.15) counteract the positive effects of confining pressure on rock strength, reducing it to the effective stress sufficient to cause rupture or runaway strain. Therefore in a stressed rock, rather than remaining *in situ* as increasingly thicker grain boundary

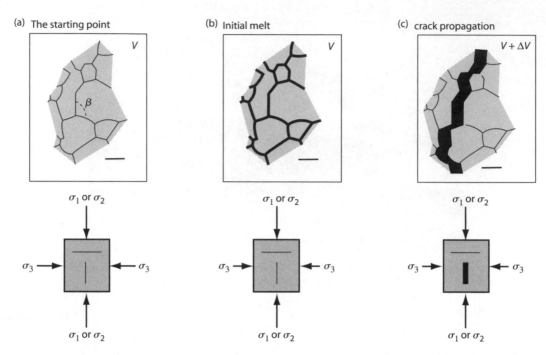

(a) The starting point   (b) Initial melt   (c) crack propagation

Fig. 5.16 (a) the starting point. Typical lower crustal source rock (a high grade metamorphic rock) for melt. Mean intracrystal face angle, β, in this case is 109°. (b) an increase in thermal energy level causes initial melt to form as rather uniform films around the constituent crystals. Melt films grow in thickness with time. (c) Critical melt film thickness reached, generates sufficient volume change, ΔV, to propagate cracks along local stress gradients enhanced by the elevated pore pressures.

films, melt fluid will tend to collect in orientations parallel to the maximum principal stress and normal to the minimum principal stress where the total fluid pressure is decreased (Fig. 5.16). This orientation is likely to be close to the vertical at depth, but as the vertical confining pressure is decreased at shallow depths, fracture-opening direction will tend to be horizontal. The resulting branching-upward dilations cause enhanced melt migration down the stress gradient, in this case toward the surface. This is analogous to the situation that is thought to occur along faults during seismic pumping, as hydrofracturing allows water migration to occur in discrete bursts. The rate of melt flow during magma-fracturing will depend upon the viscosity. From Newton's viscous flow law, $\epsilon = \sigma/\eta$, and for the low viscosities of basaltic melt pertaining close to the liquidus at Moho depths ($c.10$ kbar), very high strain rates, $c.6 \cdot 10^6$ s^{-1}, will result and the melt is expected to flow freely and instantaneously.

We thus have a picture of melt rising periodically upward through increasingly common upward-connecting channels and cracks at rates much faster than the movement of convecting mantle, perhaps at velocities of order 0.05 m a^{-1}. The rapid occurrence of fracturing and melt migration is witnessed by acoustic emissions of high frequency seismic

energy, which have been detected in the subsurface of active volcanoes undergoing melt replenishment before major eruptions. The crack (dyke) walls trend parallel to the direction of the local maximum principal stress trajectory, with the minimum principle stress normal to this. Should the dyke network be connected continuously upward to the surface, perhaps connecting crack fractures to a volcanic conduit, then the difference in lithostatic confining pressure of the ambient rock from the hydrostatic pressure of the melt "column" will ensure rapid surface eruption, the potential height that the erupting melt (now lava) can build its volcano depending upon the density difference between melt and ambient rock and the depth of hydrostatic linkage (Fig. 5.17). Recent research suggests that crack-conduits above active magma chambers may be sensitive to teleseismic waves (i.e. waves from distant earthquakes) of sufficient magnitude, causing linkage with local melt migration and volcanic eruptions.

A second possibility for melt mass transport is buoyant movement in coherent bodies, which are orders of magnitude larger than crack feeder systems. The magma rises through upper mantle and crust due to a net upward buoyancy force of magnitude $\Delta\rho g$. For typical basic melts at 20 km depth ($c.5$ kbar pressure) in mantle and

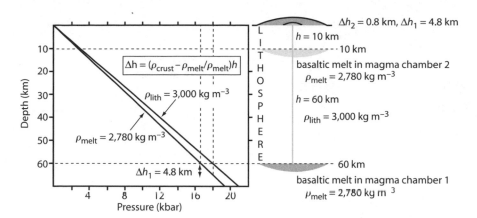

Fig. 5.17 Graph shows the two curves for variation of lithostatic pressure with depth for solid lithospheric rock of mean density 3,000 kg m^{-3} and basaltic melt of density 2,780 kg m^{-3}. For the two pressures to be equal at depth, h, the melt must rise to a height above the surface of Δh. Two examples for depth to magma chamber of 10 and 60 km are given.

continental crust of density 3,250 and 2,750 kg m^{-3}, $\Delta\rho$ is of order 550 and 50 kg m^{-3} respectively. The buoyant force per unit volume is thus of order 5,500 and 500 N. This picture of rising magma as buoyant viscous globules with low Reynolds numbers invites application of Stokes law of motion (Section 4.7), $V_p = 0.22(\sigma - \rho)r^2g/\eta$, to determine likely ascent velocities. For basic melt passing through ambient upper mantle of viscosity 10^{18} Pa s corresponding to $c.10$ km depth, V_p is $1.2 \cdot 10^{-9}$ m s^{-1} and $1.2 \cdot 10^{-7}$ m s^{-1} for globule diameters 1 and 10 km respectively. These small ascent velocities, a few centimeters a year, are comparable to the order of spreading rates at the midocean ridges, giving some credence to the crude calculations. As ascent proceeds, the combined effects of crystallization, heat, and volatile loss cause increased viscosity reducing the rate of rise until movement ceases.

Rising masses of melt (Figs 5.18 and 5.19) are termed *diapirs* – a form of mass transport by thermal plumes, involving free convective motion (Section 4.20) of an originally more-or-less continuous layer of melt. The layer undergoes an initial spatially periodic deformation, termed *Rayleigh–Taylor instability*, which amplifies into plumes. The process is analogous to the mesmeric rise of immiscible globules of oils in "lava lamps." The application of the diapir concept came about because of the large volume (10^2–10^4 km^3) of many acids to intermediate igneous intrusions revealed by deep erosion of ancient volcanic arcs (see further discussion of *magma chambers* below). Geological evidence in the form of intrusive contact relationships with sedimentary strata of known age seem to indicate that the hot melts, albeit probably partially crystallized, rose right through the cool and brittle upper continental crust on their journey upward from melt generation zones in the lower crust or upper mantle.

Numerical experiments and calculations (see above) show that silica-rich melts rise so slowly through crustal rock that conductive heat loss and embrittlement by crystallization (granular "lock-up") lead to cessation of movement well within the middle crust. Many plutons (Fig. 5.20) show evidence of these final stages of highly viscous boundary layer flow at their margins in the form of sheared crystal fabrics defining foliations that may have developed due to strain as the less viscous center of melt continued to rise buoyantly, albeit slowly.

It is thus evident that continued rise of plutons into the upper crust requires not only the outward displacement or consumption of ambient crustal rocks (for which there may be supporting geological or geochemical evidence) but also the maintenance of lubricity. Hence, the alternative concept of continued melt transport from below, of some starting plume being fed by subsequent smaller feeder plumes. These bring pulses of hotter, less crystalline melt traveling within the hotter traces of the starter plume thermal boundary layer, nourishing a large diapir at the end of their upward journeys. There is some evidence for this sustaining process from ancient plutonic bodies in the form of a myriad of minor internal contacts of small subintrusions of distinct ages and dyke-like feeder fractures. The model may also apply to magma "chamber" evolution under midocean ridges where seismic evidence disproves existence of single large melt bodies – rather than a single large space, we seem to be dealing with a number of perhaps connected small spaces; less a single magma chamber, more a magma condominium perhaps?

In addition to the cooling problem noted above, another major hitch with the whole diapiric idea is the origin of that essential prerequisite, the deep magma layer. As we have seen, the tendency at depth is for magma to be

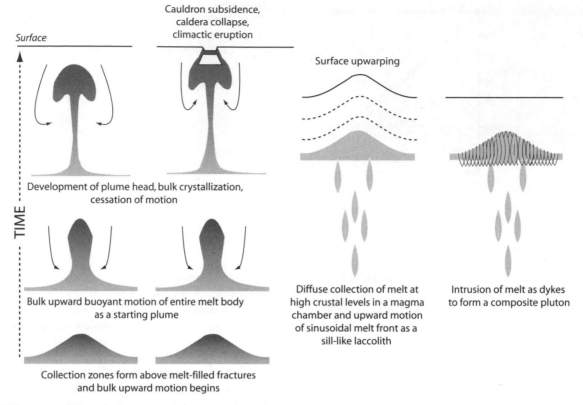

Fig. 5.18 Various possible types of magmatic diapirs or "rock mushrooms."

orientated in fractures parallel with the largely vertical axis of maximum principal stress. It is only high in the crust that the minimum stress direction deviates sufficiently from the horizontal to allow horizontal, sill-like sheets of melt to accumulate. Hence it has also been proposed that the upper crustal magma bodies grow mostly incrementally and *in situ* as sill-like blisters due to inflation by fracture-induced squirts of melt from below (Fig. 5.18). However, many ancient plutons show steeply dipping outward contacts that gravity studies reveal to persist down at least to mid-crustal depths; it seems that the predicted magma blisters were often more tumor-like in form.

Regardless of the exact mechanism involved, the accretion of magmatic material into the crust below volcanic arcs is the chief method by which continental lithospheric accretion has taken place over geological time. The essential role of water noted previously in generating this melt activity brings us back full circle to the planetary importance of water in ensuring the long-lasting dynamism of our Cybertectonic planet. More important for present purposes, magmatic accretion is also a mute witness to the volumetrically insignificant nature of volcanism compared to plutonism.

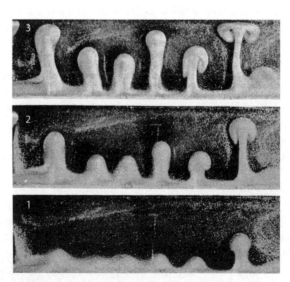

Fig. 5.19 Experimental diapirs produced as thermal plumes by heating a layer of more viscous oil uniformly from below: Sequence 1–3 in time.

Fig. 5.20 The surface trace of an igneous pluton as seen from space. This is the granitic Branberg Massif, Namibia, elevation 2.5 km. The near-circular outrop can be taken to represent a horizontal slice through the crust. Pluton long axis is 30 km. The slightly darker outer rim is the marginal zone of contact between the intrusive granite and ambient sedimentary strata into which the pluton intruded.

5.1.8 Magma bodies – crystallization in chambers and plutons

Notwithstanding doubts concerning the upward movement en masse of single large-scale magmatic diapirs, it is evident that below individual volcanic conduits and edifices, in the upper 5 km or so of crust, melt collects into small to medium scale (volume order 10^1–10^3 km³) magma chambers. These are where magma ultimately collects, being subsequently free to erupt or crystallize *in situ*; as we have noted above, only a small fraction of melt (*c*.12 percent) ever sees the light of day as volcanic lava. The direct evidence for magma chambers comes from geophysical remote sensing of active volcanoes, and indirectly from geological reconstructions of past eruptions and dimensions of solidified upper crustal magma bodies thought to represent palaeo-chambers. Occasionally, after an exceptionally powerful eruption a volcanic edifice above a magma chamber may fail along concentric fractures (Fig. 5.18) and collapse downward, leaving behind a telltale volcanic *caldera* depression bounded by high peaks of lava. The volume of the precaldera outpourings may be very large, for example, the Holocene Taupo volcanic edifice, New Zealand, erupted *c*.30 km³ of lava in a single eruption, the Pleistocene Valles caldera, New Mexico (Fig. 5.21a), an astonishing 5,000 km³.

The magma chamber is both the reservoir for volcanic eruption and also a place where crystallization can begin. We have seen that melts are mixtures of linked tetrahedral Si–O–Al groups around which metallic cations continuously diffuse. Depending upon the thermodynamic energy and concentrations of these cations, different silicate minerals are able to crystallize as the liquidus is reached during cooling. The controls upon mineral crystallization are highly complex, depending upon nucleation behavior, diffusion rate, cooling rate, and crystal growth rate, but in general we might expect an order of crystallization to proceed in the reverse order of melting as indicated by the ranking order of latent heats of fusion (Box 5.1). Note that olivine is particularly likely to be an early crystal product from basic magmas according to this crude argument, a notion given strong support by its readiness to nucleate and frequent occurrence as larger phenocryst phases in dykes (Figs 5.5 and 5.22) and lava flows. Such phenocrysts may record the beginnings of partial crystallization close to the liquidus on sparse nucleation sites before crystallization in the magma chamber was interrupted by an eruptive spasm. However, in general the interpretation of crystal size of igneous rocks is rather a complex business. As temperatures cool below the liquidus, large crystals are favored by few initial nucleation sites and subsequent size is proportional to time elapsed and inversely proportional to cooling rate. In plutonic rocks where cooling rates are very low, crystal sizes are commonly several millimeters to centimeters. In volcanic rocks, cooling rates may reach 0.1–1°C h^{-1} and crystals rarely exceed millimeter size. Crystal-free melts supercooled from above the liquidus by quenching (e.g., during subaqueous eruptions) show an amorphous or quasicrystalline texture.

In a large (>10 km) magma chamber with stationary magma, crystallization might be expected to begin at the roof and margins where conductive heat loss is greatest and where melt first cools below liquidus. As the first-formed crystal phases nucleate and grow, Stokes law determines whether the crystals sink or float through the ambient fluid. The strong pressure control upon both melt density and viscosity means that settling or floating tendencies and their rates will involve some complex feedback between ambient fluid and solid. A good example is provided by the behavior of crystals of the calcium-rich feldspar mineral *anorthosite* (Fig. 5.23). At pressures below 6 kbar, corresponding to a crustal depth of about 20 km, the mineral is denser than that of a parent basaltic melt and, together with other common minerals like pyroxene and olivine crystallizing from basaltic melt, will sink if the physical state of the magma permits (chiefly

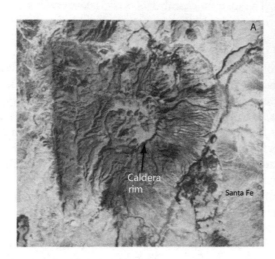

Fig. 5.21 Two expressions of magma chamber evolution. (a) Satellite image of pronounced Valles caldera in the center of the Jemez Mts, New Mexico, United States. The caldera was the source of massive outpouring of Pleistocene ignimbritic pyroclastic flows. (b) The Skaergaard intrusion in east Greenland is exposed as a deeply eroded and exhumed magma chamber exhibiting generally gabbroic rocks displaying spectacular layering (inset) of more feldspathic (light) and olivine/pyroxene (dark) rich layers formed by differential crystallization (related to double-diffusive convection?) and crystal settling in the cooling magma chamber.

overall crystallinity and state of yield stress by the time the anorthite begins to crystallize). At pressures greater than 6 kbar the anorthosite crystals are less dense than the ambient basalt melt and so the crystals will tend to float upward, segregating themselves from the other still-sinking mineral phases noted above. In both cases, the motion of crystal phases gives rise to zones of crystal concentration within a magma chamber; such zones or layers are termed *cumulates* (Fig. 5.21b). There is plenty of evidence for the existence of such cumulate layers in ancient igneous plutons. Anorthite-rich cumulates are most plentiful in plutons of great age within deeply eroded parts of the Earth's crust (in so-called PreCambrian shields), supporting the physical arguments advanced above. However, detailed calculations of crystal settling velocities are complicated by other considerations, for example, (1) crystal interactions involve the Einstein–Roscoe settling law (Section 4.7), (2) crystals increase in diameter with time, (3) settling may occur in non–Newtonian fluid whose yield strength may physically prevent settling despite the buoyancy forces being favorable, (4) the relatively low viscosity of basic melt means that Rayleigh Numbers (Section 4.20) for magma chambers are likely to exceed greatly the limit for convective stability and therefore the settling crystals will be subject to the vagaries of convective redistribution, (5) cooling of repeated injections of fresh batches of magma into a multicomponent crystallizing

Fig. 5.22 A dyke of ultrabasic composition cuts a gabbro. The dyke contact margins are well seen (arrowed), as are numerous crystal phenocrysts of olivinie set in a finely crystalline matrix. Note how the phenocrysts are more numerous and larger in the center of the dyke Rhum, scotland. This is probably due to crystal migration out from the melt boundary layer during intrusion, an effect seen during the granular shear of many fluid : grain systems. Dyke is 48 mm wide.

chamber raises the probability of occurrence of turbulent mixing, particularly effective when basic magma is rapidly jetted into preexisting basic magma (i.e. entrance Reynolds numbers for free or wall jets are large), and (6) the occurrence of double-diffusive convection (Section 4.6).

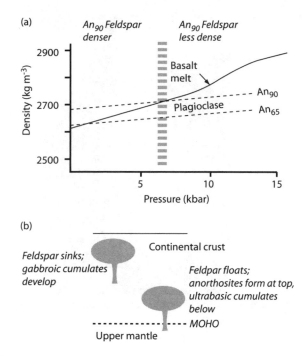

(a)

(b)

Fig. 5.23 (a) Variation of basalt melt and plagioclase crystal density with *P*. (b) Contrasts between lower and upper crustal basaltic magma chambers.

Concerning the latter, analogous experiments with cooling salt solutions show development of multiple crystal layers as the thermal gradient separates distinct compositional layers that propagate upward or sideways into the fluid body from cooling surfaces. These are considered as models for the spectacular multiple cumulates in *layered intrusions* that occur commonly in the geological record, perhaps most spectacularly in the basic plutons related to magmatism accompanying the opening of the North Atlantic ocean and currently exposed in east Greenland, most famously in the case of Skaergaard (Fig. 5.21b).

5.1.9 Magmas and volcanic eruptions

The essential fact about the majority of volcanoes is that they do not erupt continuously. Therefore, we conclude that one or more of the processes of melt formation, segregation, gathering, migration, and transport must be discontinuous. Taking a general view of the melting process as large scale, controlled by plate motions with large inertia, it would seem unlikely that the first three have the required time scale (of the order 10^2–10^4 years) of unsteadiness (this view is the equivalent of ignoring

accelerations in plate motion). The suspicion therefore emerges that the significant unsteadiness must be in the migration–transport link. Thus the deep supply of melt, say from mid-crustal depths downward, is regarded as continuous but its ultimate transport route to the near-surface magma chamber is discontinuous. From what we have seen of melt-transport mechanisms, the role of fracture mechanics in opening the transport path to magma chamber and surface vent must be key. If an eruption sequence essentially empties a chamber, then steady melt flux gradually replenishes it over some characteristic time. This replenishment is like a growing blister in that it induces a measurable swelling of the Earth's surface (see Fig. 2.15). Once full, the deviatoric stresses responsible for upper crustal fracturing build up to a critical level, ruptures reconnecting the chamber roof with the volcanic conduit, rapid upward flow of melt occurs, and the cycle begins anew.

Evidence for magma-fracturing comes not only from the seismic evidence for fracturing preceding eruptions but also from many exposed ancient plutons that were once high-level magma chambers. Around these are characteristically orientated dyke-like intrusions (Figs 5.24–5.26) whose trends either follow the predicted stresses due to bulk upward motion of the magma chamber (cone sheets) or to withdrawal and subsidence (ring dykes). When any low-viscosity fluid is forced into an elastic material, it propagates by opening narrow branching cracks in a random, fractal pattern controlled by the myriad of random defects available in the ambient material. The crack tips are locii of high pressure and spread according to the rheological properties of the host. For example, injections into low-viscosity fluid (Fig. 5.25) have an outward branching network with a multiplying number of spreading tips whereas elastic hosts concentrate their crack tips into a trilete pattern.

We conclude this section by a brief consideration of the physical processes behind flow- and blast-type volcanic eruptions. The greatest volume of volcanic eruptions (many km^3) on the planetary surface are *lava flows* represented by *flood basalts*. The coherent flow of such basic lavas, often over considerable distances, is favored when rising melt is: (1) volatile-poor, (2) of sufficiently low viscosity that any volatile phases can dissipate without disrupting the lava, and (3) erupted away from near-surface water. On the whole, silica-poor melts will tend to satisfy the first two criteria, but outside of the midocean ridge environment it is serendipity whether a rising melt finds itself interacting with surface waters of lakes, rivers, or copious artesian flows. Controls of lava flow discharge (effusion) rates are poorly known, but must depend upon the magnitude of pressure-drive from magma chamber

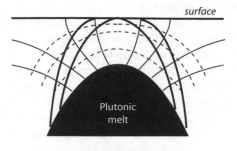

---- Minimum principal stress trajectories

___ Maximum principal stress trajectories
and melt intrusion for cone sheets

___ Ring dykes injected along planes
<90° to minimum stress trajectories

Fig. 5.24 Stress trajectories and associated intrusions above an upward-moving body of melt. Cone sheets form when magma pressure exceeds lithostatic pressure, vice versa for ring dykes.

Fig. 5.26 Vertical basic dyke cutting pillow lavas (just visible to left). Cyprus.

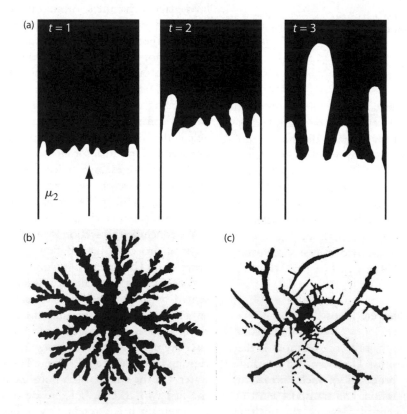

Fig. 5.25 Magma melts may intrude other melts or solids with variable viscous or elastic properties. (a) Saffman–Taylor viscous fingering results when a fluid is injected into a more viscous (and immiscible) fluid. The process is illustrated here by a time sequence of injection along a conduit. Note reduction of finger population and increased elongation with time. (b) Here the fluid is injected from the center outward in 3D – note the radial pattern, finger branching, and fractal-like nature of the process. (c) Here the fluid is injected from the center outward into an elastic material – note the reduction of branching and the many perpendicular junctions.

and friction coefficients for viscous flow through the crack-conduit system. Discharge rate and the discharge "magmograph," or time-series, will drive the subsequent course of the eruption cycle. For given discharge rates and duration, the velocity, runout length, aspect ratio (flow thickness : runout length ratio), and morphology are closely related to viscosity and development of non-Newtonian rheologies. The latter exert a major control on flow behavior through the magnitude of *flow yield strength* (Section 3.15), which must be overcome if flow is to continue; it most obviously manifests in the presence of well-developed lateral *levees* and overhanging flow snouts (Figs 5.2, 5.27, and 5.28). The viscosity control of flow morphology reaches its apotheosis in acid melts of rhyolitic composition. Here viscosity is so large and the effusion temperature relatively low that the melts erupt and flow very slowly and for short distances as bulbous *rhyolite domes* (Fig. 5.15) and *rhyolite fans* of high aspect ratio; the rapidly cooled outer glassy shells are fed from within and expand like surrealistic onions. The rapid cooling and glassy texture make the internal planes of viscous shear clearly visible in ancient examples as *flow banding*.

Rapid quenching is also the most morphologically distinctive feature of all lava flows, the *pillow lavas* of subaqueous eruptions. These are tube-like features, perhaps better described as "bolster lavas" when seen in 3D and is formed as basalt lava extrudes like toothpaste through the glassy carapaces of previous bolsters. We may regard the forms simply as extreme examples of *viscous fingering*, the result of *Saffman–Taylor instabilities* at the interface of extruding and ambient fluids of contrasting viscosity (Fig. 5.25). At the midocean ridges and in the ocean basins where hydrostatic pressures are very high (25 MPa at 2.5 km depth) there is very little obvious associated devolatilization, since sparse exsolving gases cannot enlarge into significant numbers (<5 vol.%) of visible cavities or *vesicles*. However, steam produced from seawater boiling (above 400°C at 30 MPa) below advancing lava

flows is thought to rise through the flows, lubricating their travel and forming lava pillars. The steam accumulates in cavities below upper chilled *glass rims* that later collapse into fragments as the vapor phase condenses. Such brecciated lobate flows, pillows, and bolsters form accumulations many hundreds of meters thick, thus defining the outer layer of igneous ocean crust.

Explosive eruptions (Figs 5.29 and 5.30) are favored by melts with (1) high content of volatile phases, chiefly water (up to a few wt%) and CO_2, (2) high viscosity, and (3) shallow level contact with surface water. The combination of the first and the second leads to *magmatic explosions*, which occur frequently in volcanic arcs where, as we have seen, partial melting of mantle wedge peridotite is permitted by copious dewatering of serpentinite above subducting slabs. The eruption products of such explosions tend to be dominated by magmatic materials. The third process leads to *phreatomagmatic explosions*, which occur when any magma type has high-level contact with abundant near-surface (*phreatic* – Section 6.7) waters. The explosive energy transformation involved can be very efficient in the subsurface; eruption products include shattered ambient rock (called *country rock* by geologists) mixed up with the more abundant primary magmatic products.

Magmatic explosive eruptions are marked by the production of vertical *eruption columns* (Figs 5.30–5.33) of mixed-size particulate and gas phases. Initially these *free jets* are pushed skywards at a certain muzzle velocity (of order 10^2–10^3 m s^{-1}) by the initial effects of volcanic gas

Fig. 5.28 Flow of basaltic pahoehoe lava across road, Kilauea, Hawaii, 16/7/90.

Fig. 5.27 Channelized flowing basalt a 'a' lava flow, Mauna Loa, Hawaii 30/3/84. Note older vent and onlapping lava plains (dark).

decompression; they are slowed during their subsequent rise as *buoyant plumes* (Section 4.20) by boundary layer shear mixing in *Kelvin–Helmholz vortices* (Section 4.9) and by upward linear expansion. Plume buoyancy is due to thermal energy exchange between entrained hot particulates, exsolving and expanding hot gases, and ingested air. It is the efficiency of the heat exchange, together with the sustained flux of energy that governs the overall magnitude of the buoyant plume. Ascent of the buoyant suspension is accompanied by

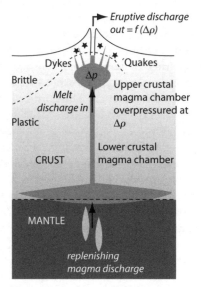

Fig. 5.29 Possible arrangement of deep and shallow magma chambers under an active volcano. The lower crustal details are largely imaginative. The brittle-ductile zone is at about 400°C.

sedimentation of increasingly finer particulates as the vertical turbulent stresses (Cookies 8 and 13) available in the column, prove incapable of holding up the clasts. Finally the plume reaches its limit and spreads laterally as a capping or *umbrella cloud* when its bulk density equals that of the local atmosphere or stratosphere. By this time, only the finest particulates remain to be entrained in zonal winds like the jet streams whereupon they may circumnavigate the globe providing a temporary but quite efficient reflector of short-wave solar radiation back to space. Lower-level ash is gradually "rained-out" or diffuses naturally back to the surface. Various scales and types of eruption column represent different levels of explosive energy (Fig. 5.30) and are named after well-known volcanoes, volcanic provinces, or their first observers, viz. Hawaiian and Strombolian fire fountains (low level; lava surface gas exsolution), Plinian (high level; gas-magma frothing, fragmentation), Vulcanian (intermediate level; sudden unblockage-related gas blasts), and Surtseyan (intermediate level; external surface water, vaporization) types. Plinean, Volcanean, and Surtseyan eruption columns also feature a very characteristic radial outward blast at the base of the erupting column (Fig. 5.3) called a *base surge*. These are actually high velocity *wall-jets* (like the turbidity currents featured in Scetion 4.12) and were first recognized as prominent features in volcanic eruptions after close observation of blast behavior during near-surface nuclear explosions. Base surges range from cool and wet to hot and dry, depending on the eruption and ambient mixing characteristics of the flows involved. They are turbulent enough to cut cone-flank channels and to produce granular bedforms like dunes during deposition. They

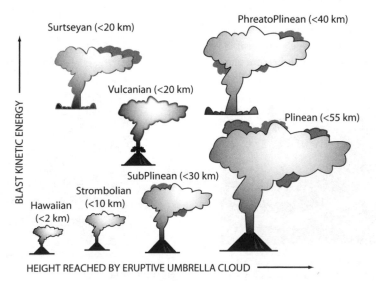

Fig. 5.30 Explosive eruptions according to energy and height of eruptive column.

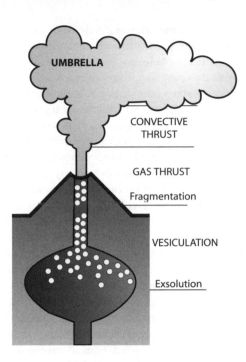

Fig. 5.31 The sequence of events thought to occur in the magma chamber and in the overlying eruption column of a Plinean eruption.

Fig. 5.32 Infamous explosive volcanoes in action. (a) Mt St Helens (Plinean phase) 1980 and (b) Vesuvius erupting 1944.

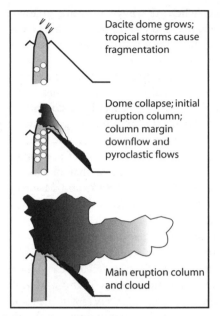

Fig. 5.33 Schematic of lava dome growth, collapse, and eruption sequence, based on Montserrate volcano.

define part of a broad spectrum of *pyroclastic density currents* (see further below).

It is necessary to consider the processes acting in Vulcanian/Plinian eruptions a little more carefully (Fig. 5.31), both on account of associated extreme hazards and for the sake of the physical exploration of the difficulties and challenges of understanding one of Earth's three-phase (solid-melt-gas) systems. Consider the general situation of an intermediate composition magma stream rising through an open volcanic conduit. Internal lithostatic pressure is continuously being reduced, which lowers volatile (chiefly water) solubility to saturation level and then permits release from solution. When this *decompression* causes local internal pressure to lower below the level of vapor pressure appropriate to the temperature involved, the disseminated volatile molecules nucleate, form discrete bubbles, and initiate melt boiling. The bubbles continue to grow by diffusion from the boiling melt, further enlarging as they rise until their internal fluid pressure equals that of the vapor phase pressure in the surrounding melt. Calculated times for bubble growth range between 16 and 70 min for major eruptions. This *first boiling phase* begins at 2–5 km depths for andesitic magma water saturation of 3–5 wt%. A second boiling phase may occur as crystallization reaches a maximum, latent heat of crystallization is released, and any remaining melt volatiles exsolve and nucleate. Eventually the rising, frothy magma reaches the surface, where the huge pressure gradients between bubbles, magma, and atmosphere triggers a rolling explosive *fragmentation front* with violent ejection of incandescent *pumice*, ambient rock, and gases in a *free jet*. This is the familiar "shaken-champagne bottle" analog for explosive eruption. The skeletal walls of pumice matrix may continue to exsolve gases in the vertical blast column, causing the more-or-less circular bubble voids to grow even larger until limited by framework cooling or volatile exhaustion.

Explosive events may also result after expulsion or fragmentation of solidified magma that blocks a vent, most famously at Mt St Pierre (1906) and Montserrat (1995, ongoing) in the Antilles volcanic arc. In the closely studied Montserrat example, cyclic behavior was evident. Degassing and crystallizing magma, evidenced by high-level precursor seismic activity, caused vent blockage with visco-plastic domes (viscosity $c.10^{14}$ Pa s) forming in the upper conduit and the whole volcano undergoing tumescence. Renewed magma ascent from a storage zone at $c.5$ km depth caused internal pressure in the rising, frothing magma stream (viscosity $c.10^6$ Pa s) below the plug to reach a critical point a few days later, causing partial ejection of the conduit-blocking plugs and violent explosive eruptions with Vulcanian columns and pyroclastic density flows (see below). There is evidence that partial dome collapse under magma pressure may be significantly triggered by the decrepitating effects of intense tropical rainfall on the hot dome surface (Fig. 5.33). Once degassed, vent

activity wanes, and the eruption cycle begins over again as another lava dome gradually constructs in the vent. The lava dome of Unzen volcano, Japan (Fig. 5.34) collapsed frequently due to toppling failure over a 4-year period as it extruded onto a steep preexisting volcanic slope.

Continued or intermittent Vulcanian/Plinian blasts are accompanied by the development of downward gravitational streaming of denser, hot pumice-rich suspensions at their outer edges (Fig. 5.33). Once descended, these form arcuate incandescent *pyroclastic density flows* (PDFs). These may also form without the aid of column collapse as frothing magma "boils over" the conduit rim, like carefully controlled emission of champagne from a bottle. In either case, such flows, originally termed *nuees ardentes* (glowing clouds), then travel outward as ground hugging, high-velocity wall jets (see Fig. 1.10), denser examples often following topographic lows like river valleys but the more energetic and dilute, perhaps driven by turbulent kinetic energy as in turbidity currents, capable of surmounting serious topography. The PDFs are sustained initially by the initial downward vertical momentum flux, then by a combination of furious intergranular contacts and continued emission of volcanic gases, which may reduce intergranular friction by fluidizing the granular flows. The exact balance between the last two mechanisms is not well known, particularly the reason behind the low-friction coefficient and mobility enhancement of granular contacts. Flow-top escape of gases and steam attest to

Mt Fugen (1359 m)

Lava dome at Mt Fugen periodically collapses downslope, generating pyroclastic flows

Mt. Fugen

Mayuyama

Mizunashi valley

East flank to show pale traces of pyroclastic flow deposits feeding lahars down the Mizunashi valley. Note arcuate scar of Mayuyama marking flank collapse of 1792

Fig. 5.34 Unzen volcano, Japan; a dacite dome fringed by pyroclastic deposits. It grew on a steep preexisting cone flank and frequently collapsed between 1991 and 1999. Thousands of pyroclastic flows and lahars were logged during this time.

gaseous emission from incandescent magmatic particles. When fluidization and drag-reduction cease, which may not be for many kilometers, the still-hot PDFs stops, cools, suffers progressive *autocompaction*, and contracts to form *columnar cooling joints* (Section 4.14). However, as with turbidity currents (Section 4.12), it is a great mistake to assume that any PDFs just represents local flow deceleration – deposition from moving flows may be cumulative at any point due to nonuniformity of overall flow. The characteristic deposits of pumice-rich pyroclastic flows are called *ignimbrites*.

We may ask why Plinian-type eruptions and pyroclastic flows involving intermediate composition melts do not occur so commonly in other basic and acidic magmas. Volatile species, concentration, and diffusion rate are the keys here: in basic melts the volatile phase is predominantly CO_2 rather than H_2O and the diffusion rate is orders of magnitude faster, so smaller and more numerous fast-growing bubbles can advect to safety at the surface through the low-viscosity (1–10 Pa s) melt much faster than in intermediate melts. At the free surface, the coalesced bubbles simply escape, contributing to fire-fountaining in Hawaiian/Strombolian eruptions, while at depth they are thought to ascend and accumulate as frothy layers at the top of magma chambers. Calculations indicate that ascent-related decompression may enhance local pressure by more than 10 MPa, triggering roof fragmentation and eruption. In acid melts of rhyolitic composition with high volatile content the diffusion coefficient is very low and bubbles have great difficulty in nucleating and growing in the rapidly chilling acidic glass (*obsidian*). However, significant volatile loss is thought to occur in Vulcanian (unblocking)

vents subject to explosive rhyolitic eruptions. Here, intense shear in conduit boundary layers is believed to induce increased permeability during vitrification/fragmentation reactions.

Finally, we must stress that our discussion of melting and volcanism has concentrated on the physical processes involved; we may have given the impression that volcanoes are rather "one-track," either acid, intermediate, or basic; effusive or explosive; gas-rich or gas poor. The immense basaltic shield volcanoes, like those of Hawaii and Iceland, are indeed mostly monogenetic, in this case made up of basalt lava flows. This is far from the general picture. Geological sections through ancient and active volcanoes reveal that their eruptive products are far from uniform. Magma-mixing and fractionation see to it that time trends exist in magma composition and a volcanic edifice or center may reveal an internal architecture made up of every combination of eruptive product; ash-fall, ash-flow, and lava flow. These strictly volcanic deposits are all mixed up with the sedimentary products of *lahar flows* that record hyperconcentrated stream flow, reworking of volcanic ash that often dominate the outer flanks and peripheries of volcanic vents (Fig 5.34). This architectural complexity is best seen in that most abundant of active volcanoes, the immense *stratovolcanoes* that dominate the volcanic arcs of the world. These are made up of intermediate composition of lava flow/pyroclastic flow/lahar/ash fall deposits built up over time, often millions of years. Their summits often show *caldera* morphology (Figs 5.18 and 5.21a), recording periods of complete or partial edifice collapse. Within the caldera walls multivent subsidiary cones may occur.

5.2 Plate tectonics

We saw in our introductory chapters that the outer 100 km or so of the solid Earth comprises a mechanical layer called the *lithosphere*. Evidence from seismology (Section 4.17) makes it clear that this more-or-less rigid layer comprises the crust and the upper part of the mantle. The minerals quartz and feldspar dominate the crust while olivine dominates the mantle. Although the lithosphere is rigid and behaves rheologically in a brittle-elastic fashion (Section 4.14), under high loads and over longer time spans it may also deform as a plastic material. What is even more remarkable is that the lithosphere is discontinuous, in the sense that it comprises a number of constituent fragments, called *plates*, which are in constant relative motion with respect to each other. There are seven major and several minor plates making up the outer solid Earth (Fig. 5.35). Plate cycling

makes Earth a highly distinctive planetary body; hence our previous references to the mode *Cybertectonica*.

5.2.1 Definitions and other facts

Plates are able to move as rigid bodies over the surface of the planet in response to driving forces (see below) because they lie over another weaker mechanical layer termed the *asthenosphere*, whose topmost part has a tiny proportion (less than 1 percent) of partial melt. The base of the oceanic lithosphere occurs at a temperature of $c.1,000°C$, just above the onset of the partial melting that characterizes the top asthenospheric low velocity zone (LVZ; see Section 4.17.4). It is this "accident" of the upper mantle geotherm

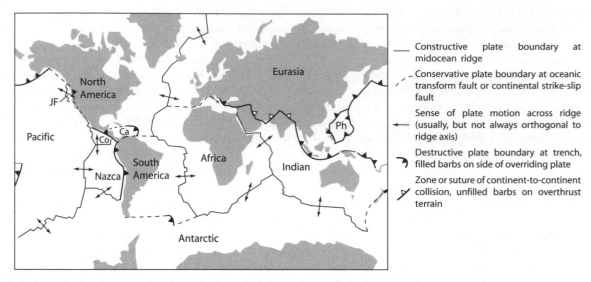

Fig. 5.35 Outline of major plates. Ca: Caribbean plate; Co: Cocos plate; Ph: Philippine plate; JF: Juan de Fuca plate.

approaching the partial melt curve for mantle rock (Section 5.1) that allows the whole process of plate tectonics to operate. The asthenosphere behaves as a high-viscosity ($c.4 \cdot 10^{19}$ Pa s) fluid in this scheme of things. Plate thickness varies according to whether continental or oceanic crust is involved in the upper layers; oceanic plate thickens laterally from zero at the ocean ridge to a maximum of 80 km, while the thickest continental lithosphere may be greater than 200 km. It is a key fact that, unlike the isostatic equilibrium of crust and mantle (Section 3.6), oceanic lithospheric is denser than the underlying asthenosphere. This inverted density stratification leads to the production of negative buoyancy forces, which drive plate destruction by subduction at the oceanic trenches.

5.2.2 A brief historic overview

It is instructive to briefly review the development of plate tectonics because the logic developed to account for various key components comes from a range of subject areas: paleontology, paleoclimatology, geology, geophysics, and geochemistry. Alfred Wegener, a meteorologist by training, developed his theory of continental drift in 1915 starting from the basis that a supercontinent called *Pangea* (Greek: "*all Earth*") progressively broke up over *c*.250 million years (My) into today's separate continental masses. Wegener and later du Toit (Wegener tragically died during an Arctic meteorological expedition),

assembled much fossil and geological evidence to support the theory of Pangea and its breakup, including the long-known jigsaw-fit of the Atlantic coastlines. Subsequently in the 1920s Holmes postulated a thermal mechanism for continental drift that involved the continents moving above convection currents in the mantle. Powerful opposition to this notion came from the geophysicist Jeffreys and others, who could not accept that the mantle could convect. This gave many skeptical and conservative geologists the excuse to ignore the theory. Major breakthroughs came with the development of paleomagnetism (study of the ancient magnetic field recorded by magnetic particles in rock) and seismic exploration of the ocean basins after World War II. The key developments were:

1 A record of diverging magnetic pole positions for different sites over Pangea indicating that continental drift had definitely occurred, though many did not believe the new science of paleomagnetism for several years after the mid-1950s.

2 A record of geomagnetic field reversals (magnetic north and south switching for long periods) in continental rocks dated precisely by radiometric dating.

3 Global mapping of midocean ridges and oceanic trenches.

4 An oceanic record of normal and reversed fields recorded in linear magnetic anomalies that lie symmetrically about the midocean ridges (Fig. 5.36): this led to the Vine–Mathews theory of sea-floor spreading in 1963.

5 The seismological recognition and significance of the LVZ (Sections 1.5 and 4.17) in defining the mechanical layers of lithosphere and asthenosphere.

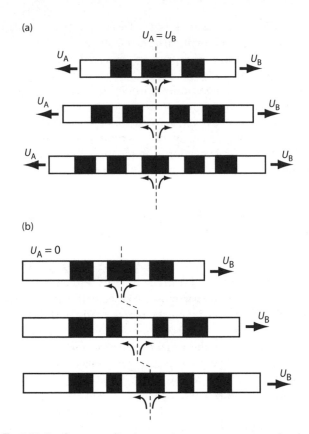

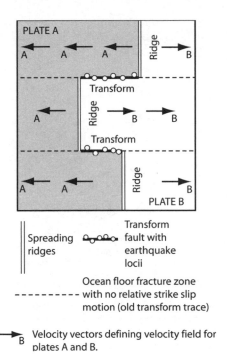

Fig. 5.37 Sketch to illustrate ridge : transform relationships between two moving plates.

Fig. 5.36 Sea floor spreading is a continuous process; magnetic minerals in the oceanic lithosphere record the orientation of the magnetic field that existed at the time of solidification. Here, the black shading depicts periods of normal magnetic polarity and the white shading reversed polarity. (a) Shows the conventional view of symmetrical spreading about a fixed midocean ridge axis, (b) shows an alternative scenario in which plate A is held fixed and the spreading ridge migrates away from it at half the spreading rate. In both cases, a symmetrical pattern of magnetic anomalies results.

6 Recognition of the particular structural features of a type of oceanic strike-slip fault, termed a transform fault (Fig. 5.37).

7 Identification of Benioff–Wadati zones of deep earthquakes along tilted interfaces under the oceanic trenches;

8 the seismological recognition of plate boundaries along (1) midocean ridges (extensional first motion earthquake mechanisms), (2) subduction zones (compressional first motion earthquake mechanisms).

9 The McKenzie–Parker kinematic theory of "tectonics on a sphere" (simply defined in Fig. 5.38) from magnetic anomaly and transform fault data, with the concept of Euler poles of rotation.

10 The parameterization of a Rayleigh Number (Section 4.20) well above critical for the existence of convection in the asthenospheric mantle.

11 Identification by Forsyth and Uyeda of the "self-propelled" theory of plate driving forces, chiefly involving slab pull.

5.2.3 Magnitude of plate motion: Rates of sea-floor spreading and other statistics

Sea-floor spreading is the evocative name given by Vine and Mathews in 1963 to the discovery that midocean ridges were the center of creation of ocean crust. They were able to say this because accurate shipboard magnetic surveying revealed geomagnetic reversals as symmetrical strips of normal and reversed ocean crust situated either side of the ridges (Fig. 5.36). The accurately dated continental record of reversals was already established and it was then possible to correlate the oceanic record with this and to establish the precise time of creation of known widths of ocean crust, something eventually traceable over 150 My. The speed of present plate motion, mostly derived from this sea-floor spreading data, varies over about an order of magnitude, from 11 to 86 mm y^{-1}. The speed of motion is related to the magnitude of the driving forces and resisting forces associated with particular plates. Table 5.1 gives relevant statistics for the major plates and some of the minor ones.

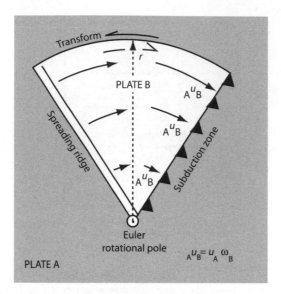

Fig. 5.38 Sketch to show Plate B rotating with respect to Plate A. Plate is generated at the spreading ridge and rotates as a solid body about circular arcs. The northern bouday to Plate B is a transform fault, an orthogonal line from which defines a great circle upon which the Euler pole lies. The linear velocity vectors are the velocities of Plate B moving with respect to Plate A. The length of the arrows is proportional to velocity magnitude which varies with the rotational radius, r, about the Euler pole shown. A typical angular velocity, ω, is about 10^{-8} radians per year.

5.2.4 Plate boundaries, earthquakes, and volcanism

Plate boundaries are described as *constructive* when new oceanic plate is being added by upwelling asthenospheric melt at the midocean ridges. Remember that this melting is due to adiabatic upwelling of mantle peridotite (Section 5.1). The volcanism is accompanied by voluminous outpourings of hot fluids along hydrothermal vents (Section 1.1.3). Shallow and relatively minor normal faulting (extensional) earthquakes accompany this plate creation along the ridge axis. *Destructive boundaries* occur at the ocean trenches where plate is lost to the deep mantle by subduction. The process is manifest by arrays of deep earthquakes (Section 4.17) along Benioff–Wadati zones and below island arcs. These latter form as water from the descending slabs dehydrate the water fluxing mantle of the overriding plate so that the mantle geotherrn intersects the peridotite solidus (Section 5.1). *Conservative boundaries* are those where no net flux of mass occurs across them, the plates simply slide past each other. In the oceans this occurs along the active parts of oceanic fracture zones, called *transform faults* (Fig. 5.37). Strike-skip faults in continental lithosphere may also mark conservative plate boundaries; examples are the San Andreas Fault, California

(Figs 1.9, 4.109, and 4.142), Dead Sea fault, Jordan (Fig. 4.110), and North Anatolian fault (Fig. 2.16).

In all the above examples, we discussed the nature of binary plate boundaries, that is, where two plates meet. However, it is theoretically possible to imagine multiple junctions meeting at topological points. In fact, points where three plates meet, termed triple junctions, are the most common. These involve various combinations of ridge, trench, and transform boundaries. The interesting thing about them is that they may migrate with time.

5.2.5 Describing the kinematics of plate motion — plate vectors, Euler poles, and rotations on a sphere

Since we all live on one of the moving plates, any statement concerning the directional vectors of the motion we undergo year upon year must be done with care. A vivid example comes from kinematic representations of the symmetrical pattern of sea-floor magnetic anomalies. Figure 5.36 shows the usual explanation for this, that of symmetrically diverging plates with equal speeds but opposite directions, that is, $u_A = -u_B$. In fact, the symmetrical spreading can equally well be achieved by motion of plate B, with plate A fixed, as long as the spreading axis also migrates in the direction of B at a velocity of $-0.5_B u$. This kind of relative motion is entirely possible since plates are self-driven entities and spreading ridges do not have to overlie upwelling limbs of convection cells fixed in asthenospheric mantle space.

We may most generally express plate velocities as relative velocities, that is with respect to adjacent plates, which have boundaries with the plate in question; Fig. 5.38 shows a simple two-plate example where the velocity of plate A with respect to plate B is minus the velocity of plate B with respect to plate A, that is, $_B u_A = -_A u_B$. To do more complicated three-plate problems, we can use the techniques of vector addition and subtraction (Fig. 5.39). Sometimes a fixed internal or external reference point is used to express the plate velocity vector. It is generally held that certain "hot-spots," the surface expressions of rising mantle plumes (the Hawaiian islands are the best known example) may approximate to such stationary points. Another trick relevant to some geographical situations is to fix one plate and relate other plate velocities with respect to that.

The linear speed, u, of any rotating plate on the spherical surface of the Earth is a function of both angular speed of the motion and the radius of the motion, r, from its rotational pole, the linear speed increasing as the length of the arc increases (Fig. 5.38). Linear speed is thus given simply by $u = r\omega$. The rotational pole is commonly termed the

Table 5.1 Plate statistics, see Fig. 5.35 for map. Asterisked plates have long trench boundaries and are fastest due to the importance of slab pull forces in generating steady plate motion (see Section 5.2.7).

Plate	Total area $\times 10^6$ km^2	Land area $\times 10^6$ km^2	Speed mm year^{-1}	Periphery $\times 10^2$ km	Ridge length $\times 10^2$ km	Trench length $\times 10^2$ km
NA	60	36	11	388	146	12
SA	41	20	13	305	87	5
PAC*	108	0	80	499	152	124
ANT	59	15	17	356	208	0
IND *	60	15	61	420	124	91
AF	79	31	6	418	230	10
EUR	69	51	7	421	90	0
NAZ*	15	0	76	187	76	53
COC*	3	0	86	88	40	25
CAR	4	0	24	88	0	0
PHIL*	5	0	64	103	0	41
ARAB	5	4	42	98	30	0
ANATOL	1	0.6	25	28	0	8

Plates: NA – North America, SA – South America, PAC – Pacific, ANT – Antarctica, IND – India, AF – Africa, EUR – Eurasia, NAZ – Nazca, COC – Cocos, CAR – Caribbean, PHIL – Phillipine, ARAB – Arabian, ANATOL – Anatolian.

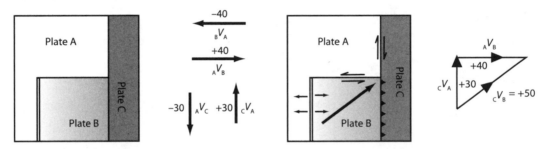

Fig. 5.39 There are three plates and alleline. The velocities of A and B and A and C with respect to each other (in millimeter year^{-1}) are known. We want to know the velocity of B with respect to C. This is given by vectorial addition, as shown on the right. Velocity vector codes like $_BV_A$ read "the velocity of A with respect to B."

Euler pole and is most easily found by drawing orthogonal lines from *transform faults* (see below), the latter being arcs of small circles on the global sphere. The Euler pole is on a great circle perpendicular to the trend of the transform fault.

5.2.6 Thermal aspects of plates and slabs

Consider first of all the likely temperature distribution in the upper 1 km of the lithosphere in a lateral transect from the mid-Atlantic ridge in Iceland to New York. At the ridge there is abundant evidence in the form of submarine volcanic activity and from heat flow measurements that temperatures in the upper crust are high and that overall heat flow is high (Fig. 5.40). In the case of offshore New York the opposite is true. Divergent plate boundaries, like this North Atlantic example, obey the simple rule that an

advecting mantle generates heat at the ridge due to adiabatic decompression. The associated melting produces new oceanic crust at the ridge axis and defines a *thermal boundary layer* in the form of cooling plate mantle that thickens away from the point of upwelling. It is thus axiomatic that lithospheric mantle above the top-asthenospheric 1,000°C isotherm must gradually thicken laterally due to conduction of the adiabatic heat released out through the upper surface of the new ocean crust into the ocean (Fig. 5.40). We ignore here the undoubted highly efficient convection witnessed at the ridge axis by hydrothermal systems responsible for "black smokers" (Fig. 3.6). In physical terms, our example means that temperature is changing with time and distance from the ridge (Fig. 5.40). However, our most complicated heat conduction scenarios to date (Section 4.18; Cookie 20) say that T only changes with distance! Advanced sums (hinted at in

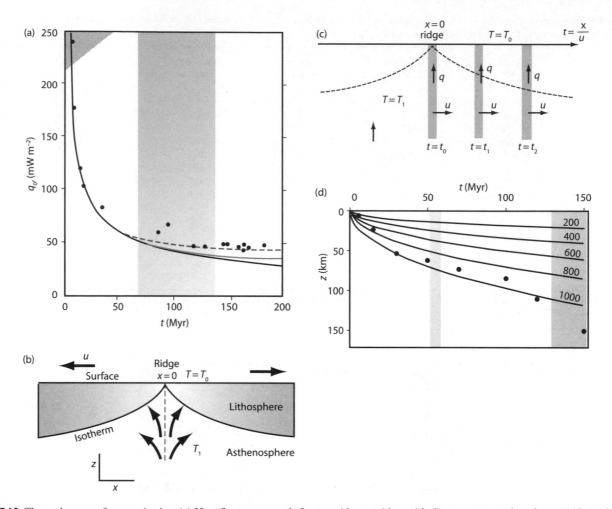

Fig. 5.40 Thermal matters for oceanic plate (a) Heat flow oceanwards from a midocean ridge, with distance expressed as plate age (derived from ocean magnetic anomalies). Dots are measurements, curves are various theoretical estimates based on the erfc argument. (b) The mechanical situation, with hot upwelling asthenosphere cooling laterally to define the plate thermal boundary layer above the isotherm at $c.1,000°C$. (c) The physical situation, with heat flow, q, conducting vertically through the ocean floor from the thickening plate above the $c.1,000°C$ isotherm. q decrease with time and the plate thickens with time according to the erfc argument (see text). (d) The proof of the pudding: data on plate thickness (dots) from seismic surveys versus estimates of plate thickness to the $1,000°C$ isotherm from heat conduction theory.

Cookie 20) tell us that the thickness, z_t, of this thermal boundary layer changes in proportion to the square root of time, t, as the simple expression $2.32\sqrt{\kappa t}$, where κ is the thermal diffusivity (Section 4.18.3). The square root term is a characteristic thermal diffusion distance and z_t refers to the thermal boundary layer thickness defined as the thickness appropriate to a base lithosphere temperature of 90 percent of steady state value $c.1,000°C$.

So, ignoring all the geological differences, we know that the lithosphere can be regarded simply as a cold, dense layer lying above warmer asthenosphere. That the situation

envisaged is buoyantly unstable is also axiomatic. Also, global continuity tells us that creation of oceanic lithosphere in one place must be accompanied by destruction elsewhere if the Earth is to maintain constant volume. The fate of oceanic plate is therefore determined; it has to be destroyed. Pushing cold slab into the hot mantle (Fig. 5.41) creates a thermal anomaly, that is, the lithosphere is cooler than it should be for the depth it has reached. This has the effect of raising the olivine : spinel transition (Section 4.17.4) by several tens of kilometers in the slab (Fig. 5.41) and creating additional negative buoyancy that adds to the slab pull force (explained

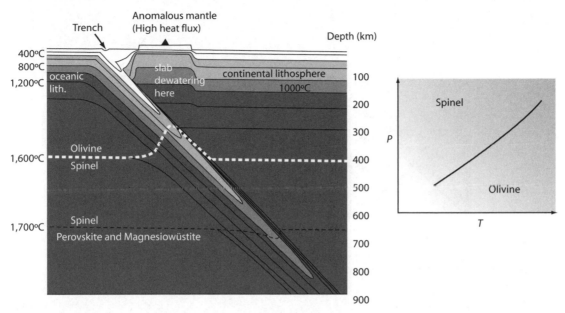

Fig. 5.41 A computed estimate of subsurface temperatures achieved as a subducting slab passes down into the lower mantle. The cold slab is denser than the ambient mantle and, despite the production of heat due to shearing along its upper interface, retains its identity to very great depths before it merges thermally with the lower mantle. Note that the olivine to spinel phase change is elevated in the descending slab and the dense spinel phase occurs at shallower depths here. The combination of cool, dense slab and elevated spinel transformation supplies the negative buoyancy necessary to drive the steady slab descent as a "slab-pull" force.

in Section 5.2.7). Also, as we have previously discussed (Section 5.1.4), volcanoes occur in volcanic arcs, not because of frictional melting, but because massive loss of water from dehydrating slab mantle serpentinite at $c.150$ km depth.

5.2.7 Why do plates move? The forces involved

Individual plates appear to be in steady, though not necessarily uniform, motion. The steadiness means that accelerations causing inertial effects are absent and therefore by Newton's Second Law all relevant forces must be in balance. The occurrence of nonuniform motion is evidenced by results from satellite GPS surveys of the continental lithosphere (see Fig. 2.16). It means that, although rigid, plates can strain internally by elastic deformation as part of the cycle of stress buildup and release associated with earthquake generation. The forces in equilibrium that drive plate motion (Fig. 5.42) may be divided into *top forces* that act because of differential topography, *edge forces* that act on peripheral plate boundaries, and *basal forces* that act on the bases of plates.

Top forces arise from the potential energy available to topography. For example, the midocean ridges lie several kilometers above the abyssal plains. A force, termed *ridge-push*, is thus pushing the plate outward from the ridge.

The topography has a thermal origin since it is due to the buoyancy of upwelling hot asthenosphere (including some partial melt), which underlies it (the Pratt-type isostatic compensation discussed in Section 3.6). Top forces are also possessed by the continental lithosphere, for the highest mountains lie up to 9 km above sea level. Potential energy possessed by such elevated terrain may be liberated as kinetic energy if the terrain in question can be decoupled from its rigid surroundings, that is, by basal sliding and along peripheral strike-slip faults. The Tibetan plateau is a case in point. Here the plateau, average elevation 5 km, lies above a very weak lower crust (probably due to a small degree of partial melting) and the whole area is collapsing outward, by basal sliding rather like a crustal glacier. At the same time it is extending by normal faulting at the surface. Another example is the Anatolian–Aegean plate (Fig. 2.16), which is being shoved outward due to the energetic impact of the Arabian plate into the Iraq/Iran part of the Asian plate along the great Zagros thrust fault. Anatolia–Aegea is decoupling ("unzipping") along the North Anatolian strike-slip fault, allowing the stored potential energy of the Anatolian Plateau, some 5 km above the deep Hellenic trench, to be released.

Edge forces result from a number of mechanisms. The chief one that seems to provide the major driver for plate motions is that of *slab-pull*. This arises as a negative buoyancy

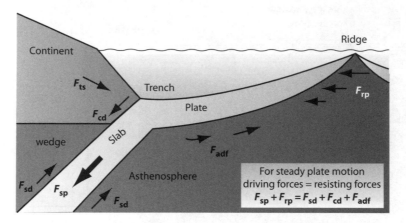

Fig. 5.42 The major forces involved in determining steady plate motion. F_{sp} – slab pull force; F_{rp} – ridge push force; F_{sd} – slab drage force; F_{cd} – collision drag force; F_{adf} – asthenospheric drag force. F_{ts} is the trench suction force, which acts to cause oceanward movement of the overriding plate if the slab should retreat oceanwards.

force because, as we mentioned before, a subducting slab of oceanic lithosphere is cooler and denser than the asthenosphere in which it finds itself. It thus sinks at a steady rate (remember Stoke's law, Sections 3.6 and 4.7) until it reaches some resisting layer within the earth or it heats up, melts, or otherwise transforms so that the buoyancy is eventually lost. Another edge force arises when a subducting plate moves oceanwards by slab collapse under an overriding plate. A *suction force* drives the overriding plate oceanwards, causing strain, stretching, and the formation of back-arc basins like the Japan Sea. *Resisting edge forces* are the frictional resisting forces that exist along plate peripheries, including *transform fault friction*, *strike-slip fault friction*, *slab drag resistance*, and for some deeply penetrating slabs, *slab end resistance*.

Basal forces have historically been the most controversial of driving mechanisms for it was the role of thermal convection (Section 4.20) in *basal traction* that was the first proposed mechanism to drive continental drift by Holmes and others. While there is little doubt that asthenospheric convection occurs, there seems little likelihood that basal traction along a convecting boundary actually drives the plate motion. This is because the almost 1 : 1 aspect ratio of Rayleigh–Benard convection cells (Section 4.20) is completely unsuitable to provide plate-wide forces of sufficient net vector: many such cells must underlie larger plates like the Pacific and their applied basal forces would largely cancel when integrated over the whole lower plate surface. *Basal resisting forces* due to asthenospheric drag are much more certain, for the motion of a plate must meet with viscous resistance over the whole lower plate surface. In this scheme, the asthenosphere passively resists motion; continuity simply requires the mass of the moving plate to be compensated by

a large-scale underlying circulation of the asthenosphere; a form of forced convection or advection (Section 4.20).

Overall, calculation of the various torques acting upon the major plates shows that the slab pull force, balanced by the basal slab resistance force, is the major control upon steady plate velocity and the slab resistance force is greater under continental plate areas than oceanic areas. This correlates with the known speeds of plates, those oceanic plates attached to subducting slabs being faster (Table 5.1).

5.2.8 Deformation of the continents

Although the oceanic and continental lithosphere are both rigid, the latter is not particularly strong; many areas are being strained due to the effects of adjacent or far-field forces.

First, we consider extensional deformation. We turn again to the eastern Mediterranean (Fig. 5.43) to illustrate this since it contains the best known and fastest extending area of continental lithosphere in the world, Anatolia–Aegea. At the leading edge of this plate we saw earlier (Fig. 2.16) that a spatial acceleration can be picked up in the Aegean region, with the largest rate across the Gulf of Corinth. You can see this in Fig. 2.16 by closely comparing the vectorial velocity field arrows to the south and north-east of the gulf, the velocity increases by more than 30 percent, about 10–15 mm yr^{-1}. Now although the whole region contains a great number of normal faults and it is evident that local strains of smaller magnitude may cause earthquakes and fault motion, the great majority of strain energy is being released along the particular array of normal faults that define the southern margin to the Gulf

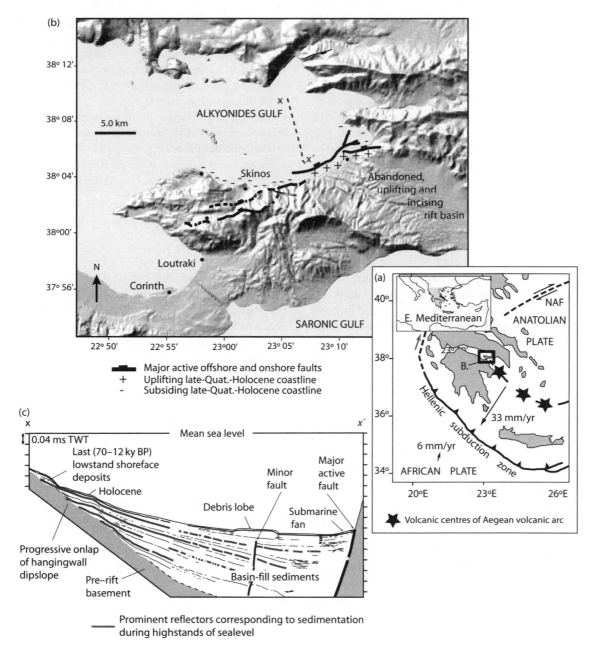

Fig. 5.43 Deformation of the continental lithosphere: Extension across the Gulf of Corinth rift, SW Anatolian–Aegea plate. (a) Context of plate, with active plate boundaries along the North Anatolian strike-slip fault (NAF) and the Hellenic subduction zone. (b) Detailed DEM to show relief and faulting associated with the active coastal fault system in the eastern rift. In this area the fault footwalls are uplifting and the hangingwalls subsiding. The dashed line *x–x'* is the line of section of C. (c) An interpreted seismic reflection survey line, *x–x'*, showing the tilted, half-graben form of the Alkyonides gulf. The Two Way Time (TWT) scale in milliseconds indicates time taken for seismic energy to pass from sea surface source to depth and back again to a receiver. Maximum water depth here is about 300 m.

of Corinth (Fig. 5.43b). The huge strains accompanying this differential motion are released periodically along powerful earthquakes on these faults (the seismogenic layer here ranges from 10 to 15 km thick). The Corinth gulf is termed a *rift* or *graben*. In the east, a *half-graben* for the normal faults that define it are only on one side, causing the prerifting crust to tilt southwards into the faults (Fig. 5.43c). Detailed GPS surveys also reveal that the southwest Greece is rotating anticlockwise with respect to the motion of the northern area. This illustrates that plates have vorticity, that is, they can spin as solid bodies about vertical axes.

The continental lithosphere may also deform under extension over vast areas, exemplified by the high (greater than 2 km) plateau of the western United States and adjacent areas of Mexico. The plateau is known as the Basin and Range on account of the myriad of individual normal fault-bounded graben and half-graben that make it up (Fig. 5.44). The individual ranges are the uplifted footwall blocks to the normal faults (Section 4.15) while the basins are the sediment-filled depressions, subsiding hangingwall ramps between the ranges. Range wavelengths are typically 5–15 km with lengths up to several times this. Today the GPS-determined velocity field in areas like Nevada is east to west at about 20 mm year^{-1} with respect to fixed eastern North America. The active normal faulting is located in distinct belts of high strain at either side of the province, chiefly associated with the Central Nevada Seismic Belt in the transition to the more rapidly northwest-moving California terrain, and to a lesser extent along the western margin to the Wasatch front in Utah. Although historic earthquakes have nucleated along steeply dipping normal faults (Fig. 5.44b) bounding individual range fronts, there is a record in the tectonic landscape of a previous phase of low-angle normal faulting (Fig. 5.44c,d). The kinematics of this kind of extension has given rise to areas of *core-complexes*, mid- to lower crustal rocks exposed in the footwalls to the low-angle faults. It is thought by some that this phase of low-angle faulting was related to a very rapid gravitational "collapse" of the over thickened Rocky Mountain crust some 20–30 Ma, with associated high-heat flow and volcano-plutonic activity.

Shortening deformation of the continents under compression occurs at plate destructive and continent–continent collision boundaries where five physical processes occur, often combined or "in-series," that cause the formation of linear mountain chains like the Pacific arcs, the Andes, and the Alpine-Himalaya (Fig. 5.45) system:

• crustal accretion by thrust faulting (see discussion of thrust duplexes in Section 4.15) of trench sediments "bulldozer-style" against the overriding plate – this results in the formation and rapid uplift of an *accretionary prism*;

• crustal thickening and buoyancy enhancement of the crust by the wholesale intrusion of lower density calc-alkaline magmas as plutonic substrates to volcanic arcs;

• whole-lithosphere thickening into a lithospheric mantle "root" by pure strain, manifest at crustal levels by shear strain along major thrusts;

• buoyant up thrust of crustal mass in thickened lithosphere resulting from the wholesale detachment of lithospheric mantle root;

• gravitational collapse of the elevated plateau with release of gravitational potential energy along active normal faults.

5.2.9   The fate of plates: Cybertectonic recycling and the "Big Picture"

Three possible scenarios concerning the large-scale recycling of plates have been envisaged at different times since the plate tectonic "revolution" in the late 1960s; they are sketched in Fig. 5.46.

1 A system of *whole-mantle convection* in which plates are carried about by applied shear stress exerted at their bases by the convecting mantle. The plates are thus part of a whole-mantle plate recycling system. The irregularity of plate areas and volumes compared to the regular system of convecting cells in Rayleigh–Benard convection (Section 4.20) is a problem with this idea. Also, the scheme requires rather wholesale mixing of slabs into the ambient mantle to prevent any lithospheric chemical signature contaminating the very uniform melt compositions represented by midocean ridge basalts (MORB).

2 This recognizes a fundamental physical discontinuity in the mantle at a depth of about 660 km due to the phase change of the mantle mineral spinel to a denser perovskite structure. A *two-tier convection/advection* system is envisaged, involving largely isolated lower mantle convection cells below the 660 km discontinuity. The upper mantle tier comprises a separate advecting system with plates driven by the edge- and top-forces discussed previously and with no slab penetration into the lower mantle. Separation of the lower and upper mantle in this way, with plate recycling restricted to the upper mantle, might be expected to gradually change the composition of the MORB through time. The scheme does not allow for the buoyant penetration of lower mantle plumes into the upper mantle and crust. The scheme was originally supported by the lack of slab-related earthquake hypocenters below 660 km.

3 This is really a hybrid scheme that has received a degree of acceptance in recent years. It involves both ongoing lower mantle convection, upper mantle advection with plates driven by edge- and top-forces and periodic slab penetration below the 660 km discontinuity. The model arose in the 1990s as advances in seismic tomographic

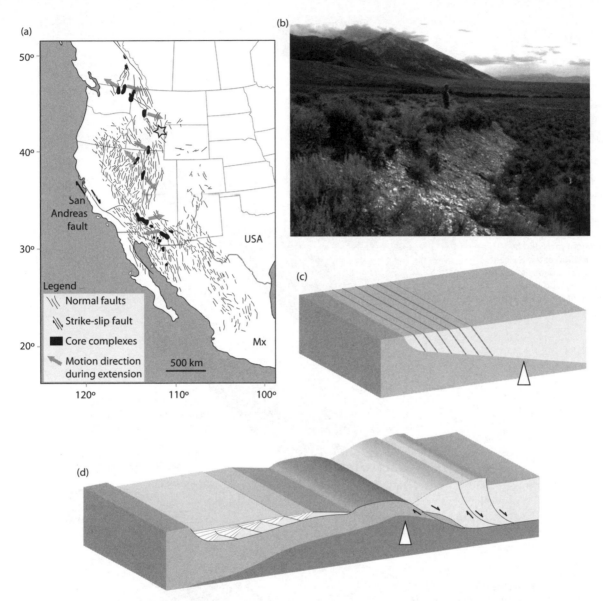

Fig. 5.44 Extensional tectonics of the western United States. (a) Map shows huge extent of the Basin and Range province with its arrays of normal faults and the location of the chief core complexes. Star indicates the site of the 1985 Borah Peak normal faulting earthquake. Large shaded arrows indicate the direction of extention revealed by the normal faults bounding the core complexes. (b) Field photo of part of the impressive surface fault break of the Borah Peak earthquake. Mike is standing on the uplifted footwall block, facing the subsided hangingwall block; total displacement is some 3.0 m. (c) and (d) show sequential development of a core complex due to a period of rapid, high extensional strain causing unroofing and uplift of mid-lower crust along a major low-angle crustal detachment normal fault system (to right of pointer).

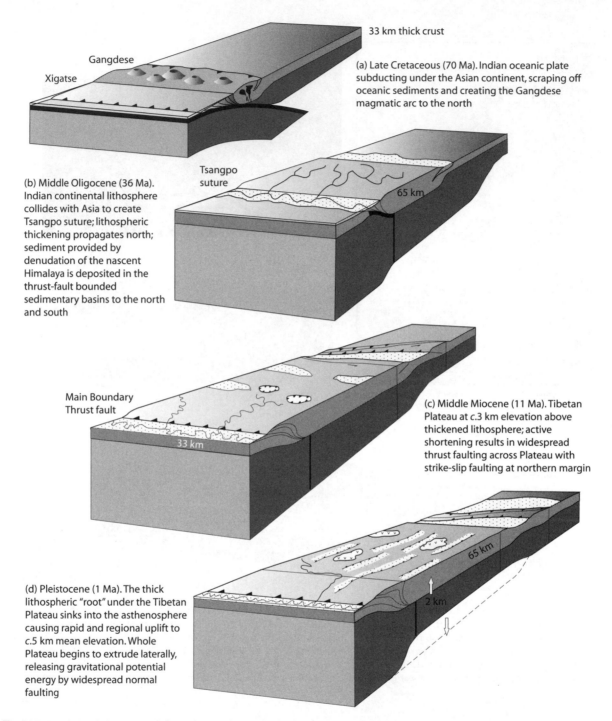

33 km thick crust

Gangdese

Xigatse

(a) Late Cretaceous (70 Ma). Indian oceanic plate subducting under the Asian continent, scraping off oceanic sediments and creating the Gangdese magmatic arc to the north

Tsangpo suture

65 km

(b) Middle Oligocene (36 Ma). Indian continental lithosphere collides with Asia to create Tsangpo suture; lithospheric thickening propagates north; sediment provided by denudation of the nascent Himalaya is deposited in the thrust-fault bounded sedimentary basins to the north and south

Main Boundary Thrust fault

33 km

(c) Middle Miocene (11 Ma). Tibetan Plateau at c.3 km elevation above thickened lithosphere; active shortening results in widespread thrust faulting across Plateau with strike-slip faulting at northern margin

65 km

2 km

(d) Pleistocene (1 Ma). The thick lithospheric "root" under the Tibetan Plateau sinks into the asthenosphere causing rapid and regional uplift to c.5 km mean elevation. Whole Plateau begins to extrude laterally, releasing gravitational potential energy by widespread normal faulting

Fig. 5.45 Continental shortening deformation on a grand scale: development of the collision of the Indian and Asian plates along the Himalayan mountain belt and the uplift and sideways collapse of the high Tibetan Plateau.

processing enabled geophysicists to track the fate of descending cool slab down to and often through the once inviolate 660 km discontinuity. This was linked to the ability of seismologists also at this time to distinguish what was interpreted as slab material at or about the core-mantle boundary, a zone termed "D". These developments suggested that mantle plumes and therefore large-scale outbursts of intraplate continental magmatism might arise from periodic "eruption" of molten slab at this boundary, though there seems little evidence that core material itself is involved in this process. It seems that *Cybertectonica* acts from surface to core/mantle boundary.

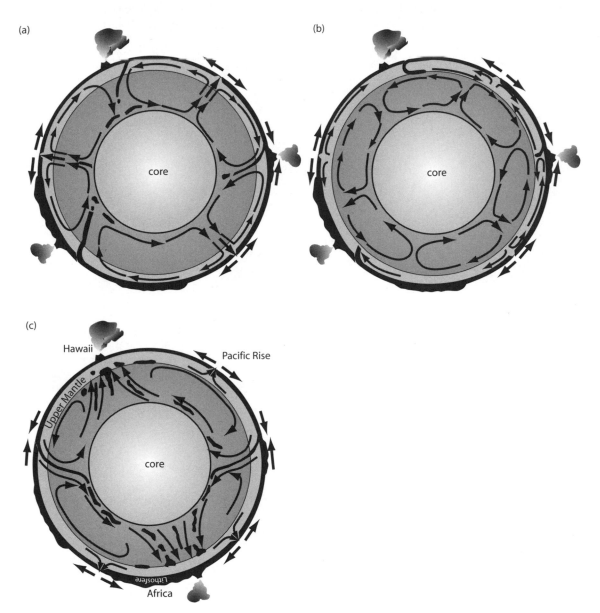

Fig. 5.46 Models for plates, plumes, and mantle convection. (a) Whole mantle convection drives and recycles plates. (b) Two-tier system with plate advection and lower mantle convection separate. (c) Upper level advection with deep slap penetration, periodic plume upwelling and lower mantle convection.

Further reading

P. Francis and C. Oppenheimer's *Volcanoes* (Oxford, 2004) is the most accessible account of volcanoes. For a comprehensive and practically orientated overview of plate tectonics from the very basics, nothing can beat A. Cox and R.B. Hart's *Plate Tectonics; How it Works* (Blackwell, 1986). Comprehensive, stimulating, and accessible accounts are P. Kearey and F. Vine's *Global Tectonics* (Blackwell, 1996) and in E.M. Moores and R.J. Twiss's *Tectonics* (Freeman & Company, 1995). The bible for advanced solid Earth studies of melting, stress, strain, and general dynamics from the point of view of mathematical physics is the unrivalled *Geodynamics* by D.L. Turcotte and G. Schubert (Cambridge, 2002). Probably the best intermediate level text is C.M.R Fowler's *The Solid Earth* (Cambridge, 2005).

6 Outer Earth processes and systems

6.1 Atmosphere

6.1.1 Radiation balance and heat transfer

It is important initially to consider the net balance of incoming solar radiation energy and outgoing reradiated energy from Earth over a long time period. We need to consider energy transformations also, like those between conductive and convective heat energy, potential and kinetic energy. In Fig. 6.1, 100 units of energy represent the magnitude of the incoming shortwave solar radiation flux (sometimes termed insolation) at the outer atmosphere; because of the Earth's planetary albedo, approximately 32 percent of this is reflected back into space. Of this total reflection, about 23 percent is from clouds behaving as perfect blackbodies and about 9 percent is directly from Earth's surface. This reflected radiation plays no part in the climate system. The remaining incoming radiation is either absorbed by the Earth's surface as direct and diffuse radiation (~49%) or absorbed by the atmosphere and clouds (~20%). Note the smaller atmospheric absorption compared with the large surface absorption. As seen previously (Section 4.19), the latter is converted into heat energy and, by Wein's law, is reradiated into the atmosphere as longwave radiation, where most of it is absorbed and then reemitted (at the same wavelengths). More longwave radiation in net terms is lost to space in this process from the troposphere (~60%), chiefly from cool cloud tops, than is absorbed (~17%). Together with the 19 percent of shortwave radiation absorbed, this means that there is a total absorption deficit of more than 35 percent.

Should the matter rest there, the Earth's troposphere and surface would cool drastically, to below 0°C. So, where does the "extra" energy come from? The deficit is provided by the transfer of heat energy from the Earth's surface by conduction and convective turbulent exchange (called *sensible heat transfer*) and as a by-product of the formation of clouds and the precipitation/evaporation of surface waters (*latent heat transfer*). Thus although the troposphere away from the tropics is in radiation deficit (Fig. 6.2), the overall positive net radiation from the Earth's surface due to the *greenhouse effect* (Section 4.19) means that the planet is in balance due to this transfer of heat from low to high latitudes by oceanographic and tropospheric circulation. To illustrate this, imagine that Earth, like its Moon, has no atmosphere. Then at any one time there would be a perfect energy balance between incoming shortwave insolation to the side of Earth facing the Sun and outgoing longwave reradiation from the whole Earth. The mean surface temperature would then be about 254 K, or −19°C. This compares to the actual mean surface temperature of around 14°C. The surplus temperature of 31°C is due to the greenhouse effect, whereby Earth's atmospheric gases absorb, reradiate, and reabsorb significant portions of the outgoing infrared radiation from the surface and make this energy available for the lateral and vertical transport of heat energy by the troposphere. It is only in dry desert areas that the Earth's climate is dominated by radiative exchanges alone, with high daily and low nightly temperatures.

The result of tropospheric heat transfer processes is a mean thermal structure shown in Fig. 6.3. The warmest temperatures, about 27°C, are at lowest latitudes, decreasing toward the poles and vertically up to the top of the tropopause at between 10 and 15 km elevation. The troposphere is thickest at low latitudes, thinning toward the poles. The greatest vertical and lateral gradients of T occur toward the top of the troposphere, a prominent boundary within the overall temperature structure of the whole atmosphere. Note that in general, particularly away from the tropics, the isotherms controlling air density diverge from the major latitudinally (zonally) averaged,

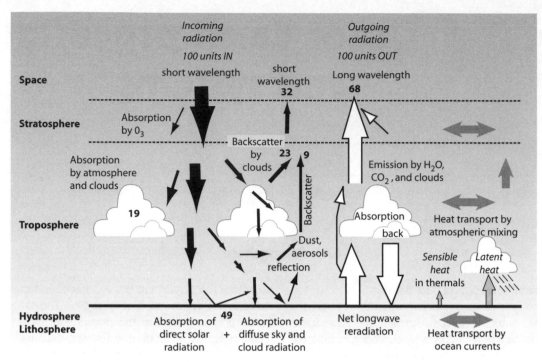

Fig. 6.1 Energy transport in and out of the atmosphere.

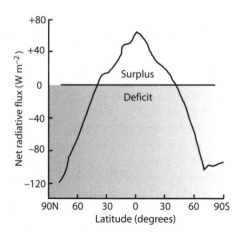

Fig. 6.2 Net annual zonally averaged radiative flux (shortwave absorbed flux minus outgoing longwave flux) from the top of the atmosphere.

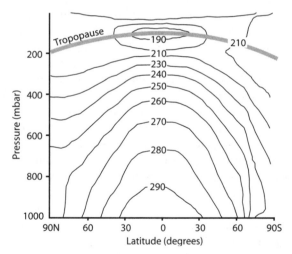

Fig. 6.3 Zonally averaged atmosphere temperature (°K) from equator to pole according to a global climate model based on sea-surface temperatures over a 10-year period. Note the vertically persistent equator-to-pole temperature gradient, its decrease with altitude, and the poleward thinning of troposphere. Also note strong baroclinicity away from equator (see discussion of "thermal wind").

equal-pressure surfaces (isobaric surfaces): in other words, the baroclinic condition (Section 3.5) dominates.

6.1.2 Thermal wind, pressure gradients, and origin of large-scale global circulation

The most fundamental features governing global atmospheric circulation (GAC) are: (i) strong negative thermal gradients from equator to pole and (ii) strong vertical density gradients set up by differential thermal heating and cooling of land and sea in equatorial latitudes. We consider feature (i) here and (ii) in a later section.

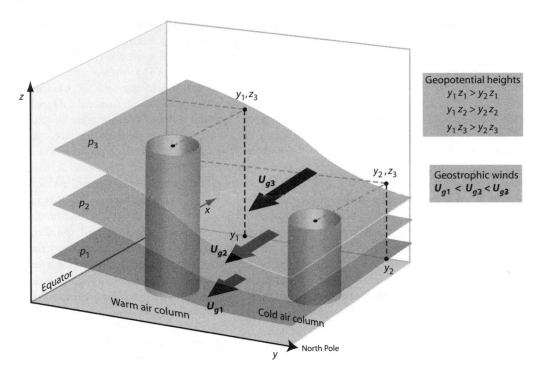

Fig. 6.4 Definition diagram to aid explanation of the thermal wind concept. p_{1-3} are pressure surfaces and their geopotential heights at positions y_1 and y_2 are $y_1 z_{1-3}$ and $y_2 z_{1-3}$ respectively.

With reference to definition diagram Fig. 6.4, the height of a particular pressure surface above sea level is termed the *geopotential height*. Differences in vertical separation (thickness) between given isobaric surfaces are due to temperature for any given pressure drop. This comes out of the hypsometric equation for layer thickness, Δz (Cookie 7). In warm air columns the layers thus have greater thickness than those in cold ones, the cumulative effect of layer-upon-layer of thickening leading to an increasing slope of the isobars with height. For the case of a negative poleward thermal gradient, we take as reference the latitudinally averaged pressure surface low in the troposphere, say at 1000 mbar, more-or-less at sea level. Measurements here (Fig. 6.5a,b) show little overall horizontal meridional pressure gradient, that is, the average poleward pressure has no large systematic changes other than those across the southern ocean and between the Azores High and Iceland Low (see weather chart of Fig. 3.21). This means that the *whole* mass of tropospheric air that exerts the near-sea-level pressure field is distributed about uniformly. Now take another pressure surface at 500 mbar in the middle of the troposphere where the air above is much less dense (Fig. 6.4). The pressure surface falls appreciably (of the order of 10–15%) due to the poleward temperature gradient through 40–60°N latitude in

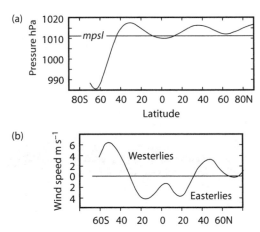

Fig. 6.5 (a) Global meridional transect of mean annual zonal pressures at sea level. Note influence of Antarctica and the southern Ocean. (b) Corresponding mean air speeds. Note the inverse relation to pressure gradient in Figure 6.4.

both summer and winter, the gradient with height decreasing further poleward in both seasons and equatorward from about 30°N in summer. Generally the air below a certain average isobar at the equator is warmer than the corresponding high-latitude air below the same isobar. The differential vertical expansion due to this means that the poleward thermal gradient is accompanied by a horizontal hydrostatic pressure gradient (Section 3.5.3) that increases vertically from virtually zero close to sea level to a maximum toward the top of the troposphere at latitudes 50–60°N. The overall concept of this thermal–wind relationship is illustrated by comparing the slopes of average isobars low down and high up in the troposphere. The increasing slope of the equal pressure surfaces and therefore the increasing strength of the resulting geostrophic wind (Fig. 6.6) is evident. The thermal–wind relationship thus refers to a rate of change of wind velocity with height; that is, it is a measure of geostrophic *wind shear*. The magnitude of the vertical wind shear is directly proportional to the horizontal temperature gradient. The high-level wind strength and its gradient can be mapped as a velocity field from pressure layer height contour maps. As a geostrophic phenomenon (Fig. 6.6) the wind travels parallel to the contours of geopotential height (Fig. 6.4) and is faster when contours are closer and vice versa. The maximum magnitude of wind shear defines the fast cores of the polar front *jet streams* that dominate the west-to-east zonal circulation of the planetary wind regime. The jets are strongest in the winter when temperature contrasts between equator and pole are greatest (Fig. 6.7).

6.1.3   Frictional wind

That part of the wind blowing well above the land or sea may be considered to flow independently of any surface

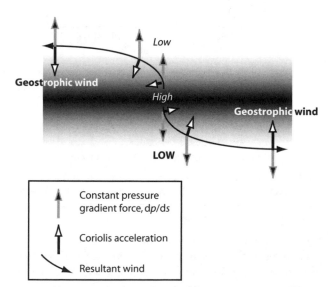

Fig. 6.6 A starting geostrophic wind begins to flow from high to low pressure along the constant pressure gradient, dp/ds, but it is progressively and increasingly turned due to the Coriolis acceleration (shown here for the southern hemisphere).

roughness, the flow approximating to that of the geostrophic wind set up due to synoptic pressure or temperature gradients. Such a wind drives flow in the atmospheric boundary layer (ABL), but because of the effects of surface friction the effectiveness of the Coriolis force in turning the geostrophic wind markedly decreases; for a northern hemisphere wind the direction of the frictional wind *backs* anticlockwise toward the surface, while in the southern hemisphere the frictional wind *veers* clockwise. In both cases the winds tend to progressively diverge from their geostrophic course parallel to isobars through the ABL. This is known as the *Ekman spiral effect* (see discussion in Section 6.4). The effect of the frictional wind on a low-pressure system is to cause inward spiralling of wind into the center; this convergence causes compensatory central upwelling of air and may cause cloudy conditions as moist air is cooled and is vice versa for high-pressure systems when divergence causes central downwelling and clear skies.

6.1.4   Energy transformation and the global atmospheric circulation (GAC)

In order to understand the principles of the wider GAC, it is now necessary to consider the role of thermal energy transfer: how energy is transported by a unit mass of moist air. In order to maintain constant total energy, E, in the face of continuous loss of longwave radiation to space (the atmospheric radiation deficit earlier discussed), it is necessary to add sensible heat from surface land and ocean and from the release of latent heat during rainfall. The general circulations of the atmosphere thus involve a poleward transport of heat energy to maintain the observed long-term temperature distributions, which are approximately constant. The simplest such arrangement possible would be a general convective upwelling of warm moist air from the equatorial regions and its transfer toward the poles, cooling by radiative heat loss as it does so, where it eventually sinks, liberating rain, snow, and latent heat. Such a simple cellular circulation, termed *Hadley circulation*, has the right principles (Fig. 6.8) but inevitably the Earth's atmosphere is more complicated, chiefly because of the effects of the Earth's rotation, and also because of pressure effects due to latitudinal variation in thickness of the troposphere. Atmospheric air masses continuously move around and at the same time energy is continuously being transformed from one form to another. Neglecting the very small kinetic energy of air masses, the following forms of energy are involved (Fig. 6.9):

1 *Latent heat energy*, E_L, arises in moist air from reversible phase changes of state between liquid, water, and gaseous

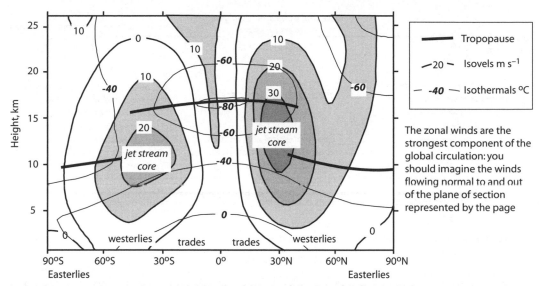

Zonal here refers to wind flow normal to a circumferential section from North pole to South pole, showing the strongest East–West components of the global circulation.

Fig. 6.7 Mean zonal winds and temperatures in January; zonal here refers to wind flow normal to a circumferential section from North pole to South pole.

Notes:

1 Jet stream with mean speeds of 30–40 ms^{-1} at limits to tropopause in regions of strong temperature/pressure gradients.

2 Discontinuities in level of troposphere at midlatitudes due to low-latitude convective upwelling and high-latitude convective sinking.

3 Greater upper troposphere wind strength in northern hemisphere winter due to enhanced seasonal contrast between the cooling northern hemisphere and equatorial regions.

4 Little variation of T with latitude in the tropical zone (*barotropic* condition), gradients increasingly diverging poleward (*baroclinic condition*)

5 Very high velocities of near-surface southern hemisphere high-latitude winds (not shown here, see Fig. 6.4) due to high-pressure gradients there.

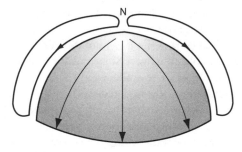

Fig. 6.8 Simple Hadley circulation on a nonrotating Earth.

water vapor (Section 3.4). It is given by the product of the latent heat of evaporation or condensation (L, $2.3 \cdot 10^5$ J kg^{-1}) times the mass (m) of water vapor involved. Evaporation of water into a parcel of air requires work to be done breaking hydrogen bonds and hence energy is taken in (from the other energy sources in the atmosphere) and cooling takes place. The opposite holds during condensa-

tion of water vapor to rain or cloud droplets and the air heats up. The large volumes of atmosphere involved mean that the process is very important in atmosphere–ocean coupling (Section 6.2).

2 *Sensible heat energy*, E_S, arises from the direct impact of radiation, the atmosphere losing or gaining sensible heat by radiation to and from space and from clouds. It is given by the product of the specific heat capacity at constant pressure, c_p (J kg^{-1} K^{-1}), and temperature, T (K). Water has a very high specific heat capacity (around 4,000 J kg^{-1} K^{-1}) compared to that of air (around 1,000 J kg^{-1} K^{-1}).

3 *Potential energy*, E_p, is that of position, y, above sea level, times gravity.

The total energy, E, present in any imaginary unit mass of moist air must remain constant; thus $E = E_L + E_S + E_p =$ constant or $E = Lm + c_pT + yg =$ constant. This *conservation of energy* equation allows us to understand energy changes in ascending or descending air masses. Such motions dominate heat exchange in the tropics and

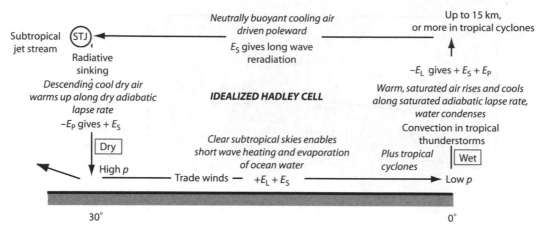

Fig. 6.9 Energy transfers in an idealized Hadley cell.

in all latitudes as air masses move over elevated topography. Consider a dry, slowly ascending mass in which we have no E_L term. The increase in E_p due to ascent must be accompanied by a fall in temperature so that E_S can decrease. We can think of this easily in physical terms with a little help from kinetic theory and the gas laws (Sections 3.4 and 4.18), since as the dry air slowly rises it moves into regimes of lower pressure where work must be done in expansion and temperature falls; vice versa for falling air. This change of T with height is known as the *dry adiabatic lapse rate*, about 1 K per 100 m. When water vapor is present in ascending air, the lapse rate is reduced because as cooling takes place the air becomes saturated, condensation occurs, and latent heat is released, becoming sensible heat. This *saturated adiabatic lapse rate* varies with T, being approximately equal to the dry rate for cold air, but much less for warm air. A typical average value would be around 0.6 K per 100 m.

6.1.5 Low-latitude circulation and climate

In the simplest representation of low-latitude circulation, termed a *Hadley cell*, between about 30° latitude north and south of the equator (Fig. 6.10), warm moist rising air originates at the equatorial *intertropical convergence zone* (ITCZ), where incoming radiative heating is at its most effective and the Coriolis force is minimal. This warm air uplifts the higher atmosphere air column, increasing the pressure at any height and causing an outflow into higher latitude areas. The warm saturated equatorial air ascends mostly in convective thunderstorms. As upflowing air is cooled, precipitation releases latent heat energy. Periodicity of deep convective rainfall on a roughly

monthly timescale in the ITCZ in the Indian Ocean, southeast Asia, and southwest Pacific is due to a curious linked atmosphere–ocean phenomenon known as the *Madden-Julian Oscillation*. Recent decadal warming of the tropical oceans, particularly the Indo-Pacific, is thought to be the teleclimatic link that has forced changes in higher-latitude climate systems (discussed later for North Atlantic).

As it is cooled by radiative heat loss due to emission of longwave radiation gained by its high sensible heat content the equatorial air sinks, with a transfer of potential energy to sensible heat according to the dry adiabatic lapse rate. Thus the dry cloudless subtropical deserts have their radiation deficits, due to high infrared radiative heat loss from the high-albedo/low-cloud-cover environment, compensated by the lateral transfer of heat from the equator. Over the subtropical oceans the low-humidity air and cloudless conditions enable shortwave solar radiation to warm and evaporate seawater. The trade winds then complete the Hadley circuit, transporting latent and sensible heat in the form of moist, near-saturated winds returning once more to the ITCZ at low levels as the generally steady and dependable easterly trade winds. The convergence or confluence of these then causes general upwelling in the equatorial troposphere. Winds in the ITCZ itself are usually light: the area of the doldrums.

The general Hadley cell circulation is thus clockwise in the northern hemisphere and anticlockwise in the southern hemisphere. The warm low-density equatorial air creates a low-pressure zone at the tropics, which shifts north and south of the equator with the ITCZ during the course of the year. The descending and cooling high-density subtropical air creates the zonal high-pressure belts marked on the continents by the great trade wind deserts.

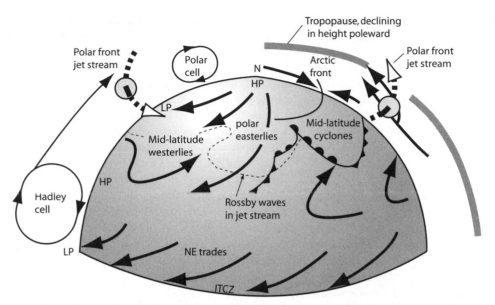

Fig. 6.10 The observed General Atmospheric Circulation, involving convective cells with Coriolis turning, jet stream, Rossby waves, and associated frontal systems.

6.1.6 Mid-latitude circulation and climates

So far we have implied that the equatorial air masses that cool-convect down to the subtropical surface flow back west to form the trade winds. In fact, the poleward horizontal pressure gradient and the resulting thermal wind ensures that a substantial poleward-moving component comes into contact with equator-moving cool polar air, the polar easterlies, at mid-latitudes 40–55°. The two masses meet along what is known as the *polar front*, where Coriolis deflection leads to the formation of a zone of westerly winds these latitudes. The westerly winds so characteristic of mid-latitude climates are really the low-altitude remnants of the much stronger *polar front jet stream* wind (see Section 6.1.2). Observations indicate not only that the jet streams encircle the globe, but that the seasonal-averaged winds vary in strength and direction because of two to four wave-like billows that occur with wavelengths of several thousands of kilometers. These *planetary long waves* are often seen as seasonal-permanent features of the atmospheric circulation on mean pressure maps. They are termed *Rossby long waves* (Fig. 6.10) and serve to transfer momentum and heat across the mid-latitudes. Rossby waves owe their origin to differential heating of major continental land masses and sea surfaces in the lower atmosphere region below the jet stream. This engenders a wave-like diversion of jet stream flow around the isobars of the resulting pressure anomalies, the waves themselves traveling much more slowly than the air in the jet stream itself.

Higher-frequency Rossby waves superimposed on the long-Rossby waves represent junctions between warm equatorial air and cold polar air with large temperature gradients across them, known as *fronts* (Fig. 6.10). Fronts slope gently upward from low to high latitudes and are the sites of what is termed "slantwise convection," that is, the forced upwelling of warm low-latitude air over sinking cold polar air; they are a form of rotating *density current* (Section 4.12). The intersection of a front with the earth's surface is not simple for there are often smaller "parasitic" waves and fronts superimposed that are shed off by the vorticity of the major Rossby waves. These moving air masses comprise stable air of contrasting temperature and pressure separated by the frontal surfaces. They dominate the weather and climate of mid-latitudes, giving rise to a more-or-less predictable sequence of weather, but which travel at more-or-less unpredictable rates, much to the chagrin of forecasters.

Important climatic variability in the northern hemisphere mid-latitudes (and probably elsewhere through some teleclimatic connection, probably from warming of the tropical Indo-Pacific oceans) seems to be correlated with what has become known as the *Northern Hemisphere Annular Mode* (*NAM*), also known as the *North Atlantic Oscillation* (*NAO*). It is represented as the difference in sea level pressure between the Azores High (descending low-latitude air) and the Iceland Low (see weather chart of Fig. 3.21). High-index time periods (large statistically significant variations in pressure) are marked by anomalously strong subpolar westerlies, which reduce the severity of winter weather over much of mid- to high-latitude continental regions. The recent decadal trend in this direction is suspected as having a partly human cause (though as we write this, in winter 2005, the huge blocking high from Azores to Iceland has been around for over a month and it hardly feels like it).

6.1.7 Monsoonal circulations

These can be most simply regarded as continent-scale sea/land breezes. They result from seasonal variations in the Hadley circulation because of changing thermal gradients, notably the trans-equatorial migration of the ITCZ and substantial enhancement of subtropical highs in winter over the cool continents and their diminution and change to deep lows in summer as the land warms up. Generally (Fig. 6.11), as land warms up the overlying air is heated, expands, and ascends, and hence creates higher pressures above. The high level land-to-seaward pressure gradient causes divergence or diffluence and net outflow of air in the upper levels, causing a low pressure at the land surface. Air now flows in from the ocean along the resulting sea level pressure gradient. The overall effect is the formation of a top to bottom convective circulation of air. The effect in the subtropical maritime zones of Asia and Africa, together with smaller regional zones like the southwestern USA, is noteworthy because it leads to seasonal reversals (remember that the Coriolis force is low at low latitudes) of the trade winds that lead to the inflow of moist warm maritime air onto the continents. It is the ascent and cooling of these *monsoonal air masses* onto the Himalayas that causes some of Earth's highest and most intense rainfall.

In recent years the role of large continental plateaux, notably the Tibetan Plateau and, to a lesser extent, the western USA, has come center-stage in efforts to explain not only regional but also world-wide climate. Tibet in particular seems to play a major role in enhancing the Indian Ocean monsoon. It is a site of high sensible heat transfer from the atmosphere in late spring and summer and a ground-level low-pressure system results, with upward-flowing air defining an upper-atmosphere Tibetan anticyclone. The area thus contrasts greatly with the high-pressure trade wind deserts of Africa at similar latitudes, which lie in the continental interior at much lower ground elevations. A strong ($>110 \text{ km h}^{-1}$) *easterly subtropical jet stream* forms south of the Tibetan Plateau during the monsoon in response to strong temperature and pressure gradients between the warm rising Tibetan Plateau air and the cooler Indian Ocean air (Fig. 6.11). The magnitude of

this vortex (formed as southward- and downward-flowing air along the Tibetan anticyclonic north to south pressure gradient and turned westward by Coriolis force) enhances the downward flow to add to the strength of lower-level summer monsoonal winds. Thus there is a correlation between strength of the easterly subtropical jet stream and the amount of monsoonal precipitation. Such a correlation is further evidence of the enhancement role of the Tibetan Plateau in monsoon development. Thus the monsoonal winds from the Indian Ocean (which ought to be flowing, like all trade winds in the northern hemisphere, to the southwest) turn to penetrate high into the foothills (particularly the eastern foothills) and ranges of the Himalayas, contributing additional latent heat to the rising dry air on the Tibetan Plateau to the north, which is itself markedly arid during the summer.

6.1.8 High-latitude climates

The polar and subpolar ($>70°$ latitude) regions of Earth are influenced by highly seasonal shortwave insolation that varies from nonexistent to low. Longwave diffuse reradiation predominates in the totally dark winters. Polar skies are often cloudy and it might be thought that these clouds should trap more reradiated energy than they actually do. The clouds are in fact very "thin," with a sparse content of water droplets present in the very cold and undersaturated air. Albedo is high all year round over permanent ice- and snowfields and high in winter everywhere because of seasonal snowfall. The net radiation deficit is partly filled by the poleward atmospheric transport of heat described above and partly by oceanic transport. In northern latitudes there is a major radiation deficit over the Norwegian Sea associated with the formation and descent of deep water, one of Earth's major heat sinks. Over the poles themselves, radiative sinking dominates, there existing more-or-less permanent but weak high-pressure systems of descending air that diverge surfaceward to form the polar easterly air masses (Fig. 6.10) that help define the Arctic and Antarctic fronts.

In the southern hemisphere the climate over the high latitudes is dominated by the very strong circumpolar westerlies acting about the polar vortex. Like the northern hemisphere *NAM* (see Section 6.1.6), strengthening of the *Southern Hemisphere Annular Mode* (*SAM*, also known as the *Antarctic Oscillation*) seems to have occurred in recent decades, interpreted as due to the effects of Antarctic ozone depletion. This is because in the stratosphere, observed ozone depletion has induced cooling over the polar icecap as a result of reduced absorption of springtime

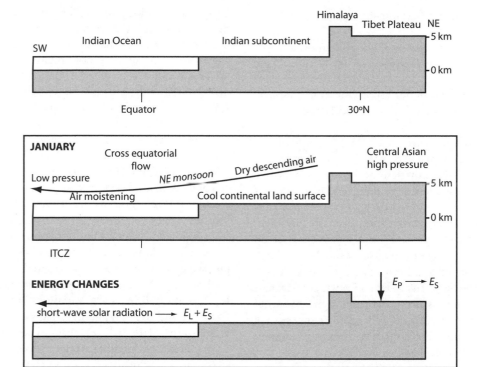

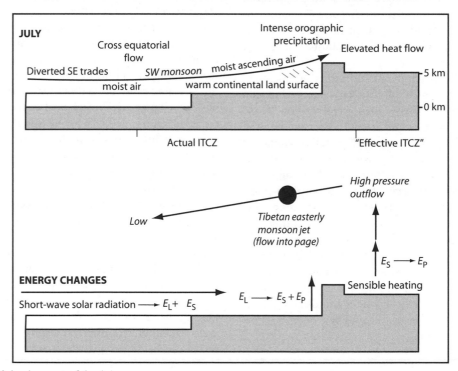

Fig. 6.11 The seasonal development of the Asian monsoon.

ultraviolet radiation. The stratospheric cooling induces increased circumpolar flow between latitudes 50 and 60°S, which is transferred to the troposphere in the summer months after a lag of a month or so. However, recent results also point to a natural period of unusually high positive variation in the SAM-index 50 years ago, before the advent of human-induced ozone changes. The debate on the forcing factors responsible (solar variability, etc.) continues.

6.1.9 Global climates: A summary

1 Equatorial latitudes are dominated by high insolation, low pressures, light to variable winds (these are the latitudes of the "Doldrums"), high mean annual temperatures, low daily temperature changes, high water vapor saturation pressures, and copious convective precipitation, particularly in summer months.

2 Trade wind latitudes extend 15–35°N and S of the equatorial low-pressure belt. Over land, air masses are dry and skies clear over the hot deserts, with high radiative heat transfer and consequent high daily temperature variations from hot to cold. Over the oceans, evaporation rates into the initially unsaturated advecting air masses are high. Hurricanes and typhoons result from instabilities set up during convective oceanic heating (Section 6.2). The large and high landmass of southern Asia (also east Africa, Southeast Asia, northern Australia, and northern Mexico) develops its own low-pressure system and high latitude jet stream in summer that attracts and reverses the normal northeast trade winds, thus initiating torrential summer monsoonal precipitation.

3 The subtropical high-pressure belt at about latitudes 35–40°N and S is characterized by descending unsaturated air masses and generally light winds, resulting in Mediterranean climates with hot dry summers and cool wetter winters.

4 The mid-latitude to temperate-latitude maritime low-pressure zones are dominated by the movement of frontal systems that form at the polar–subtropical transition. These sweep warmer saturated air north and cooler unsaturated air south as wave-like intrusions. Plentiful precipitation results in the cool to cold spring, winter, and autumn seasons. Oceanic currents like the warm north-flowing Gulf Stream lead to highly important contributions to air temperatures by latent heat released in the rainstorms associated with frontal systems blowing over them. Continental areas at these latitudes (40–60°N and S) suffer much larger temperature extremes and generally lower precipitation.

5 The polar anticyclone presides over a stable regime of cold to very cold descending dry air masses with very high albedos over snow- and ice fields under cloudless or pervasive "thin" cloudy skies in summer giving high radiative heat losses to the atmosphere.

6.1.10 Milankovich mechanisms for long-term climate change

We accept the notion of a mean climate for particular regions and a net global balance between incoming and outgoing radiation, yet looking back through recorded history and into the young and then older geological record, it is obvious that significant and sometimes major changes have occurred in both regional and global climate. We first enquire as to whether the global amount, surface distribution, and greenhouse entrapment of incoming solar energy have remained constant through geological time. Given the great importance of the oceans in the climate machine, we must also enquire as to how known changes in continent–ocean distributions may have affected global and regional climate.

There is little direct evidence that the solar constant changes in response to short-term solar activities like sunspot cycles but there is circumstantial evidence that over periods of several hundred years the decreased activity of sunspots may be reflected in lower solar energy output since such periods are associated with severe global cooling (e.g. Maunder sunspot minimum and the Little Ice Age of *c.*350 years BP). Variations in the orbital path of Earth around the Sun and in Earth's own rotation induce longer-term (10^4–10^5 yr) changes in the *relative* solar energy flux to particular parts of the planetary surface. We stress the term "relative" because small changes in orbital parameters lead to no net increase or decrease in solar radiation received: the changes simply tend to apportion the radiation at different times of the solar cycle in particular hemispheres. It is this cyclical preferred apportionment that is thought to lead to longer-term climate change and the accumulation or melting of great ice-sheets. These physical changes in the seasonal distribution of incoming energy cannot of course be measured, but, in one of the great scientific breakthroughs of the twentieth century, their indirect climatic effects have been carefully ascertained by sophisticated geochemical studies of Quaternary marine fossils.

Following the lead of the nineteenth-century amateur scientist James Croll, Milankovitch in the 1920s and 1930s calculated how variations in the three orbital parameters (Fig. 6.12) – eccentricity, wobble, and tilt – would lead to different amounts of radiation being received at different latitudes. The key was found in the variation of radiation received at temperate latitudes during summer. During winter we know that the polar and high latitudes are cold enough to form snow and ice; it is the survival of these seasonal features that will determine whether the ice-sheets can expand below the Arctic or Antarctic circles. Thus any orbitally induced changes that encourage summer cooling by decreasing received radiation should lead to climate change sufficient to trigger an Ice Age.

The first orbital mechanism is based on the fact that Earth's rotation around the Sun is elliptical and not

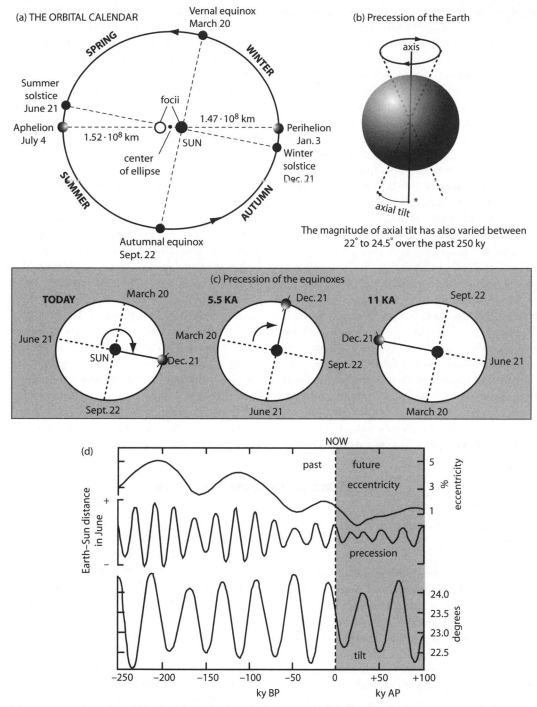

Fig. 6.12 (a), (b), and (c) are the various Milankovich mechanisms for long-term global climate change; (d) Computed changes in eccentricity, tilt, and precession for the past 0.25 my and for the future 0.1 my.

circular, a fact known since the work of Kepler in the seventeenth century. The very nature of an elliptical path, with the radiating Sun at one focus, means that there are seasonal variations in the amount of radiation received by Earth at aphelion and perihelion. This is based on the premise that the intensity of solar radiation is reduced with distance from the Sun. At present, Earth is nearest the Sun, by about $4.6 \cdot 10^6$ km, on 2–3 January and furthest away on 5–6 July (note that these dates are not the same as the times of solstice, see below). As a consequence the solar radiation received varies by about ±3.5 percent from the mean value. Although these figures are not appreciable compared to the other effects noted below, exact calculations of the gravitational effects of the other planets in the Solar System on Earth's elliptical orbit led to the later theory (due to Leverrier in 1843) of time-variable eccentricity, whereby the yearly orbit becomes more and less eccentric on the rather long timescale of around 10^5 yr. At the present time and for the foreseeable future (Fig. 6.12) we are in a period of average to low eccentricity; at times of highest eccentricity it is calculated (originally by Croll) that the change of radiation may be greater than 5 percent.

A second orbital mechanism is based on the regular "wobble" of Earth's inclined spin axis relative to the plane of rotation around the Sun or to some point fixed in space. This wobble is due to the gravitational attraction of the Sun and Moon upon Earth's own equatorial bulge. The practical effect of this leads to the "precession of the equinoxes," a phenomenon discovered in about 120 BC by Hipparchos of Alexandria (see Section 1.4), whose own observations of star clusters taken at fixed yearly times and positions compared to those observed by earlier Egyptian and Babylonian astronomers (going back to about 4,000 BC) led him to note the gradual shift of familiar star clusters around the Earth's ecliptic, the plane of the solar orbit, at times of solstice. One complete wobble involves a circuit of the spin axis about a circle, causing the northern and southern hemispheres to change their times of closest and furthest approach, at aphelion and perihelion respectively, approximately every $2.2 \cdot 10^4$ yr. This is the explanation for the fact noted above that the solstices (times of maximum tilt of the Earth away from and toward the Sun) do not have to coincide with aphelion and perihelion. In terms of solar radiation received at Earth's surface the Lambert–Bouguer law (Section 4.19) makes it clear that the effects of precession are greatest at the equator, decreasing toward the poles. Minimum levels of radiation for either hemisphere away from the polar circles occur when perihelion corresponds to winter. Today we are close to the situation of southern hemisphere summer at perihelion: about 11 ka any Palaeolithic astronomers would have experienced northern hemisphere summers at perihelion.

A final orbital mechanism depends upon the angle of the inclined spin axis changing relative to the ecliptic. Calculations and observations indicate that this is currently changing by about $1 \cdot 10^{-4}$ deg yr^{-1}. Over a $4 \cdot 10^4$ yr period the axis varies about extreme values of 21.8° and 24.4°. The current value is around 23.44°. In terms of solar radiation received, again according to the Lambert–Bouguer law, minimum levels are to be expected in winter when tilt is maximum, but the effect makes no difference to high polar latitudes since these are in darkness anyway. The effect has greatest influence in moderate to high latitudes.

6.2   Atmosphere–ocean interface

6.2.1   Atmospheric boundary layer: Momentum exchange over the ocean

Having considered general atmospheric circulation we now consider the behavior of the *atmospheric boundary layer* (ABL), part of the frictional wind driven over the ocean. Momentum exchange occurs in the lower parts of the ABL close to the ocean–atmosphere interface; offshore platforms are used to determine this (Fig. 6.13). We know from Chapter 3 that fluid in any boundary layer transmits a stress to its bounding medium through the transfer of momentum, the rate of transfer being proportional to the overall rate of fluid flow as measured by some local time-mean velocity. In the very high Reynolds number situations that exist in air flows we may completely ignore viscous stresses and ascribe the fluid momentum transfer and mixing processes in the interface region to the turbulence. We have seen previously (Section 4.5) that this region defines what is known as the "wall-layer" or "logarithmic zone" of turbulent shear flows (Fig. 6.14). Here we can relate the rate of increase of mean velocity with height above the water or land interface, $d\bar{u}/dz$, to a shear or friction velocity, u_*, which defines the degree of turbulent momentum exchange or momentum flux associated with any wind. This causes a surface shear stress, $\tau_{zx} = C_d \rho (u_*)^2$, where C_d is a drag (friction) coefficient and ρ is the density of air. The practical problems of measuring C_d are numerous and beyond the scope of this text; however, they provide the key to many practical ocean forecast models. The great importance of estimates of surface wind shear lies in its key role in determining coupled ocean–atmosphere interactions, especially their role in determining surface flow direction in climate models. For this purpose longer-term time-mean values, such as monthly means, $\bar{\tau}$ must be taken from areas of the ocean corresponding to required model grid spacing. However, computations of shear must also take into account shorter-term fluctuations, τ', about $\bar{\tau}$ over periods of days or even hours. A key application is in the field of tropical cyclone modeling

Fig. 6.13 The stable FLIP platform for measurement of wind velocity/boundary layer data in the offshore environment 50 km west of Monterey, California.

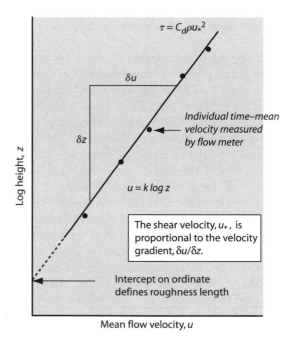

Fig. 6.14 Schematic of typical ABL velocity distribution with height to show the logarithmic relationship and the computation of shear velocity and hence shear stress, τ, as a measure of flow turbulent momentum exchange.

and forecasting; the speed of storm advance is sensitive to degree of surface friction at the ocean–atmosphere boundary (Section 6.2.4).

An important roughness feature on the sea surface is water waves. These occur on a variety of scales. The waves take the form of smooth sinusoids through to breaking waves (whitecaps). Wave formation from a flat sea surface is due to velocity and pressure fluctuations in the wind set up during wind–wave coupling over initially small waves that then grow to equilibrium. They are broadly analogous to the Kelvin–Helmholz waves described in Section 4.9. The growing waves then strongly influence momentum, energy, and mass exchanges at the interface. Any overall wind velocity, u, is the sum of a long-term mean velocity, $\bar{u}$, a turbulent fluctuation from the mean, u', and a periodic velocity associated with the waves, $\tilde{u}$ so that $u = \bar{u} + u' + \tilde{u}$. The same applies to Bernoulli pressure fluctuations induced by the wind: when the wave pressure is less on the lee side than the stoss side of the wave then energy is transferred from air to wave and the wave grows.

We might ask ourselves how far upward the log layer extends in natural ABLs which may be in motion for many hundreds of meters to kilometers upward above any measuring platform. The answer comes from data collected in connection with hurricane studies in which radio sondes, remotely-tracked by Global Positioning System (GPS) receivers, were released into the ABL from aircraft (Fig. 6.15). The frictional influence of the flow boundary (ocean water in this case) extends upward as a log layer $c.200$ m. Maximum flow velocities were reached at about 500 m, gradually weakening upward to a height of 3 km. Surprisingly, the most energetic storm winds lead to a reduction of surface roughness friction, despite the production of larger surface waves. This is thought to be the result of the production of abundant surface foam, which somehow acts as a *drag-reduction agent*. The reduced friction enables tropical cyclonic storms to move faster than predicted from conventional considerations of boundary drag. Such momentum-exchange processes obviously lead to gaseous exchange from the ABL into the ocean, but also vice versa, in surface ocean layer gases from planktonic photosynthesis. Also, both sensible and latent heat are exchanged via radiation gains and losses, evaporation of surface waters due to forced convection, and from the condensation of water vapor.

6.2.2 Dynamic ocean topography: Atmospheric wind forcing of surface ocean currents and circulation

The low resistance of ocean water to surface shear by the blowing wind leads to a net mean motion of the

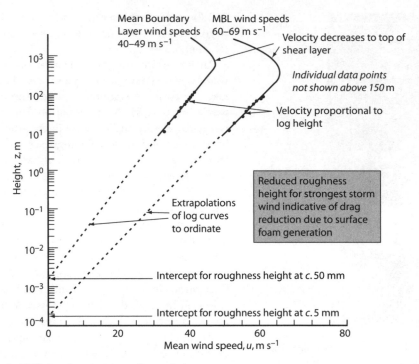

Fig. 6.15 Radio-sonde/GPS data from hurricane force winds.

water. However, in the surface layers of the deep ocean this motion is not in the downwind direction, as common sense might predict. In fact, the combined effects of surface shear and the Coriolis force (see Cookie 21) leads to the surface flow being turned clockwise to the wind direction, by 30° or so in the northern hemisphere and anticlockwise in the southern hemisphere. This deviation of mass transport is termed *Ekman transport* (we noted the phenomenon of frictional flow deviation in connection with the frictional wind in Section 6.1.3). Below the water surface an intricate spiral pattern of flow is set up that is progressively more deviated from the surface wind direction with water depth; this is called the *Ekman spiral*.

The major surface currents of the ocean are caused by the effects of wind shear due to general atmospheric circulation, combined with the Ekman transport. The omnipresent trade winds begin the process of wind shear in low latitudes, creating the north and south Equatorial Currents whose warm waters journey into higher latitudes on the western margins of the oceans. Between 25° and 30° latitudes they are further urged on by coupling with strong westerly winds. In high latitudes a return flow of cool waters is initiated along the eastern sides of the oceans. The resulting large-scale motions are termed *subtropical gyres*. The rotary anticyclonic motion of the

wind shear causes Ekman transport of surface waters toward the centers of the gyres (Fig. 6.16), resulting in an upward slope of the ocean surface toward their centers.

An important consequence of wind-driven Ekman transport is the phenomenon of *upwelling*. Convergent winds or coast-parallel winds cause surface water flow divergence away from the line of wind convergence or coastline respectively (Fig. 6.16). This forcing away of surface waters is necessarily accompanied by upwelling of deeper waters to take their place. Upwelling is particularly important, for example, off the coasts of Peru, northwest and southwest Africa, Galicia, and California. Upwelling waters bring with them nutrients such as phosphorus and nitrogen, which cause a greatly increased plankton biomass. Along the intertropical zone of convergent trade winds, divergent transport poleward (at about $0.1 \, \text{ms}^{-1}$) by the westward-flowing equatorial currents causes steady equatorial upwelling of about 1 m per day. The north and south equatorial currents are bisected by a prominent narrow eastward-flowing surface counter-current and a spectacular shallow subsurface counter-counter-current toward the west at up to $1 \, \text{ms}^{-1}$. Shear interactions in these near-equatorial zones cause powerful convergences and divergences, upwelling of cold nutrient-rich waters, and marked eddy mixing.

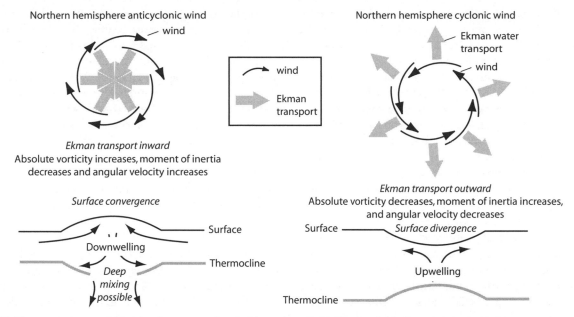

Fig. 6.16 Ekman transport and vorticity changes associated with northern hemisphere anticyclonic and cyclonic winds.

6.2.3 Atmospheric boundary layers and heat exchange: General

The conventional view of the Oceanic Boundary Layer (OBL) thermal reservoir is that it gives up its thermal energy in a one-way heat transfer to the overlying ABL, mostly in the form of latent heat of evaporation. The evaporated seawater in air thus carries most of the transferred heat energy, linking the ocean thermal system directly with the atmosphere. We may apply this concept of an ocean–atmosphere heat engine between the limits of the ocean surface and the tropopause. We have noted previously (Section 6.1) that there exists strong evidence that warming of the Indo-Pacific oceans might be responsible, through a teleclimatic forcing connection, for recent decadal change in the *NAM* and *SAM*. At the same time we must also stress the role of atmospheric flow and forced convection on cooling ocean water in polar latitudes when cold polar air jets chilled by passage over snow or ice or cooled by descent then pass over the ocean. The resulting loss of heat from the surface ocean by conductive transfer and forced convection is a major process in the production of unstable surface cold water that sinks to form Arctic and Antarctic deepwater.

6.2.4 The tropical ocean–atmosphere heat engine: Tropical cyclones

Many thousands of tropical thunderstorms are generated each year in the intertropical "heat engine" but only a few undergo extreme development to tropical cyclonic storms, variously called hurricanes or typhoons. For example, between 5 and 10 hurricanes typically develop in the southern North Atlantic each year. Hurricanes are revolving storms (i.e. vortices) of great ferocity (surface windspeeds 33–70 m s^{-1}) sourced over the global tropical oceans. Because of their danger to humans and despite their infrequence, of all meteorological phenomena their correct forecast is probably of the greatest importance.

Cyclones grow from spatially concentrated *seed banks* of cumulonimbus clouds on the downflow margins of the trade wind belts, where sufficient passage has occurred over warm tropical ocean so that saturation vapor pressures are high. The position of the late summer Bermuda High plays an important role in guiding the storm tracks from east to west and hence northwest landwards toward the Caribbean Islands and southeast North America. During strong El Nino Southern Oscillation (ENSO) (El Niño, see Section 6.2.5) years the Bermuda High is forced eastward and storm tracks rarely impact land; vice versa for weak ENSO years (El Niña). Tropical cyclones cannot be generated within about 5° or so of the Equator because here the Coriolis force is insignificantly small to zero and a geostrophic balance cannot be reached. Their energy is derived from latent heat transfer above the very warmest tropical oceans in late summertime and thus any cyclic perturbations to seasonal Sea Surface Temperatures (SSTs), like those associated with ENSO oscillations, have an important influence on the frequency of hurricane genesis. Cyclones grow upward above the very warm ocean water

($T > 26$–$27°C$) into a moist atmosphere (without temperature inversions that might prevent high ascent) and where the generally converging flow of easterly waves in the trade winds causes upward motion. Very low core pressures (~950 mbar) attract adjoining trade winds over a large area, for about 80–90 percent of hurricane motion depends upon flow and pressure conditions in the *steering flow* of the adjacent undisturbed atmosphere. The cyclone's vorticity comes from shallow atmospheric conditions that cause cyclonic (anticlockwise in northern hemisphere) shear enhancement of positive vorticity (Section 3.8) – this is commonly due to convergent flow causing rising air masses to form.

Cyclone morphology is distinctive (Figs 6.17 and 6.18), comprising a mass of high velocity anticlockwise-revolving clouds rimming a clear and relatively still core. Incoming winds are forced to turn anticlockwise by the Coriolis force. The rapidity of this process around the hurricane eye means that a narrow diameter "solid" mass of still air blocks an outer mass of spinning air. In terms of force balance, the pressure gradient from vortex to center is balanced by the inward centripetal acceleration and thus an outward centrifugal force (Section 3.7). Once inside the developing *hurricane vortex* the moist winds are forced to spiral upward, further warming up by the latent heat released from condensing water vapor as they do so. Once at high levels (*c.*12 km) divergent flow occurs outward and downward above the troposphere and this is turned by the Coriolis force to assume a clockwise rotation that accelerates the risen air far outward as cirrus clouds. Thus a vertical energy transfer cycle from ocean surface to tropopause is set in progress.

At this stage the nascent hurricane is sensitive to the temperature difference between its core and the wider surrounding ocean waters; perhaps a difference of just a few degrees celsius or so increase enables the central hurricane eye with intense downdrafts to form. Certainly it is well documented that hurricanes traveling onto warmer ocean current gyres (if only differing by 2–3°C) may undergo rapid pressure intensification ("rapid deepening") and wind acceleration, as observed with Hurricane Opal in October 1995. Rapid deepening is believed to be the mechanism whereby an average hurricane is transformed into a very dangerous storm: it signifies the very great importance of heat interchange between surface ocean and atmosphere under such conditions. Successful hurricane prediction models treat the phenomenon thermodynamically, with the hurricane as a *heat engine* running between the warm ocean and the cool troposphere, with a ΔT of some 100°C. However, corrections must be made for the effect of increased hurricane winds causing surface ocean layer turbulent mixing and therefore cooling due to momentum and heat transport in the dynamic models.

In addition to their obvious role in increasing wind shear and forward momentum transport in associated waves, tropical cyclones have two major effects on the ocean itself. First, their cyclonic flow sets up a divergence of water (Fig. 6.16) in the top 100 m or so of the oceanic surface boundary layer (Section 6.4). This leads to upwelling and mixing of cooler waters over a large area in the track of the storm (an effective "fingerprint" when seen using thermal imagery). Second, the extremely low pressures associated with the center of the cyclone cause a rise in local mean sea level independent of any wind shear effects or state of the local tidal wave. This effect is termed a *storm surge*, with a possible rise in sea level of over 5 m, in the most intense storms.

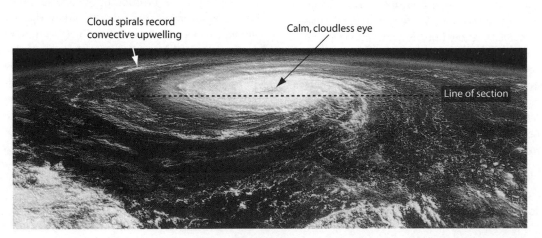

Fig. 6.17 Tropical cyclone from space.

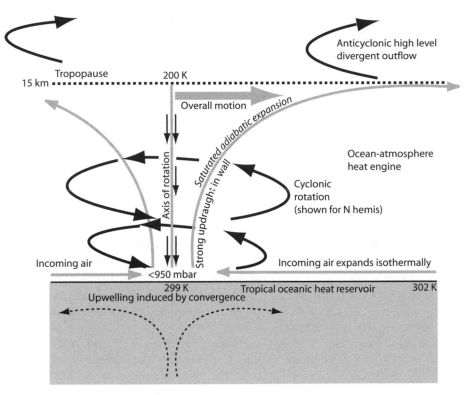

Fig. 6.18 Schematic section through tropical cyclone.

6.2.5 The tropical ocean–atmosphere heat engine: El Niño Southern Oscillation as a teleconnector

The El Niño Southern Oscillation is a periodic aspect of the tropical ocean–atmosphere heat engine that initially affects the equatorial Pacific Ocean but once in action seems to have global teleconnections. In fact, early research into the phenomenon took place because of the disastrous failure of the Asian monsoon in 1877 and subsequent periods of high mortalities from flooding, landslides, fishing failures, and droughts in the wider Indo-Pacific equatorial region. ENSO events occur every 3–5 years and are characterized by warmer than average (by up to 2°C) SST in the south equatorial east Pacific that peak around Christmas, hence the name El Niño (the little boy). During the peak of the southern hemisphere summer the warm SSTs in the east Pacific off central South America are accompanied by weakened southeast trade winds as a consequence of the easterly migration and weakening of the Indonesian tropical low-pressure center and weakening of the South Pacific High. The periodic swings in the magnitude of these pressure centers are the "oscillation" inherent to the ENSO

phenomenon. The El Niño warm-phase is characterized by easterly spread of the Indonesian equatorial "hot pool" waters (SSTs > 28°C), abnormally wet weather in the equatorial Andes, depression of the eastern Pacific Ocean thermocline, and cessation of the upwelling that usually characterizes the Pacific coast of Peru. Between many El Niño events there occur anomalously cool (usually <1.2°C) sea surface conditions, termed La Niña cold-phase events. Dynamically, the El Niño part of the ENSO leads to collapse of the elevated west Pacific waters previously kept higher (by up to 2 m) than the eastern Pacific by strong wind shear from the southeast trades. The elevated warm water migrates east rapidly as a Kelvin wave (with no Coriolis turning at these equatorial latitudes), depressing the thermocline as it does so. Once it reaches the South American shelf the wave spreads laterally north and south along the coast and partly reflects back as a Rossby wave of deformation. The teleconnection involves simultaneous migration and modulation of other major pressure centers and attendant weather systems, including the winter North American High Pressure and summer Bermuda High Pressure cells, the latter having a strong influence as a "path guide" for tropical cyclones (Section 6.2.4).

6.3   Atmosphere–land interface

As with the ocean surface, the interaction between the ABL and the land surface involves the transfer of momentum, mass, and heat. Mass transfer of water vapor and photosynthetic gases arises as the ABL interacts with surface water in lakes, soil substrates, and vegetation canopies. Heat transfer from land surfaces to the ABL occurs via reradiated long-wave thermal energy and may be viewed as a type of forced convection. In the summary in Sections 6.3.1 and 6.3.2, we focus on sediment mass transport and associated exchanges of momentum.

6.3.1   Peculiarities of the ABL over land

In detail, the flow behavior and characteristics of the local or regional ABL over land is more variable than that over oceans in particular because of:

• the existence of commoner stable seasonal atmospheric stratification or inversions, which cause cooler and denser air masses to "buffer" the transmission of momentum from any overlying gradient wind, leading to often temporary "calm" conditions and the undesirable affects of urban pollution

• local flow due to motions arising from differential land surface heating and cooling as typified by land and sea breezes, by the generation of thermals over deserts and other hot plains, and by the outflow of cold dense *katabatic* winds from ice, mountains, and plateaux downslope into surrounding plains and ocean

• topographic enhancement of local or synoptic wind speed causing acceleration or decelerations as land surfaces intersect higher up or lower down the gradient wind column

• formation of lee-side effects like the well-known undulating waves of cloud formed downwind from linear mountain ranges and the tip jets characteristic of mountainous capes bordering oceans, for example, Cape Farewell, S. Greenland etc.

• downslope flow of dry air driven by a pressure reduction on the lee side of mountains, the flow accompanied by adiabatic warming at the dry air lapse rate – the *foehn* effect

6.3.2   Synoptic aeolian sand and dust transport in the atmospheric boundary layer

In the great trade wind sand seas (*ergs*) of the Sahara and central Australia, there is a close correspondence between dominant wind flow and sand transport. Meteorological observations, bedform orientations, and the trend of erosional lineations (*yardangs*) enable sand-flow distributions to be mapped out regionally (Fig. 6.19). Ideally, a *sand-flow map* should show resultant directions as flowlines and resultant magnitudes as contours, analogous to a combined wind direction and pressure map. Sand-flow maps are also analogous to drainage maps in that they show divides separating distinct "drainage" basins: peaks in fixed high-pressure areas and saddles in between them. Unlike water drainage, there is little direct relation between sand flow and topography since winds and their sandy bedload may blow uphill. For example, the sand-flow lines for North Africa extend from erg to erg, implying very long transport distances downwind. Evidence for this is provided by satellite images showing yardangs linking ergs along sand-flow lines. All the sand-flow lines arise within the desert itself, with the main clockwise circulatory cell roughly corresponding to the subtropical high-pressure zone.

Notice from Figs 6.19 and 6.20 that the Saharan sediment-flow lines eventually lead to the ocean, coincident with the great plume of Saharan dust that extends out for thousands of kilometers into the Atlantic Ocean. A significant part reaches as far as the Lesser Antilles, providing a steady flux of fine silt particles to the deep ocean. Such mineral dust is a potent light-scattering aerosol and has an important cooling role to play in climate change as well as a source of nutrients (chiefly labile Fe^{2+}) to the oceans. Current estimates of total global dust flux to the atmosphere vary between 1 and $3 \cdot 10^9$ ton yr^{-1}. It has been estimated that the total flux of Saharan mineral dust is some $2.6 \cdot 10^8$ ton yr^{-1}, the magnitude having varied throughout the Quaternary in response to cycles of climate change. The evidence for this comes from deep-sea cores, which enable dust deposition rates to be calculated. Present-day fluxes are low compared to those at the last glacial maximum. A worrying recent trend has seen marked increases in levels of wind-blown dust in the troposphere worldwide, particularly in winds sourced from the expanding deserts and dried-up lake basins of northern China and Mongolia. It is not known exactly whether it is recent climate change to aridification or the poor maintenance of land use that is the cause of this trend.

In mountainous terrains and over the high latitude ice sheets, storm-force winds play a major role in the transport of particulate ice and snow, driving it into major accumulations. As with the transport of sand grains in deserts, the transporting particles exert an important feedback as flow and solid momentum is exchanged.

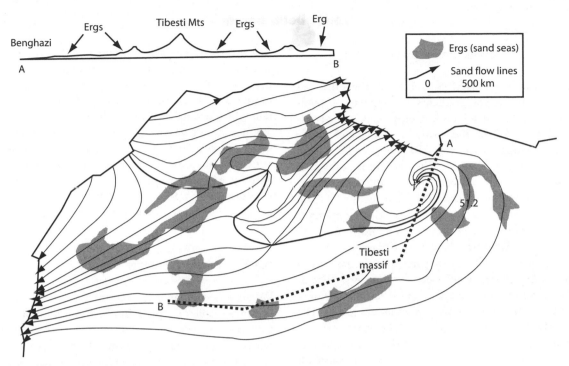

Fig. 6.19 Regional sand-flow lines reflecting the yearly mean trade winds, Sahara desert.

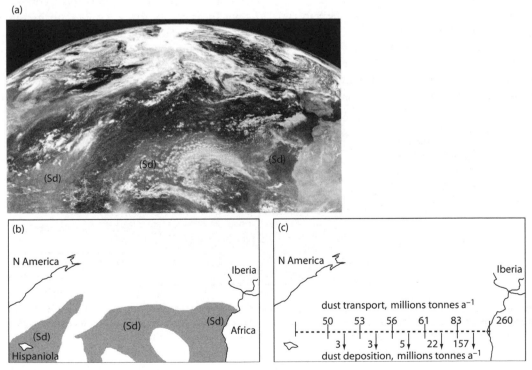

Fig. 6.20 (a) Satellite image showing intermingling of Saharan dust (Sd) in the ABL with normal cloud and frontal systems in the central Atlantic ocean; (b) shows extent of dust in sketch form; (c) shows estimated annual transport and depositional fluxes and their variation with distance from the Saharan source, based on studies of Atlantic ocean floor sediment cores. Such long distance transport has important repurcussions for reflectivity of incoming short-wave radiation, rain droplet nuleation, cloud formation, and precipitation.

6.4 Deep ocean

6.4.1 General: Oceanic boundary layer

Below a shallow well-mixed surface layer, the OBL has strong vertical and lateral gradients of temperature, salinity, velocity, and turbulent kinetic energy. As discussed previously (Section 6.2.3) the OBL is a locus of active interaction between the moving ABL above and the main mass of ocean waters beneath. The occurrence of a relatively slow-moving (in comparison with the ABL) large-scale surface circulation of the OBL, first recognized and systematically recorded as a practical exercise by mariners over hundreds of years, has become well known in detail over the past 20 years since the advent of precise geodetic observations from satellites. In particular, it has drawn attention to the existence of oceanic "weather," that is, regional variations in surface flow vectors, defining unsteady and nonuniform flows, in *mesoscale eddies*. The OBL may also have a surprisingly deep signature.

The major forces available to drive OBL circulation are:
1 *Lateral gradients in hydrostatic pressure*. Newtonian fluids cannot resist lateral pressure gradients, geostrophic or gradient flow results, the latter influenced by centrifugal effects as seen in the flow of water round bends (Section 3.7).
2 *Interplanetary gravity forces*. Chiefly from the Sun and the Moon, they cause the production of long-period oceanic waves responsible for tides. These are generally of very low surface amplitude in the open oceans but are considerably amplified on shelves and around coastlines. We consider these further in a subsequent section.

3 *Atmospheric pressure gradients*. Under intense cyclonic storms (Section 6.2) these drive strong local currents into the lowered pressure zone at the storm centers where mean sea level may be up to several decimeters higher than that which normal tide-producing forces would produce.

Once in motion three major modifying forces arise:
• *Coriolis force* due to planetary vorticity (Section 3.8).
• *Lateral boundary friction force* due to boundary layer current shear and momentum flux (Sections 4.5 and 6.2).
• *Internal waves* (Section 4.10) break under critical vertical gradients in imposed shear thereby creating turbulence.

6.4.2 Gradient currents

We have seen that in a well-mixed surface ocean layer, pressure gradients, dp/ds, can be brought about purely by the action of surface forces, chiefly *wind shear*. In cases where temperature and density do not vary laterally the subsurface isobars are parallel to the sea surface; this is the *barotropic condition* (Fig. 6.21). However, the existence of *thermohaline gradients* in seawater density caused by differential means that the barotropic condition may not apply. Instead, subsurface *isopycnal surfaces* are not parallel to the sea surface and the isobars become increasingly divergent from the overlying sea surface slope with depth. This is the *baroclinic condition* (Fig. 6.22). In the open oceans, away from frictional influences at solid boundaries (see below), the resulting flow in both baroclinic and

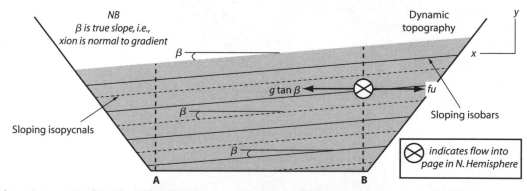

Hydrostatic pressures above **B** > hydrostatic pressures at all equivalent heights above **A**, by a constant gradient given by the water surface slope, tan β. The pressure gradient, dp/dx is ρg tan β per unit volume, or g tan β per unit mass. The pressure gradient and hence magnitude of the gradient current are equal at all depths

In the absence of friction the horizontal pressure force is balanced by the Coriolis force so that
$$g \tan \beta = fu$$
or
$$u = (g/f) \tan \beta$$

Fig.6.21 Barotropic conditions: isobaric and isopycnal surfaces are parallel in well-mixed water bodies (assuming atmospheric pressure is equal across line of section).

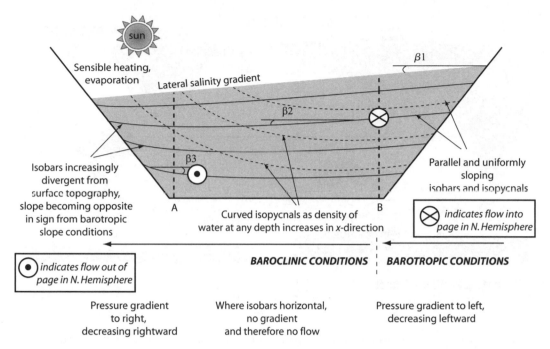

Fig. 6.22 Baroclinic conditions: isobaric and isopycnal surfaces are NOT parallel in water bodies with laterally varying density. Illustrated here by a salinity gradient caused by solar evaporation similar to the situation in the eastern Mediterranean.

barotropic cases is balanced by the local Coriolis force to define a geostrophic flow of magnitude $u = (g/f)\tan\theta$ along a path parallel to the isobars.

Large-scale oceanic circulation in the subtropical gyres is due to pressure-driven flow outward from the centers of ocean surface topography caused by Ekman transport (Section 6.2.2). Surface flow is then set up, initially as gradient currents that run down the surface slope and then is turned by the action of the Coriolis force to run parallel to the dynamic sea surface topography and to the regional ABL flow (Fig. 6.23). Like their atmospheric equivalents (Section 6.1) such surface currents are also geostrophic.

6.4.3 Western amplification of geostrophic currents

The broad pattern of global surface ocean topography and currents is shown in Figs 6.24 and 6.25. The circumpolar *Antarctic Current* is strong because of the extreme degree of wind forcing in response to great lateral pressure gradients at these high polar latitudes (Section 6.1). But how can we explain the intensification of surface flow on the western borders of the oceans, manifest in strong western boundary currents such as the North Atlantic *Gulf Stream* and Pacific *Kuroshio*? It is common to measure speeds of over $1\,\mathrm{m\,s^{-1}}$ ($3.6\,\mathrm{km\,h^{-1}}$) in these currents. The Gulf Stream, for example, is usually a continuous,

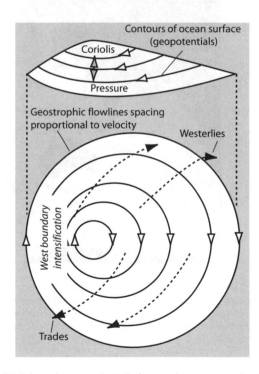

Fig. 6.23 Balance between Coriolis force and pressure gradient causing geostrophic flow around an ocean "bump."

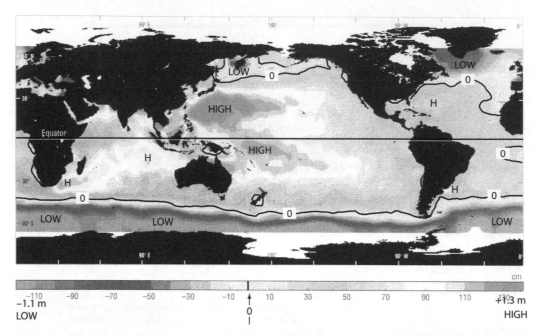

Fig. 6.24 The remarkable satellite-measured topography of the mean sea surface (with wave and tidal wave effects subtracted).

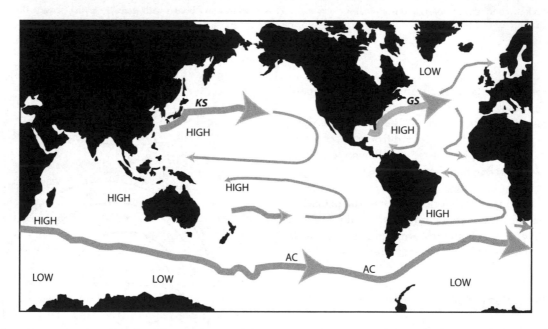

Fig. 6.25 The major pattern of gradient flow from the computed dynamic sea surface. Note the control of current vectors (both magnitude and direction) by the magnitude of the spatial gradients in water topography, that is, OBL flow is parallel to the gradient lines, with an intensity proportional to grayscale thickness. Note western intensification of Pacific Kuroshio (KS) and Atlantic Gulf Stream (GS) currents and the strong circumpolar Antarctic current (AC).

though complex, meandering filament of warm Caribbean water in transit to the shores of northwest Europe (Figs 6.26 and 6.27). In the mid-twentieth century, Stommel explained these most striking features of the general oceanic circulation by a consideration of both lateral friction and conservation of angular momentum.

We have seen that all moving fluid masses possess vorticity appropriate to the latitude in which they find themselves (Section 3.8) and that the total, or absolute, vorticity $(f + \zeta)$ must be conserved. Thus a northward-moving mass of water, impelled by wind shear to spin clockwise, will gain planetary vorticity as it moves. In order to keep

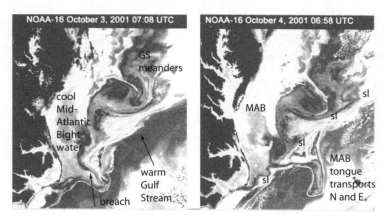

Fig. 6.26 The Gulf Stream is usually a continuous, though complex, meandering filament of warm Caribbean water in transit to the shores of northwestern Europe. In these satellite images an unusually strong north wind has driven cool waters from the Mid Atlantic Bight across the track of the Gulf Stream, breaching it as a cool tongue that is eventually itself transported north and east in the main current. Main northern margin to Gulf Stream is a boundary shear layer (sl).

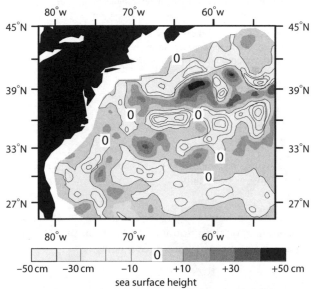

sea surface height
(1 m of topography over a typical eddy length of c. 250 km
gives a mean slope of 1:250,000: note the asymetric slopes caused by
radial flow around the meandering Stream)

Fig. 6.27 Map of northwest Atlantic sea surface topography as measured by remote sensing from altimetric satellite Jason-1. The map shows strong topographic features (mesoscale eddies) associated with meanders of the surface Gulf Stream current. Geostrophic theory (Fig. 6.5) says that flow should parallel the topography, defining in this case the sinuous flow around a compex series of warm and cold core eddies.

absolute vorticity constant it must therefore lose relative vorticity. As the *major* part of the flow away from the ocean bottom boundary layer is deemed frictionless the external flow lags rotation of Earth and therefore loses positive relative vorticity, that is, gains negative relative vorticity. In other words the flow rotates clockwise (i.e. to the right) in

the northern hemisphere and anticlockwise (to the left) in the southern hemisphere.

Let us apply these simple notions of conservation of angular momentum to real-world oceanic gyres by a *vorticity balance*, taking into account the action of wind shear, the change of f with latitude, and the effects of boundary layer friction at the ocean edges. The simplest physical model for a symmetrical wind-driven gyre would be in 2D and have westerlies and trades blowing opposite in a clockwise circulation, both declining to zero at the horse latitudes (Fig. 6.28). One can see immediately that the wind velocity gradients will cause a clockwise angular velocity of rotation (i.e. addition of negative vorticity to the water) and that the magnitude of the pressure gradients due to Ekman transport will determine the strength of the resulting water flow. We must also take into account the linear rate of change of the planetary vorticity, f, with latitude, as this also determines the transport vector. Finally, since we are concerned with solving the problem of western intensification against the solid boundary of the continental rise, we recognize that the sense of boundary layer friction will cause the addition of positive vorticity on both western and eastern boundaries. The combined effect of wind and f on the western side enhances the negative vorticity. On eastern margins the two effects roughly cancel out. For the western current to remain steady and in balance the frictional addition of positive vorticity must be made more intense. This can only be done by increasing the current velocity, since the braking action provided by boundary layer friction is proportional to velocity squared. The warm western currents are thus extremely strong, up to ten times the strength of the cool eastern currents.

It should not be thought that strong western boundary currents have no effect at oceanic depths. Direct current

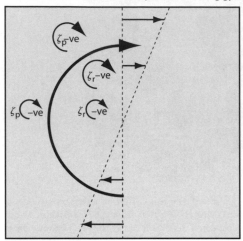

Western half of northern hemisphere circulating gyre

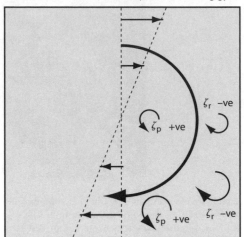

Eastern half of northern hemisphere circulating gyre

W-side story: *f* increases N and so ζ_p more negative N
ζ_r from wind stress is negative
Overall on this westward leg a net decrease of relative
vorticity $(-\zeta_p - \zeta_r < 0)$

E-side story: *f* decreases S and so ζ_p more positive S
ζ_r from wind stress is still negative
Overall on this eastward leg a net balance of relative
vorticity $(+\zeta_p - \zeta_r \sim 0)$

Overall, across the whole circuit (west and east combined) there is net loss of vorticity. This is not allowed
because the total vorticity must be kept constant. Extra relative vorticity must be generated by either
pronounced western lateral boundary shear or by western bottom shear, or a combination of both.
The eastern flow needs no such enhancement and is thus weaker and more spatially uniform.

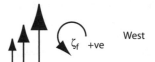 West

 East

Fig. 6.28 Sketches to show that conservation of vorticity requires western boundary currents to be stronger than eastern ones. ζ_p is planetary
vorticity (or *f*), ζ_r is relative vorticity due to wind shear, and ζ_f is relative vorticity due to lateral friction.

measurements and bottom scour features indicate that
strong vortex motions are sometimes able to propagate turbulent energy all the way (i.e. >4 km) down to the ocean
floor, where they cause unsteadiness in the deep thermohaline current flow (see Section 6.4.5; so-called *deep-sea
storms*), enhanced resuspension of bottom sediment, and
nutrient mixing. Also, the currents are unsteady with time,
both on the longer time scale, for example, major erosive
events on the Blake Plateau have been attributed to Gulf
Stream flow during glacial epochs when the current was
thought to be at its strongest, and on a subyearly basis as
spectacular eddy motions, meanders and cutoffs of cooler
waters form *cold-core mesoscale eddies* (Figs 6.26 and 6.27).
Notions that the Gulf Stream circulation might "fail" due to
global warming and a shutoff in the deep circulation (see
below) are erroneous: in the words of one oceanographer,
"As long as the wind blows and the Earth turns then the
surface current will exist." The one thing that will change is
the junction between the warm surface current and the cold
southerly flows from the Arctic Ocean along the Polar

Front: this is known to shift zonally by large amounts
depending upon the amount of cold but buoyant freshwater
issuing out of the Arctic from ice melting.

6.4.4 Internal waves and overturning: "Mixing with latitude"

Internal waves (Section 4.10) of much longer period than
normal wind-driven surface waves have recently been discovered to be a major source of turbulent mixing in the
deep oceans. The internal wave field arises due to wave-like
disturbances of the density stratification that occurs at various depths, but particularly within the deep-ocean water
column. The disturbances or forcing occurs due to:
1 *Internal tides* formed when the main ocean tidal currents
flow over rough sea-floor topography and act upon the internal stratification to form tidal period internal waves.
2 A response of the stratification to *inertial surface waves*
piled up by wind shear during storms, the internal waves

have periods relating to the Coriolis force and thus are a strong function of increasing latitude.

In both cases it is the property of vertical propagation of the internal waves that makes them so effective in spreading momentum; unlike surface ocean waves which only propagate horizontally. The internal waves cause vertical internal shear as $(du/dz)^2$ along their wavy interface (cf. Prandtl's mixing layer theory for turbulent shear flows; Cookie 12) and it is postulated that such shear zones act as in any turbulent boundary layer to transfer turbulent kinetic energy to shorter period eddies as the waves progressively break up. The mixing process is much more effective at higher rates of shear and thus the resultant mixing is more efficacious at higher latitudes where the Coriolis force, f, is greatest.

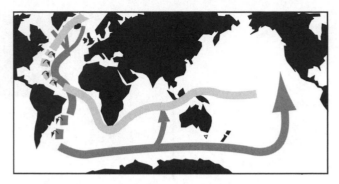

Fig. 6.29 The general ocean bottom (darker shading) and surface return legs of the global thermohaline system. Both surface and deep currents show periodic breakup into spectacular rotating warm-core eddies, shown here for the surface north Brazilian and Gulf Stream currents and the deep thermohaline North Atlantic Deepwater in the South Atlantic.

6.4.5 Benthic oceanic boundary layer: Deep ocean currents and circulation

We have seen that motion of the upper ocean reflects momentum exchange across the atmosphere–ocean interface as modified by vorticity gradients from equator to pole. But what of the deeper ocean? We still know very little of the benthic oceanic boundary layer, as problems of logistics and instrumentation have prevented progress in the area until quite recently. Radioisotope tracers indicate that all deep waters must reestablish contact with the atmosphere on a 500 year timescale. This requires a system of circulation that allows such links. In the last 40 years, theoretical results and detailed temperature, density, and isotopic studies worldwide have revealed a system of deep (1500–4000 m), dense currents (Fig. 6.29), termed *thermohaline currents* from the dual role that temperature and salinity have in producing them. Thus at low latitudes the upper ocean is heated by solar radiation (density decreases), but also loses water by evaporation (density increases). At high latitudes the upper ocean is cooled by contact with a very cold lower atmosphere during winter (density increases), but freshened by precipitation, river runoff, and inflows of polar glacial meltwater (density decreases). At the same time the production of sea ice leads to saltier residual seawater (density increases). Thermohaline circulation can thus have several causes, most varying seasonally, favored by destabilizing processes that lead to density inversions due to increased surface water density and the production of negative buoyancy. There is also a vital role played in cold water formation by atmospheric wind forcing and Ekman suction/pumping (Section 6.2), chiefly by regional gyres of high vorticity like the Irminger Sea tip jet to the east of Greenland

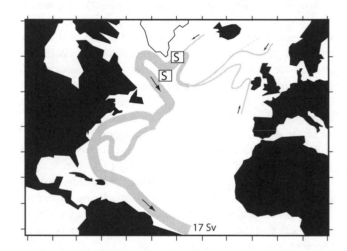

Fig. 6.30 Cold water sources and generalized flow of North Atlantic deepwater ($T = 1.8$–4°C). S – major sources of downwelling in the Labrador and Greeenland seas, the latter due to wind shear by the Irminger tip jet.

(Fig. 6.30), and by more local shear producing mixing gyres, as in the *mistral* wind in the West Mediterranean and the *bora* of the Adriatic.

Thermohaline currents are linked to compensatory intermediate and shallow warmer currents in a complicated pattern of downwelling and upwelling, whose detailed paths in the Pacific and Indian Oceans are still uncertain. The amount of water discharged by the currents is staggering, one estimate for deepwater being some $5 \cdot 10^7$ m^3 s^{-1} (50 *Sv* [*Sverdrup units*: each 10^6 m^3 s^{-1}]). This is about 50 times the flow of the world's rivers; about half of the total ocean volume is sourced from the cooled

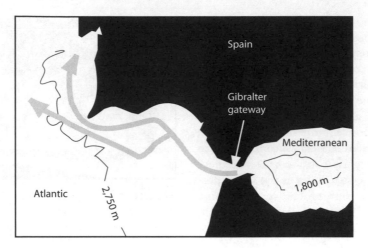

Fig. 6.31 Deep outflow of dense Mediterranean water through the Gibralter gateway.

sinking waters of the polar oceans (Fig. 6.30). The nature of the oceanic circulation, with its links from surface to depth, and its role in heat transport and redistribution, has led to its description as a *global conveyor belt* of both heat and kinetic energy. The consequences of this deep circulation are profound, since steady current velocities of up to 0.25 m s^{-1} have been recorded in some areas where the normally slow ($c.0.05 \text{ m s}^{-1}$) thermohaline currents are accelerated on the western sides of oceans (for the same vorticity reasons as discussed earlier for surface currents) and in *topographic constrictions* like gaps between mid-ocean ridges, oceanic fracture zones, and oceanic island chains and plateaux margins. In all these case *turbulent mixing* is accentuated due to the rough topography, a phenomenon that occurs at all scales from laboratory flows (Section 4.5) to the oceans.

Dense water masses from the Antarctic and Arctic seas sink to become the Antarctic Bottom Water (ABW) and North Atlantic Deep Water (NADW); total discharge in the range of 10–40 Sv respectively. ABW forms the majority of the bottom flow around the Antarctic as a circumpolar current, receiving NADW from the western South Atlantic in a series of huge migrating warm-core eddies and in turn leaking large discharges northward from the Weddell sea and other sources into the South Atlantic (under and alongside the NADW), Indian and Pacific Oceans. Intra-ocean transfers occur in the winter as evaporative fluxes from the Mediterranean to the Atlantic and from the Red Sea/Arabian Gulf to the Indian Ocean. The Mediterranean example is a classic case of flow forced to intensify through the narrow constriction at the Straits of Gibraltar (Fig. 6.31), at velocities that exceed 3 m s^{-1}, then decelerating out into the Gulf of Cadiz, but is still as

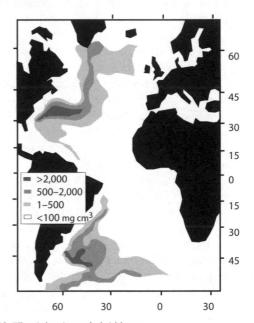

Fig. 6.32 The Atlantic nepheloid layer.

high as 0.2 m s^{-1} at Cape St Vincent. The Mediterranean Outflow Water (MOW) is warm (13°C) and saline ($>37 \text{ g l}^{-1}$) and spreads out to mid-depth (800–1200 m) in the North Atlantic. The MOW is compensated by an inflow of Atlantic water: the combined circulations being described as *anti-estuarine*, that is, salty dense water out and fresher less-dense water in.

Recent results also confirm earlier observations that there is significant flux of deepwater through fracture zones across and along mid-ocean ridges. Thus transfer of ABW from the western to the eastern side of the South Atlantic occurs through the larger silled fracture zones of the mid-Atlantic Ridge, with intense turbulent mixing along the upper interface. Tracer studies at the interface of other shallow water masses reveal a low value of the mixing rate, about 10^{-5} m^{-2} s^{-1}. This implies a low rate of turbulent mixing along density interfaces relative to lateral spread, a conclusion also established by turbulent stress calculations. However, it is likely that other mixing mechanisms exist, for example breaking internal waves generated during ocean tides, which will lead to much larger turbulent dissipation.

A feature of deep ocean waters is attributed in part to the action of thermohaline currents and in part to the occurrence of deep-sea storms (see discussion in Section 6.4.5). This is the phenomenon of increased suspended material, revealed by light-scattering techniques (Fig. 6.32). The source of the suspended sediment in these bottom *nepheloid layers* is variable: distant sourcing from polar regions, local erosional resuspension of ocean-floor muds by "storms" and enhanced thermohaline currents, windblown dust, and dilute distal turbidity current flows probably all have a role. Some nepheloid layers may be up to 2 km thick, although 100–200 m is a more usual figure. Sediment in nepheloid layers is usually <2 μm in size although fine silt up to 12 μm may be suspended, normally at concentrations of up to 500 mgl^{-1} rising to 5000 mg l^{-1} a few meters off the bottom during deep-sea "storms." Nepheloid layers are also known in many areas from intermediate depths, often at the junction between different water masses. These are thought to arise through the erosion of bottom sediments by internal waves (Section 4.13) and tides, amplified on certain critical bottom slopes. The layers, once formed, intrude laterally into the adjacent open ocean as layers many tens of meters thick.

6.5 Shallow ocean

Shallow (<200 m depth) ocean dynamics (Fig. 6.33) are more complicated than the open ocean both because of the effects of the shallow water on wave and tide and proximity to land. A generalized physical description of the shelf boundary layer (Fig. 6.34) defines an inner shelf mixed layer where frictional effects of wave and tide are dominant in the less than 60 m shallow waters. In the deepening mid- to outer shelf there is differentiation into surface and bottom boundary layers separated by a "core" zone. The shallow water enables waves to directly influence the bottom and for the longer-period tidal wave to amplify as it is forced shelfward from the open ocean. Proximity to land causes interactions of wave and tide with effluent plumes sourced from river estuaries and delta distributaries (Fig. 6.35). Coastal geometry also has a strong local influence upon water dynamics. Shelves have been classified into tide- and weather-dominated, but most shelves show a mixture of processes over both time and space. The majority of shelves have a tidal range less than 2 m but this may be amplified several times around their margins.

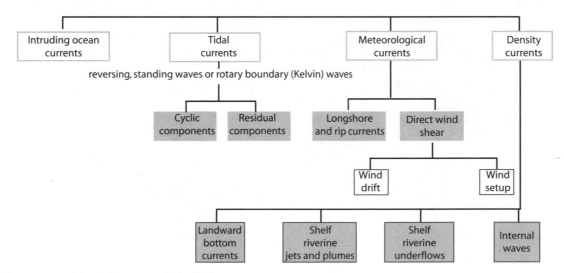

Fig. 6.33 Components of the shelf current velocity field.

6.5.1 Shelf tides

In the oceans the twice-daily tidal wavelength, λ, is very large (about 10^4 km) compared with water depth, h (say 5 km), and is thus still of shallow-water (long-wave) type (i.e. $h/\lambda \ll 0.1$). From Section 4.9 the maximum tidal wave velocity in the open oceans is thus given approximately by $u = (gh)^{0.5}$, about 220 m s^{-1}. The open ocean tidal wave decelerates as it crosses the shallowing waters of the shelf edge. This causes wave refraction of obliquely incident waves into parallelism with the shelf break and partial reflection of normally incident waves. At the same time the wave amplitude, a, of the transmitted tidal wave is enhanced. This follows from the energy equation for gravity waves $E = 0.5\rho g a^2 (gh)^{0.5}$ (Section 4.9); the supremacy of the square versus the square root terms means that the overall wave amplitude must increase. The tidal current velocity of a water particle (as distinct from the tidal wavelength) also increases because this depends upon the instantaneous amplitude of the wave.

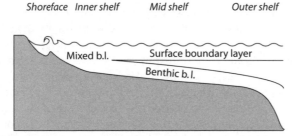

Shoreface Inner shelf   Mid shelf   Outer shelf

Mixed b.l.   Surface boundary layer
                     Benthic b. l.

Fig. 6.34 Simple division of shelf waters into mixed, surface, and bottom boundary layers. Inner shelf mixed b.l. has tide and wave mixing, though the degree of mixing is seasonally variable. Outer shelf is often stratified into a surface b.l. with geostrophic flows and a friction-dominated benthic boundary layer.

Tidal strength may also vary because of the nature of the connection between the shelf or sea and the open ocean. In the case of the Mediterranean Sea, for example, the connection with the Atlantic has become so narrow and restricted that the Atlantic tide cannot reach any significant range over most of its area. Locally, in the Straits of Gibraltar, the Straits of Messina, and the Venetian Adriatic, for example, the tidal currents (but not necessarily the tidal range) may be greatly amplified when water levels between unrelated tidal gyres or standing waves interrelate.

Another cause of spatially varying tidal strength is the *resonant* effect (Section 4.9) of the shelf acting upon the open oceanic tide (Fig. 6.36) and creating *standing waves*. Resonance greatly increases the oceanic tidal range in nearshore environments and leads to the establishment of very strong tidal currents. Most shelves are too narrow and deep (Fig. 6.36) to show significant resonance across them, that is, $L < 0.25\lambda$. In most cases, for example in the shelf of the eastern USA, a simple slow linear increase of tidal amplitude and currents occurs across the shelf. Open coastal basins like estuaries, bays, and lagoons must receive the 12-hourly oceanic tidal wave and a standing wave (of period 12 h) may be set up, with a node at the mouth and an antinode at the end (by no means the only resonant possibility). In the limiting scenario, with $L = 0.25\lambda$, we have $T = 4L/\sqrt{gh}$. The Bay of Fundy, Maritime Canada, is the world's most spectacular example of a gulf that resonates with the *c*.12 h period of the semidiurnal ocean tide. The gulf has a length of about 270 km (calculated from the gulf head to the major change of slope at the shelf edge) and is about 70 m deep on average, giving the required approximately 12 h characteristic resonant period. The standing resonant oscillation has a node at its entrance, which causes the tidal range to increase from

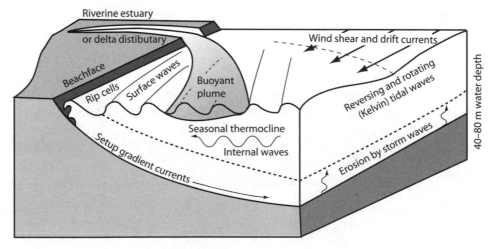

Fig. 6.35 Major controls on cross-shelf water and sediment transport.

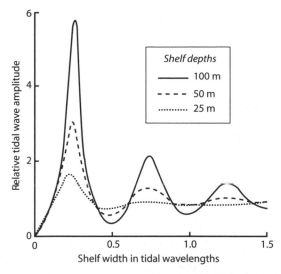

Fig. 6.36 Tidal wave resonance across shelves of different width and water depth.

3 m to a spring maximum of some 15.6 m along its length to the antinode.

The Coriolis force acts as a moderating influence on tidal streams in semi-enclosed large shelves, like the north-western European shelf, the Yellow Sea, and the Gulf of St. Lawrence. In the former, the progressive anticlockwise tidal wave of the North Atlantic enters first into the Irish Sea and the English Channel and then several hours later it veers down into the North Sea proper through the Norway–Shetland gap in a great anticlockwise *rotary wave* (whose passage north to south was noted by the monk Bede in the eighth century). Why should such rotary motions occur? The answer is that the tidal gravity wave, unlike normal surface gravity waves due to wind shear or swell (Section 4.9), has a sufficiently long period that it must be deflected by the Coriolis force. Since the water on continental shelf embayments like the North Sea is bounded by solid coastlines, often on two or three sides, the deflected tide rotates against the sides (Figs 6.37 and 6.38) as a *boundary wave*. Such waves of rotation against solid boundaries are termed *Kelvin waves*, the propagating wave being forced against the solid boundaries by the effects of the Coriolis parameter, f. The water builds up as a wave whose radial slope exerts a pressure gradient that exactly balances the Coriolis effect at equilibrium (Fig. 6.39). Tidal currents due to the wave are coast parallel at the coast (Fig. 6.40a) with velocities at maximum in the crest or trough (reverse) and minimum at the half-wave height. The wave decays in height exponentially seaward toward an *amphidromic node* of zero displacement. The resonant period in the North Sea is around 40 h, a figure large enough to support three multinodal standing waves (Fig. 6.41). The crest of the tidal Kelvin wave is a

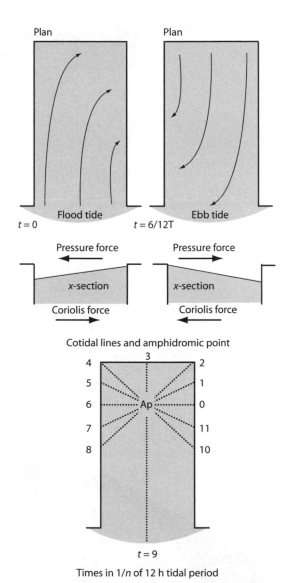

Fig. 6.37 The development of amphidromic circulation within a partly enclosed shelf sea by Coriolis turning of the tidal wave into a Kelvin wave of circulation.

radius of the roughly circular basin and is also a *cotidal line* along which tidal minima and maxima coincide. Concentric circles drawn about the node are lines of equal tidal displacement. Tidal range is thus increased outward from the amphidromic node by the rotary action. Further resonant and funnelling amplification may of course take place at the coastline, particularly in estuaries (see Section 6.6.3). Not all basins can develop a rotary tidal wave: there must be sufficient width, since the wave decays away exponentially with distance. The critical width is termed the *Rossby radius of deformation*, R, given by the ratio of the velocity of a shallow-water wave to the magnitude of the Coriolis parameter, that is, $R = \sqrt{gh}/f$. At

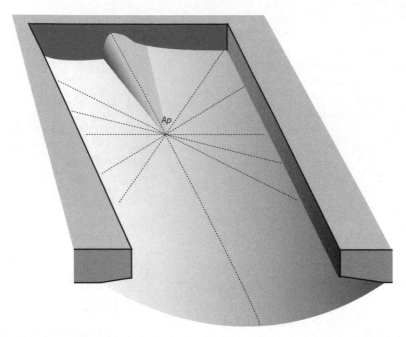

Fig. 6.38 The Kelvin rotating tidal wave travels anticlockwise in the northern hemisphere, decreasing in amplitude inward toward the amphodromic point, *Ap*, of zero displacement.

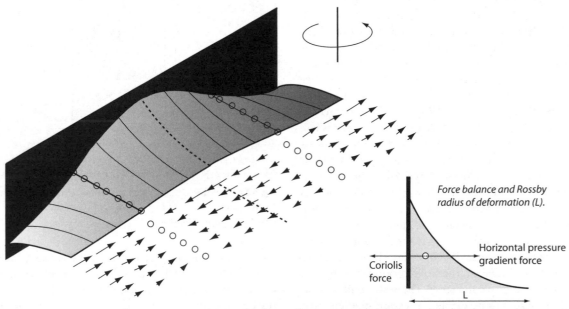

Fig. 6.39 Topography and bottom flow associated with the edge of an anticlockwise-rotating Kelvin tidal wave. The rotary component is neglected for clarity.

this distance the amplitude of any Kelvin wave has reduced to $1/e$, 0.37 of its initial value.

We may usefully summarize the vector variation of tidal currents by means of *tidal current ellipses* whose ellipticity is a direct function of tidal current type and vector asymmetry

(Fig. 6.40). For example, the inequality between ebb and flow on the northwest European continental shelf is largely determined by a harmonic of the main lunar tide. Since sediment transport is a cubic function of current velocity it can be appreciated that quite small residual tidal currents can

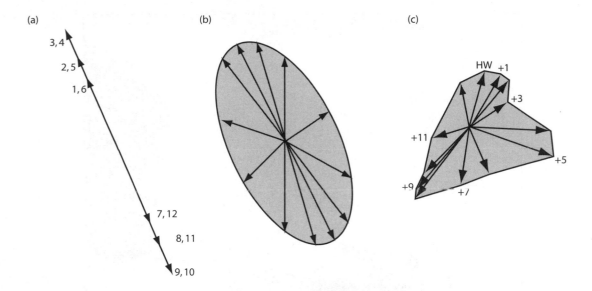

Fig. 6.40 Tidal current variations with time. (a) Linear symmetrical ebb-flood with zero residual; (b) symmetrical tidal ellipse with zero residual current; (c) Irregular tidal ellipse with complex residuals.

cause appreciable net sediment transport in the direction of the residual current. The turbulent stresses of the residual currents will be further enhanced should there be a superimposed wave oscillatory flow close to the bed (Section 4.10). A further consideration arises from the fact that turbulence intensities are higher during decelerating tidal flow than during accelerating tidal flow, due to unfavorable pressure gradients. Increased bed shear stress during deceleration thus causes increased sediment transport compared to that during acceleration, so that the net transport direction of sediment will lie at an angle to the long axis of the tidal ellipse.

A final point concerns the importance of internal tides and other internal waves (Section 4.9), particularly in the outer shelf region. These are common in summer months when the outer-shelf water body is at its most density-stratified, with a stable warm surface layer of thickness h' and density ρ_1 overlying a denser layer, ρ_2. They are also common in fjords. If a wave motion is set up at the stable density interface (due to storm-induced wind stress or the incoming tide), the restoring force of *reduced gravity*, is much smaller than at the surface and so the internal waves cannot be damped quickly; they provide important mixing mechanisms when they break at external boundaries.

6.5.2 Wind drift shelf currents

Although all continental shelves suffer the action of storms, weather-dominated shelves are those that also show low tidal ranges (<1 m) and correspondingly weak tidal currents ($<0.3 \text{ m s}^{-1}$). Also it is not uncommon for the inner shelf to shoreface to be tide-dominated during the summer months but wave-dominated during the winter. In any case, tidal currents and wave currents are progressively less important offshore, so that at the outer shelf margin it is only the largest storms that affect the bottom boundary layer. In these areas it is common to find a multilayer water system, with a surface boundary layer dominated by wind shear effects, a middle "core" layer, and a basal boundary layer dominated by upwelling, downwelling, or intruding ocean currents (Fig. 6.34). Winter wind systems assume an overriding dominance on most shelves, causing net residual currents arising from wind drift, wind set-up, and storm surge. Wind shear causes water and sediment mass transport at an angle to the dominant wind direction because of the Ekman effect arising from the influence of the Coriolis force (see Section 6.2). For example, southward-blowing, coast-parallel winds with the coast to the left in the northern hemisphere will cause net offshore transport of surface waters and the occurrence of compensatory upwelling.

From all this the reader can appreciate that outer-shelf dynamics are extremely sensitive to the magnitude of shelf wind systems. Depending upon dominant wind regime, either import or export is possible: for example, cool shelf waters can be driven far oceanward as intruding tongues that may interfere with ocean currents like the Gulf Stream (Section 6.4).

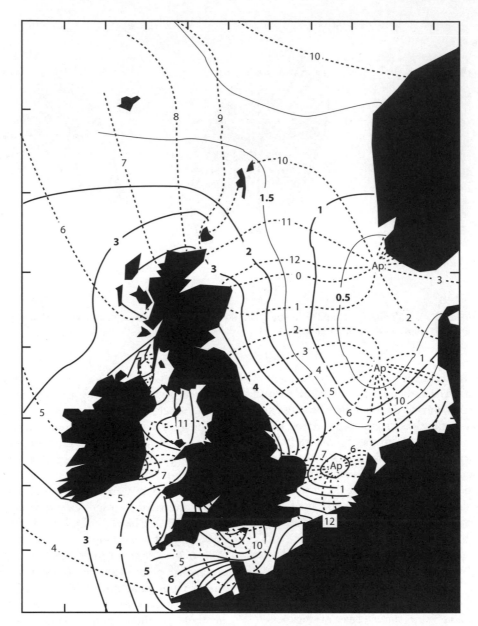

Fig. 6.41 Amphidromic tidal gyres of the North Sea and surrounding areas. Each of the the three systems has anticlockwise sense of rotation. Full lines are co-range lines with tidal range in meters. Dotted lines are cotidal lines indicating the level of high water at the stated number of hours lapsed since the Moon passed over the Greenwich meridian.

6.5.3 Storm set-up and wind-forced geostrophic currents

Let us examine the effects of storm winds in more detail, for, as we shall see later, major shelf erosion and deposition result during such episodes. As in lakes, wind shear drift causes set-up of coastal waters; should this coincide with a spring high tide, then major coastal flooding results. The effects are well known in the southern North Sea (where the Thames Barrage now protects low-lying London), in the Bay of Bengal, and in the Venetian Adriatic (where in both places the inhabitants are not so lucky). The very low barometric pressures during storms cause a sea-level rise under the storm pressure minimum. The magnitude of this effect is about 1 cm rise per millibar decrease of pressure. So passage of the eye of a tropical storm of pressure 960 mbar might cause a few tens of centimeters of sea-level rise. The very low core pressures of coastal tornadoes are particularly effective at raising the setup of shelf waters, sometimes up to 4 m or more above mean high-water level, as in Hurricane Carla on Padre Island, Gulf of Mexico.

The magnitude of wind shear setup can be roughly estimated by assuming that the shearing stress, τ, due to the wind balances the pressure gradient due to the sloping sea surface, $\partial p/\partial x$, that is $\tau = \rho g h \partial p/\partial x$, where h is water depth and ρ is water density. Solving for the slope term for storm winds of 30 m s^{-1} acting on 40 m water depth yields about $2.2 \cdot 10^{-6}$ for the 600 km long North Sea, leading to a superelevation of about 1.3 m. This is 50 percent or so less than the observed surge height because we have neglected important effects due to the Coriolis force, which pushes the current against adjacent shorelines where it is further amplified by resonance and funneling. In the case of the major southern North Sea storm of 1953, the southerly directed wind drift was first forced westward onto the Scottish coast with the southward traveling (anticlockwise) Kelvin tidal wave, where it ultimately gave rise, some 18 h later to a +3.0 m superelevated surge along the Dutch and Belgian coasts. The Kelvin wave nature of storm surges enables prediction for vulnerable areas like the North Sea and the Adriatic by reference to monitored upcurrent changes in sea level during storm development. Offshore, the large wave setup during storms means that a compensatory bottom flow occurs out to sea, driven by the onshore to offshore pressure gradient. Such geostrophic or gradient currents (which are also turned by Coriolis forcing; Fig. 6.42) have been proven by measurements during storms to reach over 1 m s^{-1}, running for several hours (a fact suspected by submariners since 1914, see Fig. 6.42). They are a major means of offshore transport from coast to shelf.

6.5.4 Shelf density currents

Density currents are also important in shelf transport. *Hypopycnal* (positively buoyant) jets of fresh to brackish water with some suspended sediment issue from most estuaries and delta distributary mouths. In higher

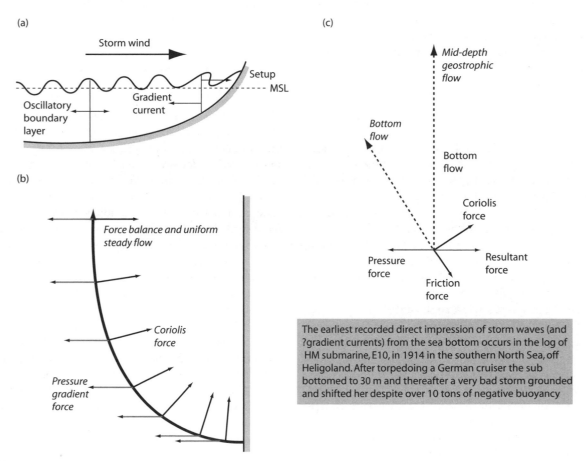

The earliest recorded direct impression of storm waves (and ?gradient currents) from the sea bottom occurs in the log of HM submarine, E10, in 1914 in the southern North Sea, off Heligoland. After torpedoing a German cruiser the sub bottomed to 30 m and thereafter a very bad storm grounded and shifted her despite over 10 tons of negative buoyancy

Fig. 6.42 Shoreface to shelf geostrophic gradient currents. (a) Section; (b) force balance; (c) plan.

latitudes, small to moderate buoyancy fluxes are soon turned by the Coriolis force, and they may be trapped along-source in the mid- to inner shelf where they form coastal currents or linear fronts. Mixing vortices develop along the free shear layer of the fronts and offshore circulating shelf waters. Plumes are very sensitive to the effects of coastal upwelling or downwelling currents caused by winds. They may reach some way out into the mid-shelf or right across the shelf break, depending upon their dynamic characteristics and those of the shelf. Low slopes encourage long passage, whilst the development of vorticity on steeper slopes encourages turning and termination. The large buoyancy flux of many late spring and summer Arctic rivers, for example, causes plumes to extend for up to 500 km offshore, well into the Arctic Ocean.

6.6 Ocean–land interface: coasts

Coasts are dynamic interfaces between land and sea where energy is continuously being transferred by the action of traveling waves, including the tide. This incoming wave energy flux also interacts with energy inputs from the land, in the form of river flows. The nature of any coastal interface varies according to the type and magnitude of these various energy fluxes and also to the geological situation determined by bedrock type (more or less resistant). Like any interface the coast may be largely static in time and space or it may be highly mobile, either advancing seaward when sedimentary deposition dominates, a *prograding coastline*, or retreating landward when erosion and net transport outward to the shelf dominates, a *retreating coastline*.

6.6.1 Nearshore wave behavior

As the typical sinusoidal swell of the deep ocean passes landward over the continental shelf the dispersive wave groups (Section 4.9) undergo a transformation as they react to the bottom at values of between about 0.5 and 0.25 of wavelength, λ. In this transformation to shallow-water waves, wave speed and wavelength decrease whilst wave height, H, increases. Peaked crests and flat troughs develop as the waves become more solitary in behavior until oversteepening causes wave breakage. Waves break when the water velocity at the crest is equal to the wave speed. This occurs as the apical angle of the wave reaches a value of about 120°. In deepwater the tendency toward breaking may be expressed in terms of a limiting wave steepness given by $H/\lambda \approx 0.14$. Breaking waves spill, plunge, or surge (Figs 6.43 and 6.44); the behavior varies according to steepness of the beach face. Steep beaches possess a narrow surf zone in which the waves steepen rapidly and show high orbital velocities. Wave collapse is dominated by the plunging mechanism and there is much interaction on the breaking waves by backwash from a previous wave-collapse cycle. Gently sloping beaches show a wide surf zone in which the waves steepen slowly, show low orbital velocities, and surge up the beach with very minor backwash effects.

The shallow water nature of incoming coastal waves means that the wave trains are no longer made up of dispersive waveforms, as for deepwater waves (Section 4.9). Instead, the speed depends only upon water depth and so the impact of waves upon shallow topography leads to a number of interesting features, chiefly the familiar curvature or *refraction* of approaching oblique wave crests as they "feel bottom" at different times (Fig. 6.45).

6.6.2 Waves arriving at coasts: The role of radiation stress

The forward energy flux or power associated with waves approaching a shoreline (Section 4.9) is, Ecn, where E is the wave energy per unit area, c is the local wave velocity, and $n = 0.5$ in deepwater and 1 in shallow water. Because of this forward energy flux there exists a shoreward-directed momentum flux or *radiation stress* outside the zone of breaking waves. This radiation stress is the excess shoreward flux of momentum due to the presence of groups of water waves, the waves outside the breaker zone exerting a thrust on the water inside the breaker zone. This thrust arises because the forward velocity associated with the arrival of groups of shallow-water waves gives rise to a net flux of wave momentum (Fig. 6.46). For wave crests advancing toward a beach there are two relevant components of the stress, τ_{ij}. One is τ_{xx}, with the x-axis in the direction of wave advance and the other, τ_{yy}, with the y-axis parallel to the wave crest. These components are $\tau_{xx} = E/2$ for deepwater or $3E/2$ for shallow water, and $\tau_{yy} = 0$ for deepwater or $E/2$ for shallow water. Radiation stress plays an important role in the origin of a number of coastal processes, including wave setup and setdown, generation of longshore currents, and the origin of rip currents (Fig. 6.47).

The nearshore current system may include a remarkable cellular system of circulation comprising *rip currents*. The narrow zones of rip currents make up the powerful "undertow" on many steep beaches and are potentially hazardous to swimmers because of their high velocities (several meters per second). Rip currents arise because of variations in *wave setup* along steep beaches. Wave setup is the small (centimeter to meter) rise of mean water level above still water level caused by the presence of shallow-water waves. It originates from that portion of the

Energy flux of swell wave = Energy flux of shoreface wave
or
$$E_1c_1 = E_2c_2$$

Fig. 6.43 A familiar sight on the sea or lake coast; swell waves slowing down ($c_1 > c_2$) and amplifying over the shelving coast, increasing in height and steepness until they break on the beachface. Energy flux (power) is conserved throughout until finally dissipated in the turbulence, cavitation, and sediment transport of the swash zone.

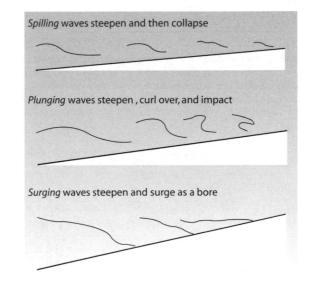

Spilling waves steepen and then collapse

Plunging waves steepen , curl over, and impact

Surging waves steepen and surge as a bore

Fig. 6.44 Types of breaking waves.

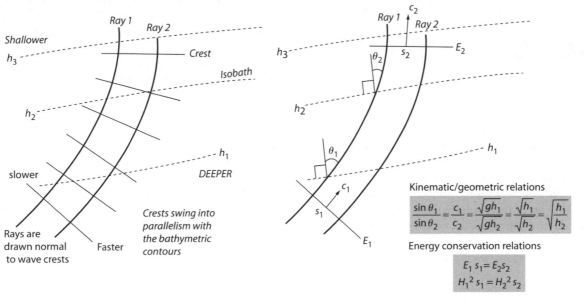

Kinematic/geometric relations

$$\frac{\sin \theta_1}{\sin \theta_2} = \frac{c_1}{c_2} = \frac{\sqrt{gh_1}}{\sqrt{gh_2}} = \frac{\sqrt{h_1}}{\sqrt{h_2}} = \sqrt{\frac{h_1}{h_2}}$$

Energy conservation relations

$$E_1 s_1 = E_2 s_2$$
$$H_1^2 s_1 = H_2^2 s_2$$

Fig. 6.45 Wave refraction from deeper to shallower water by shallow water waves of height H whose speed is purely a function of water depth.

Sign convention

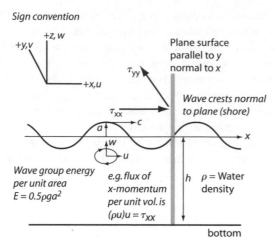

Fig. 6.46 Definition diagram for the radiation stress, τ, exerted on the positive side of the *xy* plane by wave groups approaching from the left hand side. The radiation stress is the momentum flux (i.e. pressure) due to the waves.

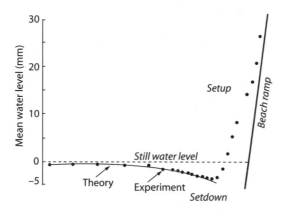

Fig. 6.47 Wave setup and setdown as produced by radiation stress caused by incoming waves in an experimental tank.

radiation stress τ_{xx} remaining after wave reflection and bottom drag and is balanced close inshore by a pressure gradient due to the sloping water surface (Fig. 6.48). In the breaker zone the setup is greater shoreward of large breaking waves than smaller waves, so that a longshore pressure gradient causes longshore currents to move from areas of high to low breaking waves. These currents turn seaward where setup is lowest and where adjacent currents converge.

What mechanism(s) can produce variations in wave height parallel to the shore in the breaker zone? Wave refraction is one mechanism; some rip current cells are closely related to offshore variations in topography. Since

rip cells also exist on long straight beaches with little variation in offshore topography, another mechanism must also act to provide lateral variations in wave height. This is thought to be that of standing *edge waves* (Fig. 6.48), which form as trapped waveforms due to refraction and refracting wave interactions with strong backflowing wave swash on relatively steep beaches. Edge waves were first detected on natural beaches as short-period waves acting at the first subharmonic of the incident wave frequency, decaying rapidly in amplitude offshore. The addition of incoming waves to edge waves give marked longshore variations in breaker height, the summed height being greatest where the two wave systems are in phase. It is thought that trapped edge waves may be connected with the formation of the common cuspate form of many beaches; these have wavelengths of a few to tens of meters, approximately equal to the known wavelengths of measured edge waves. Results concerning the effects of edge waves and "leaky" mode standing waves (where some proportion of energy is reflected seaward as long waves at infragravity frequency, 0.03–0.003 Hz) indicate that both shoreward and seaward transport may result, dependent on conditions. Usually, water entrained under groups of large waves in arriving wavepackets is preferentially transported seaward under the trough of the bound long period group wave.

The familiar longshore currents are produced by oblique wave attack upon the shoreline; these may be superimposed upon the rip cells described earlier. Such currents, which give a lateral thrust in the surf zone, are caused by τ_{xy}, the flux toward the shoreline (*x*-direction) of momentum directed parallel to the shoreline (*y*-direction). This is given by $\tau_{xy} = 0.25 E \sin 2\alpha$, where α is the angle between wave crest and shore (shore-parallel crests = 0°; shore-normal = 90°). The τ_{xy} value reaches a maximum when $\sin 2\alpha = 1$, or when the angle of wave incidence is 45°. Field data give the longshore velocity component, u_l, as $2.7 u_{max} \sin \alpha \cos \alpha$.

6.6.3 Estuarine circulation dynamics

Water and sediment dynamics in estuaries are closely dependent upon the relative magnitude of tide, river, and wave processes. The incoming progressive tidal wave is modified as it travels along a funnel-shaped estuary whose width and depth steadily decrease upstream. For a 2D wave that suffers little energy loss due to friction or reflection (a severe simplification), the wave energy flux will remain constant, causing the wave to amplify and shorten as it passes upstream into narrower reaches. This is the

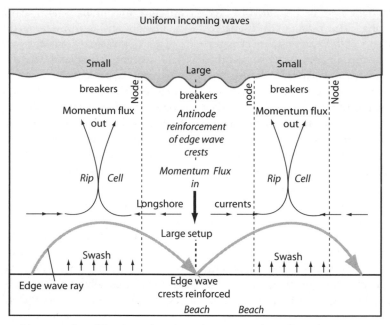

Fig. 6.48 Rip current cells located in areas of small breakers where incoming waves and standing edge waves are out of phase.

convergence effect. Thus for wave energy, E, per unit length of an estuary, Eb is the energy per unit length, where b is total estuary width. Multiplying by the wave speed, c, gives the energy flux up the estuary as Ebc = constant. Writing $E = (\rho g a^2)/2$ and the wave equation for shallow water waves as $c = (gh)^{0.5}$, we have $0.5(\rho g a^2)b\sqrt{gh}$ 5 constant, or, $a \propto b^{-0.5}h^{-0.25}$. We can see that narrowing has more effect on changing wave amplitude than shallowing. Shallowing also causes the wave speed to decrease and, since wave frequency is constant, the wavelength must decrease by the argument $c = f\lambda$. Since $\lambda = c/f = \sqrt{gh}/f$, we have $\lambda \propto h^{0.5}$. Thus tidal waves increase in amplitude and decrease in wavelength up many estuaries. But we cannot ignore frictional retardation of the tidal wave in this discussion; this causes a reduction in amplitude of the tide upstream and is greatest when channel depth decreases rapidly. In some estuaries the tidal wave changes little in amplitude since the convergence effect is balanced by frictional retardation. Resonant effects with tide or wave may also affect currents in estuaries (Section 6.6).

The most fundamental way of considering estuarine dynamics is through the principle of mass conservation, which states that the time rate of change of salinity or suspended sediment concentration at a fixed point is caused by two contrasting processes: turbulent diffusion and circulatory advection. Viewed in this way, water dynamics in estuaries may be conveniently represented by four major end-members (Fig. 6.49). However, it is important to realize that a single estuary may change its hydrodynamic character with time according to changing river, tidal, and wave conditions.

Type A well-stratified estuaries are those river-dominated estuaries where tidal and wave mixing processes are permanently or temporarily at a minimum. The stratified system is dominated by river discharge, with the tidal : river discharge ratio being low, less than 20. An upstream tapering salt wedge occurs, over which the fresh river water flows as a buoyant plume (Fig. 6.50). Shear waves of Kelvin–Helmholtz type may occur at the *halocline* interface, the waves cause upward advective mixing of salt water with fresh water. Should flow occur over topography then internal solitary wave trains may be triggered at the interface. A prominent zone of deposition and shoaling at the tip of the salt wedge arises when sediment deposition from bedload occurs in both fresh water and seawater. This zone of deposition shifts upstream and downstream in response to changes in river discharge and, to a much lesser extent, to tidal oscillation.

Type B partially stratified estuaries are those in which turbulence destroys the upper salt–wedge interface, producing a more gradual salinity gradient from bed to surface water by intense turbulent mixing. The tidal : river discharge ratio is between about 20 and 200. Down-estuary changes in the salinity gradient at the mixing zone occur so that the zone moves upward toward higher salinities. Earth rotational effects cause the mixing surface to be slightly tilted so that in the northern hemisphere the tidal flow up the estuary is nearer the surface and strongest to the right. Sediment dynamics is strongly influenced by the upstream and downstream movement of salt water over the various phases of the tidal cycle. The resulting *turbidity maximum* is particularly prominent in the upper estuary (around 1–5 ppt salinity) on spring and large neap ebb and flood tidal phases, and less prominent at slackwater periods due to settling and deposition. Turbidity maxima are affected by the magnitude of freshwater runoff. A seasonal cycle of dry-season upstream migration of the turbidity maximum and locus of maximum deposition is followed by wet-season downstream migration and resuspension by erosion. The

turbidity maximum is also acted on by gravity-induced circulations arising from excess density.

Type C well-mixed estuaries are those in which strong tidal currents completely destroy the salt-wedge/freshwater interface over the entire estuarine cross-section. The ratio of tide : river discharge is greater than 200. Longitudinal and lateral advection processes dominate. Vertical salinity gradients no longer exist but there is a steady downstream increase in overall salinity. In addition, the rotational effect of the Earth may still cause a pronounced lateral salinity gradient, as in Type B estuaries. Transport dynamics are dominated by strong tidal flow, with estuarine circulation gyres produced by the lateral salinity gradient. Extremely high suspended sediment concentrations may occur close to the bed in the inner reaches of some tidally dominated estuaries. Sediment particles of river origin, some flocculated, will undergo various transport paths, usually of a "closed loop" kind (Fig. 6.51), in response to settling into the salt layer and subsequent

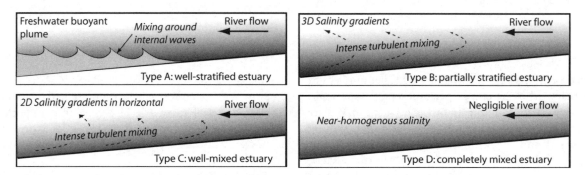

Fig. 6.49 A useful classification of estuaries according to the dynamic processes of mixing and salinity gradients.

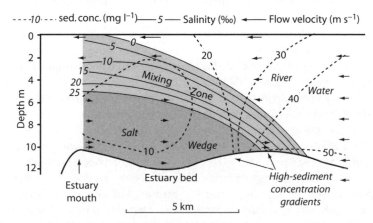

Fig. 6.50 Salinity, velocity, and suspended sediment profiles taken during high tide along transect of the well-stratified (salt wedge) Fraser River estuary.

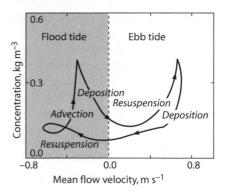

Fig. 6.51 Variation of estuarine suspended sediment concentration over several tidal cycles. Velocities are negative for the flood (incoming) tide and positive for the ebb.

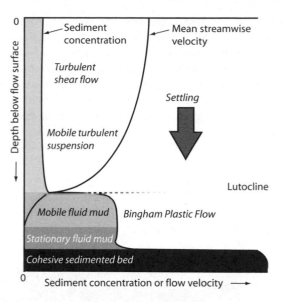

Fig. 6.52 To illustrate the process of fluid mud formation.

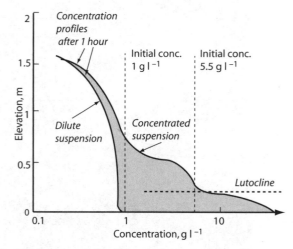

Fig. 6.53 Experimental data to contrast the behavior of dilute and concentrated settling sediment suspensions. Note the stepped profile that forms in the latter case with the formation of a lutocline as hindered settling and flocculation delay fall.

transport by the net upstream tidal flow. Settling of bound aggregates of silt- and sand-sized particles creates large areas of stationary and moving mud suspensions (Figs 6.52 and 6.53), loosely termed *fluid mud*, that characterize the outer estuarine reaches of tide-dominant estuaries. This may be mobile or fixed, the latter grading into areas of more-or-less settled mud. Stationary suspensions up to 3 m thick can show sharp upper surfaces on sonar records and may deposit very quickly. Such suspensions form during slackwater periods, progressively thickening during the spring to neap transition. They are easily eroded, to be taken up in suspension once more by the accelerating phases of spring tidal cycles.

Type D estuaries are theoretical end-members of the estuarine continuum in that they show both lateral and vertical homogeneity of salinity. Such conditions apply only in the outer parts of many type B and C estuaries; they are clearly transitional to open shelf conditions. Under equilibrium conditions, saline water is diffused upstream to replace that lost by advective mixing. Sediment movement is dominated entirely by tidal motions, again with no internal sediment trap.

6.6.4 Estuarine sedimentation

The mixing of fresh and salt water causes estuarine circulation in response to density gradients. Sedimentary particles may be of both marine and river origin, with *flocculation* and floc destruction by turbulent shear and resuspension of bed material as important controls upon particle size. Flocculation is a process whereby the usually repulsive van der Waals electrostatic forces present between closely located clay particles is made positive by absorption of abundant cations from salt water. Also, the higher the amount of suspended clay, the more likely particle collisions will occur, leading to flocculation of aggregates whose settling velocity is enhanced. At the same time, the higher the particle concentration, the lower will be the rate of settling as a result of the effects of particle hindrance (Section 4.7). These two effects, agglomeration and hindrance, lead to the formation of distinct layers of suspended material during the period of relatively slack water in estuaries where tidal currents are important (Figs 6.52 and 6.53). The net accumulation of sediment in the water column due to tidal pumping arises because of inequality in the local magnitude of the ebb and flood tides. If the flood is dominant in the upper estuary, as is often the case, then more sediment enters the upper estuary than leaves, and hence a *turbidity maximum* occurs.

6.6.5 Delta distributaries

Consider the nature of the combined discharge of sediment and fresh water issuing from the mouth of a major delta distributary (Fig. 6.54). This occurs as a jet, analogous to the expanding flow of fluid issuing from any nozzle or opening (Section 4.1). The nature of the discharge, the physiography of the receiving basin, and the degree to which the discharge is modified by wave and tide will control the gross morphology of a delta and the distribution of sediment. Bates first considered the role of jets as relevant and essential to the theory of delta formation. As effluent fluid moves into the marine basin it has the possibility to expand in both horizontal and vertical directions. *Plane jets* just expand horizontally while *axial jets* expand in all directions. Gently sloping coasts restrict vertical expansion and cause plane jet formation. Buoyant effects between effluent and ambient fluids can give rise to

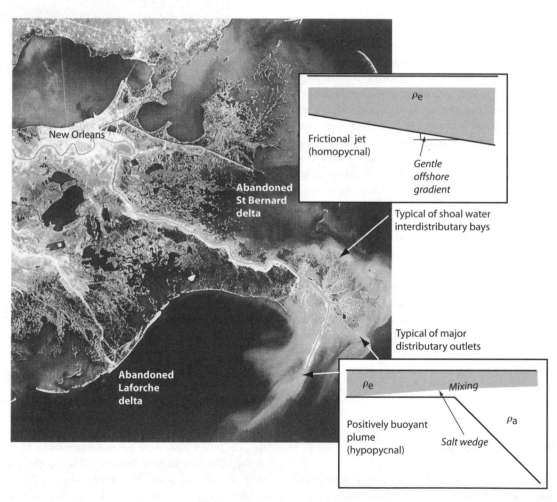

Fig.6. 54 Coastal jets illustrated from the Mississippi "birdsfoot" delta. Effluent jets and plumes rich in suspended sediment appear gray in this satellite image. Note the form of this river-dominated delta, with its numerous distributaries issuing from the seaward extension of the main river channel. The pattern of these gives rise to the term "birds-foot" delta. Most sediment deposition occurs during high river flow close to the mouths of the distributaries, forming accumulations of sediment called "mouth bars." Note the abandoned older Holocene deltas to the southwest and northeast, which are now being reworked by wave action under conditions of rising local relative and absolute sea level: the city of New Orleans is immensely vulnerable to both river flooding and marine inundation during major hurricane impact, as events of summer 2005 have proved.

significant gravitational body forces of the form $[(\rho_a - \rho_e)/\rho_a]g$ per unit volume of effluent fluid, where ρ_a is ambient density and ρ_e is effluent fluid density. The behavior of the plume thus depends upon the resultant of the various buoyancy contributions due to temperature, salinity, and suspended sediment concentration. For example, negative buoyancy acts when sediment-laden effluent jets of cool river water enter into marine basins at delta fronts. The extent of influence of buoyancy on jet behavior is expressed by the *densimetric Froude number*, $Fr' = \bar{u}/\sqrt{gh'\gamma}$ where $\bar{u}$ is the mean effluent velocity, h' is the depth of the density interface from the surface of the jet, and γ is the density ratio $1-(\rho_e/\rho_a)$. For values of $Fr' > 1$, waves form at the effluent ambient interface; these cause enhanced mixing, increased friction, and greater deceleration of the buoyant fluid. The spreading and expansion of a buoyant jet is best considered by reference to the production of superelevation of the effluent arising from its buoyancy: the jet floats with its surface at some small height (δh) above the ambient fluid.

In summary, three factors may influence the nature of the sediment-laden freshwater jet itself: (i) the inertial and turbulent diffusional interactions between the jet and the ambient fluid; (ii) frictional drag exerted on the base of the jet by the delta front slope; and (iii) any buoyant force due to the jet's density contrast with the ambient fluid.

Jets dominated by their own inertia and by turbulent diffusion are said to be *homopycnal*, with virtually the same density for jet and ambient fluid. The majority of such jets are dominated by turbulent effects. This is clear from a simple calculation of an outlet Reynolds number of the form $Re_o = u_o[h_o(b_o/2)]^{0.5}/v$ where u_o is the mean centerline outlet velocity, h_o and b_o are the depth and width of the outlet, respectively, and v is the effluent kinematic viscosity. Most deltas show outlet Reynolds numbers greater than 3,000, indicating the dominance of turbulent mixing. A turbulent jet will expand linearly with distance from the outlet as the homopycnal jet expands laterally and vertically. Delta fronts dominated by homopycnal flows are commonest in lakes.

When the shoreface of the subaqueous delta slopes quite gently and water depth is shallow relative to the magnitude of the incoming effluent jet (Fig. 6.54), frictional effects arising from bottom drag on the jet become very important. Such plumes experience rapid seaward spreading, deceleration, and hence deposition of bedload sediment. Such friction-dominated jets quickly deposit sediment as a *distributary mouth bar*.

Low values of Fr' (<1) suggest dominance by buoyant forces whereby the outflow spreads as a narrow expanding jet above a salt wedge (see Section 6.5.5) that may extend

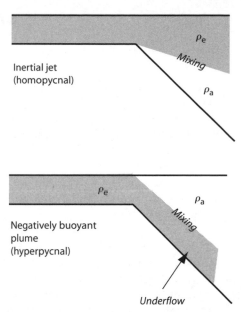

Fig. 6.55 Other kinds of coastal jets and plumes.

for a considerable distance up the distributary channel. Such jets are termed *hypopycnal*. As was discussed in the context of estuary behavior, salt wedges are best developed in deep channels with low tidal ranges. In large river deltas like the Amazon the effluent jets remain dominant far onto the shelf.

When the combined density of effluent jet water and its suspended solids exceed that of the basin ambient fluid ($\rho_e/\rho_a > 1$), the conditions are set for the jet to *underflow* in a state known as *hyperpycnal* (Fig. 6.55). This is more likely to occur in lake waters since a suspended load of at least or greater than $28\ kg\,m^{-3}$ must be present just to counteract the density of normal seawater. Perhaps the most spectacular underflowing delta system is that of the Huang Hue, whose colossal suspended load picked up on its passage through the central China loess belt enables it to sink without trace in the offshore region.

Waves and tides have a great effect on these simple jet models of delta front dynamics. Wave power is substantially reduced as waves pass from offshore areas over very gently sloping nearshore zones; indeed some extremely gentle slopes may cause almost complete dissipation of wave energy. In coastal areas of high wave power relative to river discharge, effluent jets may be completely disrupted by wave reworking. The coastlines of such deltas tend to be very much more linear in plan view than those of more moderate to low wave power.

6.7 Land surface

6.7.1 Hydrology

The *hydrological cycle* on land (Fig. 6.56) involves consideration of:
- Interception of precipitation by vegetation
- Utilization of water by vegetation in the photosynthetic cycle through evapotranspiration
- Surface runoff as overland flow
- Subsurface percolation and soil water throughflow
- Groundwater flow and groundwater seepage to river channels to make up streamflow

All this takes place within the spatial entity known as a *drainage catchment*, countless of which cover the entire land surface of the Earth. However, it is a grave mistake to assume that the hydrological cycle in a catchment is simply a kinematic concept. Although it is a balancing budget exercise for water where *Input = Output + Storage*, it is also highly dynamic, with both potential, kinetic, and thermodynamic energy transfers and transformations taking place constantly within the system (Fig. 6.57). Thus within each catchment the balance of water flux and storage is determined by a unique and self-sustaining combination of:
- Ambient temperature from solar radiation balance
- Magnitude of incoming water supply determined by climatic/meteorological conditions
- Fertility and permeability of bedrock
- Bedrock mineral alteration by percolating groundwater
- Production of surface biomass through ecological energetics of plant productivity
- Breakdown of dead plant biomass through respiration
- Gravitational force components available to water fluxes down hillside slopes

Thus, in a way, the catchment creates the landscape from a number of prior conditions, rather analogous to the "Nature versus Nurture" concept for individual animal development. The genetic makeup of an individual (nature providing) is acted upon by external circumstances (nurture modifying). Tectonics, climate, and geology are any given landscape's "genes," while water : rock and water : organic reactions, groundwater throughflow, surface runoff, gravity slope, mass movements, and sediment transport are the environmental variables that nurture and modify.

First let us consider the nature of the aerated soil water that lies in partially filled pore spaces above the water table. This may reside in soil, sediment, or chemically altered bedrock termed *saprolite*. A mature, well-developed soil with plentiful *in situ* organic and clay fractions and a natural

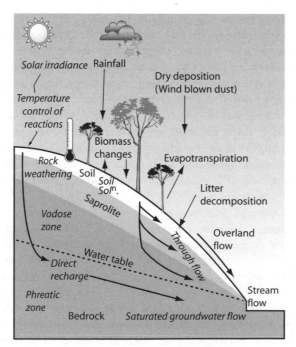

Fig. 6.56 The hydrological variables of a hillslope system.

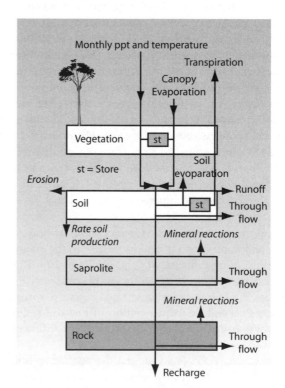

Fig. 6.57 Flow chart for use in water modeling.

open framework acts as a buffer or valve, holding moisture and protecting the easily eroded subsoil from direct rain drop impact. The soil zone acts as an important reservoir of water during dry periods since capillary uprise of water through soil pores by the soil water's *osmotic potential* creates a flux that can take the place of water evaporated at or near the surface. A *zero-flux* plane may be defined that varies in depth seasonally according to land use and thermal conduction; it separates upward-moving capillary soil water from downward-moving recharge water. It usually lies at a depth of a few decimeters to a meter or so. The soil zone also allows a proportion of intercepted rain to naturally *throughflow* at a rate that is directly dependant upon the *infiltration capacity* of the soil in question. This is initially very high after a prolonged drought but measurements suggest it eventually settles down to an equilibrium value, the *saturated hydraulic conductivity*, controlled by gravity. The rate of throughflow will thus depend upon both the hydraulic gradient defined by the hillslope gradient and saturated hydraulic conductivity as expressed in a form of Darcy's Law (Section 4.13) modified for flow through partially saturated media. Areas of throughflow *convergence*, associated for example with slope concavities, may cause significantly higher soil water saturation levels and lead to *overland flow*. Such flows are extremely aggressive since the tractive bed stresses imposed by turbulent flow on steep hillside slopes, especially as the flow aggregates into rivulets and minor channels, may cause extensive erosion, rilling, and general environmental degradation. However overland flow is usually only significant over immature, severely desiccated, or disturbed soils, particularly artificially indurated examples.

Concerning runoff becoming streamflow, the components of the time-series of runoff are called a *hydrograph* (Fig. 6.58). We usually measure runoff in a channel as discharge in cumecs ($m^3 s^{-1}$). It is important to be able to understand the sequence of events that starts with background river discharge, involves a precipitation event in the catchment, and ends with the passage of a flood peak passing down the channel. We talk of *baseflow* to describe the discharge that varies very slowly or not at all, due mostly to near-constant supply from groundwater (see the following sections). *Quickflow* is the contribution from rainfall events. The shape of the flood hydrograph depends on the many variables in a catchment that control the rate of quickflow discharge, chiefly the infiltration capacity and its controls like prior degree of soil saturation, desiccation, occurrence of frozen ground, nature of natural vegetation or land use, urbanization, and contribution of drains etc. Human modifications are very significant in this respect, for example, the USA has an area the size of the state of Ohio under buildings or road surfaces. Hydrograph shape also depends on the measurement site, for as the flood wave passes downstream it may be heightened or diminished due to tributary effects, floodplain storage, and so on.

Concerning *groundwater flow*, we have previously seen how Darcy's Law (Sections 4.13) provides us with a simple but general approach via energy gradient to the flow of fluid through porous media. Consider now the energetics of slow and continuous groundwater flow through a porous and permeable rock (defining an *aquifer*) in the saturated zone beneath a water table (Fig. 6.59). At any point along the aquifer the groundwater flow possesses energy sufficient to keep up a certain height of water to a measurable level in a manometer (piezometer) tube or well drilled down to intersect the aquifer water table. Note first of all that the water table is not a flat planar surface as its name seems to imply. In fact it follows the general shape of the topography, although in a "subdued" manner, rising beneath hills and falling toward valley bottoms, intersecting the surface at spring lines, lakes, or at river level. This is because groundwater that flows away is usually being replenished from above by infiltrating soil water. The rates of the two fluxes determine the depth and slopes of the water table. As shown in Fig. 6.59, groundwater flow is along streamlines, always acting normal to and down the maximum gradient of the equipotential surfaces (Section 4.13). If there was no replenishment of groundwater then the end result of potential flow would be a perfectly horizontal water table of minimum potential energy where pore pressures beneath were all hydrostatic.

Catchment *hillslopes* are the feeder systems, not just for water but for eroded soil, sediment, and rockfall to rivers, lakes, and the ocean. Hillslope processes work mostly under the influence of gravity. Thus sediment and soil particles are moved by combinations of kinetic energy transfer during rain splash, by slow surface, and subsurface mass

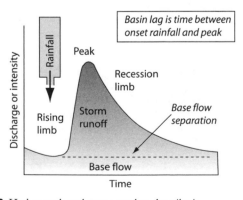

Fig. 6.58 Hydrograph and terms used to describe it.

flow termed *soil creep*, turbulent transport in overland flow, by mass wasting from rockfall avalanches and, after saprolite shear failure on steeper slopes, in slides, slumps, and debris flows. Since these processes are all driven by gravitational forces it is a truism to state that they are more important in mountainous terrains. Large volume mass failures are especially important in areas associated with rapid active tectonic uplift of basically weak rock in orogenic belts. These are normally triggered by excess pore pressures associated with either infiltration and through-flow after abnormal rainfall events or by the effects of seismic shock and fabric rearrangement associated with major earthquakes.

6.7.2 Standing water: Lakes

Lakes are sinks for both water and sediment, cover about 2 percent of the Earth's surface and contain about 0.02 percent by volume of the biosphere's water. They form when runoff or river flow is interrupted, usually because a depression causes water build-up that cannot be neutralized by seepage or evaporation. The commonest causes of large lakes are tectonic subsidence and glacial erosion. Lakes have great environmental and economic importance, for example their sink-like properties make them highly important repositories of evidence for past climate change, but, unfortunately, also for pollutants. Climate is the chief modulator of physical lake dynamics; even the world's largest lakes are too small to exhibit more than minute tidal oscillations. Solar radiation provides energy transfer through its control of surface water temperature and hence density, giving rise to *thermal density stratification*, the distinct layers differing not only in their density, but also in chemical makeup. A temperate lake in summertime (Fig. 6.60) will show well-marked thermal stratification, with an upper warm layer, the *epilimnion*, separated from deeper, cold water that makes up the *hypolimnion* by a layer of water exhibiting a changing temperature, the *metalimnion*. The *thermocline* defines a surface of

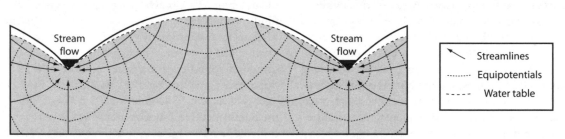

In 2D cross section the groundwater flow is always down the total energy gradient, $d\phi/dx$, determined by the slope of the line joining the points of intersection of the water surface . This line is merely the intersection in 2D of a 3D potential surface that maps out the elevation of the energy available to drive the groundwater flow. In fact, flow is always down the maximum gradient in ϕ. This is written as $\nabla\phi$ for any surface. From Darcy's Law the rate of flow is then $q = -K\,\nabla\phi$, where K is the hydraulic conductivity. It can be seen that the expression for this groundwater potential flow is mathematically identical and physically analogous to that for the flow of heat by molecular conduction and for the diffusion of ionic species along molecular concentration gradients (Section 4.18). All the expressions relate to the mass movement of some quantity from higher to lower potential surfaces. Since we are assuming that the conservation of mass applies then $\nabla q = 0$ and $\nabla \cdot \nabla \phi = 0$ (Cookie 2). The latter expression is the celebrated Laplace's equation which allows us to mathematically determine the variation of potential flow fields in space.

Fig. 6.59 Computed groundwater flow net for symmetrical topography with an underlying mirror image subdued water table. Net comprises equipotential lines and normal flow streamlines.

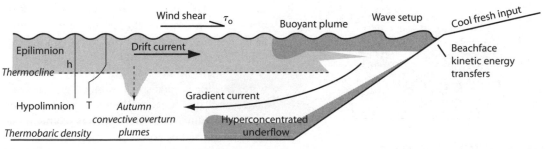

Fig. 6.60 Generalised summary of physical processes affecting lakes.

maximum temperature gradient. Most heat is trapped in the surface epilimnion until, in autumn, cooling from the water surface downward causes density inversions and mixing of the epilimnion with the deep hypolimnion. Melting of winter ice causes wholesale sinking of cold surface water, giving rise to the *spring overturn*. In early spring the water of a moderately deep lake will all be at a temperature of approximately 4°C. The topmost waters will be gradually warmed by solar radiation and mixed downward by wind action. As heating continues the warm surface water become buoyant, resisting wholesale mixing to remain above colder deeper water. The process of overturn in thermally stratified lakes causes the production of alternating annual sediment- and organic-rich laminae; these are termed *varves*.

There are a large number of variations in lake circulation and stratification recognized by limnologists. Some of these, applied to lakes deep enough to form a hypolimnion, are summarized below:

- *Amictic* – lakes permanently isolated from the atmosphere by ice cover.
- Cold *monomictic* – ≤4°C; one period of circulation in the summer.
- Cool *dimictic* – Lake-water freely circulates twice yearly in spring and autumn (described earlier).
- Warm *monomictic* – greater than 4°C; freely circulating in the winter and stratifying directly in the summer.
- *Oligomictic* – rare circulation; greater than 4°C; stable stratification with small temperature versus depth variations.
- *Polymictic* – common circulation due to strong winds and/or strong short-term temperature variations.
- *Meromictic* – a *pycnocline* separates near-permanent saltier bottom water from the main water mass.

Concerning input fluxes, water and sediment come at point sources via river outlets. Only a small proportion of surface runoff enters a lake as surface jets. Due to the generally higher density of cooler inflowing water, *hyperpycnal underflows* (Sections 4.12, 6.6.5) and turbidity currents are very common in lakes (Fig. 6.61). The underflows bring oxygenated water into the deep hypolimnion and prevent permanent stagnation in deep lakes. Proximity to source is a fundamental control on the nature of lake mixing with external sources. Successively finer sediment will be deposited outward from the point source, although this regular pattern is affected by surface currents due to direct *wind shear* (Figs 6.61 and 6.62; Box 6.1). Density current development is hindered by turbulent dissipation in very shallow, well-mixed lakes with gently sloping margins. In thermally stratified lakes the density of the inflowing water may be greater than that of the lake epilimnion but less than that of the hypolimnion, so that the density current moves along the top of the metalimnion as an *interflow*. High concentrations of suspended sediment at this level may then be dispersed over the lake by wind-driven circulation.

Away from river influxes, water movement in lakes is controlled entirely by wind-driven progressive waves and gradient currents (Fig. 6.62). Gradient currents have the ability to interact with turbidity undercurrents in an interesting way (Fig. 6.61). Wind-driven surface waves effectively mix the upper levels of lake-water and give rise to wave currents along shallow lake margins. The size and effectiveness of lake waves depend upon the square root of the fetch of the lake winds and therefore on the physical size of the lake itself. The energy associated with traveling waves is dissipated along the shoreline as the waves break. Internal waves (Section 4.9.6) may also form at the epilimnion–metalimnion interface. A steady wind causes a

Fig. 6.61 The steady underflow of cold, sediment-rich glacial meltwater and its interaction with a gradient current produces intense turbulent mixing and sedimentation. Example from Peyto Lake (Alberta, Canada).

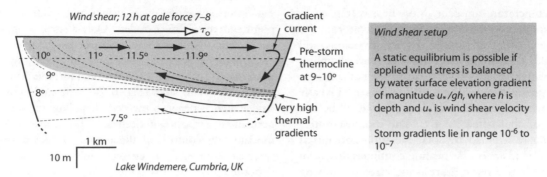

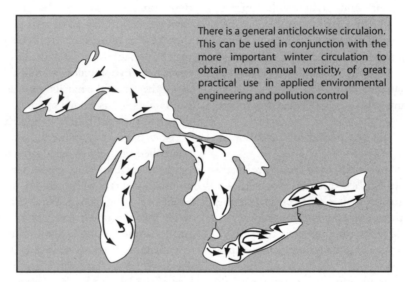

Wind shear; 12 h at gale force 7–8

Gradient current

Pre-storm thermocline at 9–10°

Very high thermal gradients

Wind shear setup

A static equilibrium is possible if applied wind stress is balanced by water surface elevation gradient of magnitude u_*/gh, where h is depth and u_* is wind shear velocity

Storm gradients lie in range 10^{-6} to 10^{-7}

Fig. 6.62 Extreme wind shear causes isothermal displacement, density inversion, windward wave setup, and gradient currents.

There is a general anticlockwise circulaion. This can be used in conjunction with the more important winter circulation to obtain mean annual vorticity, of great practical use in applied environmental engineering and pollution control

Fig. 6.63 Mean summer surface circulation in the Great Lakes of North America.

mass transport of surface water by wind shear, most effective in very large and deep lakes (Fig. 6.63). The application or disappearance of wind stress causes lake-surface and internal oscillations known as *seiches*, which may further mix surface waters or subsurface stratified layers and cause erosion and entrainment along shorelines.

6.7.3 Rivers and the hydraulic boundary layer

The hydrological cycle ensures that some part of the precipitation that falls on the Earth's surface eventually finds itself flowing as channelized runoff: river channels are conduits for the dispersal of weathering products derived from their catchments and are highly sensitive indicators of tectonic slope changes, sourceland geology, and climate. River channels vary greatly in size, over more than four orders of magnitude, from mere ditches to greater than 20 km wide lower reaches of the Brahmaputra and the

Ganges. The magnitude of any channel may be described in terms of its width, w, and depth, h, for bankfull flow. The bigger the channel, the more water it can carry, so we must also characterize channels according to the magnitude of the mean annual discharge, Q. Since the mean flow velocity, u, in any channel is Q/wh, we have $Q = whu$. Expressing width, depth, and mean velocity of flow as functions of the mean discharge, we can derive the basic expressions of *hydraulic geometry* as $w = aQ^d$, $h = bQ^e$, $u = cQ^f$, where $a + b + c = 1$ and $d + e + f = 1$. The magnitudes of the exponents and constants vary according to different stream types and climatic conditions.

The remarkably smooth, concave longitudinal profiles of most channels once they emerge from their bedrock valleys reflect a long-term ability of rivers to overcome initial or imposed gradient irregularities. This can be readily understood by noting that the downstream increase in discharge associated with the stream network must be accompanied by a downstream decrease in slope if equilibrium, that is, neither erosion nor deposition, is to be maintained. If slope stayed constant or increased, then erosion would result in a lowering of the bed and the production of a concavity. Profile concavity may be described by an equation of the form $H = H_0 e^{-kL}$, where H is local profile height with respect to an initial height H_0, k is an erosional or depositional constant, and L is distance downstream. The downstream concavity of profile in most aggrading rivers (ignoring the effects of tributaries) leads to a diminution of bed shear stress and hence to a general downstream fining of bed sediment grain size, although abrasion due to transport also undoubtedly contributes to this.

Channels possess form as well as magnitude and are best described by a combination of:
- Planform deviation from a straight path (*sinuosity, P*)
- Degree of channel subdivision by mobile bars (*braiding*)
- Degree of channel subdivision by permanent islands (*anastomosing*)

No rigid classification of any single channel on anything longer than a reach level (a few channel widths) is universally practicable since many rivers show combinations of sinuosity and braiding.

Concerning river channel sediment transport processes, the largest river bedforms have dimensions that bear some constant relationship with the dimensions of the containing channel. The bedforms, henceforth referred to as *bars*, are said to scale with the channel size, usually expressed as total active channel width, and are frequently referred to as *macroforms*. These include point, lateral, and in-channel bars (Figs 6.64–6.66). The former positions of barforms

may be studied from aerial photographs since slow migration of bar and/or channel leads to partial preservation of the barform as topographic features (*swale-and-ridge* topography) in plan view. These are the surface expressions of periodic *lateral accretion* of sediment on to the bar, particularly due to flow expansion downstream.

Helical flow due to radial forcing (Section 3.7) is the common denominator to each kind of sinuous channel reach. It causes downstream bar migration to be accompanied by bar lateral accretion. Why is this? Owing to helical flow, the bed shear stress vector deviates by a small angle δ from the mean flow direction, so that near-bed sediment grains tend to move inward (Fig. 6.65). From theory and experiment $\tan \delta = 11 h/r$, where h is local flow depth and r is the local radius of curvature of the bend. The actual path of a moving bedload grain depends upon the balance of the frictional, fluid, and gravity forces acting on it. For steady, uniform flow the net rate of transverse sediment transport must be zero, or a stationary bed cannot be attained. To counterbalance the inward transport, a transverse bed slope develops. This is the point bar or channel bar lateral surface where grains travel parallel to the mean flow direction, around the "contours" of equal bed shear stress. Given a wide range of available grain sizes, δ will vary directly with bed shear over any bar surface, the general decline of shear as the water shallows causing a general lateral fining from bar toe to top. At the same time as this inner bend deposition, erosion of the outside bend *cut-bank* takes place. The rate of this depends upon the structure of the bank and the properties of its constituents – whether

Fig. 6.64 Meander bends of a sinuous reach ($P > 2$), South Angling Channel, Saskatchewan River, Canada. The light grey areas on the inside bends are exposed parts of sandy point bars revealed at low river flow. These extend below water into the deepest part of the channel. The dark wooded areas inboard of each point bar are older active bar surfaces recording the gradual outward migration of the channel bends with time. Faint concentric lines are *swale-and-ridge topography* (SRT); this marks former inner banks to old active point bar surface margins.

they are noncohesive, cohesive, or composite. Bank undercutting by the river leads to collapse along planes formed by soil shrinkage. Clay banks may fail along rotational slides. In both cases, failure and collapse occur preferentially during falling river stage in response to changing porewater levels and pore pressures. The combined processes of outer bank erosion and inner bank accretion over the surface of a bar lead to whole channel migration. Channel bends may grow, migrate, and rotate with time, frequently leading to the formation of *cutoffs* as the channels take shortcut routes across meander necks during high discharges. Periodic bend migration and bar accretion in response to major floods lead to the production of a characteristic arcuate topography of *scroll bars* on bar surfaces (Fig. 6.65).

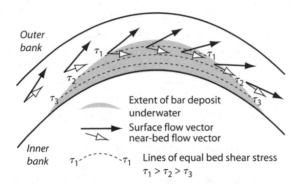

Fig. 6.65 Helical flow cells due to radial acceleration occur in *all* channel bends. They explain the inward movement of sediment to form the point or channel bar. The lessening magnitude inward of the flow vectors explains the observed inward trend to finer sediment grain size.

At the zone of river *confluences*, experimental and field studies reveal the existence of strong vertical axis vortical structures of Kelvin–Helmholtz type (Fig. 6.67; Section 4.9.6), which scour deeply into the underlying substrate. The depth of scour may reach several times that of the bankfull depths of the contributing tributaries, so that in the case of major rivers, the highly mobile scours may reach up to 30 m below low water level.

Adjacent to river channels are swathes of periodically flooded *wetlands* whose environmental well-being has often been neglected in the cause of floodplain cultivation and habitation. Data on the magnitude of deposition during flooding events are sparse, but careful trapping experiments, comparisons of upstream and downstream gauging station records, and radiometric dating of floodplain cores reveal that a surprisingly high proportion (30–70%) of upstream suspended load may be deposited on the floodplain reach during flooding. Most sand is dumped on the *levees* very close to the channel margins (Fig. 6.68). The net effect of repeated flooding is the production of an *alluvial ridge*, whose topography of levees and active and abandoned meander loops stand above the general floodplain level. The observed falloff in mean net deposition rate, r, at any distance, z, from the edge of the channel over the levee to the floodplain margin is most simply given by power-law expressions like $r = a(z + 1)^{-b}$, where a is the maximum net deposition rate at the edge of the channel belt, and b is an exponent that describes the rapidity with which the rate of deposition decreases with distance from the meander belt. The coefficients vary according to factors such as climate, river size, timing of flood, and sediment load.

Fig. 6.66 Braided reach of the Platte River, Nebraska during low flow. The wide, shallow channel divides and rejoins around channel bars. Note vegetated bar to left indicative of short term stability and the active, migrating bars under shallow water to the right.

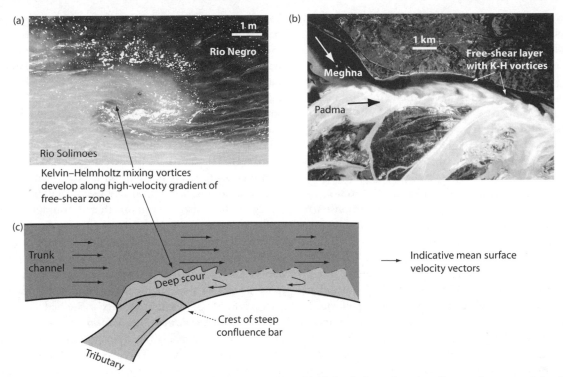

Kelvin–Helmholtz mixing vortices develop along high-velocity gradient of free-shear zone

Fig. 6.67 Turbulent mixing and scour at river channel tributary junctions highlighted when tributaries of contrasting suspended sediment content meet. Both the Negro (Amazon; a) and Meghna (b) rivers have low suspended sediment and high organic content. (c) Plan view schematic.

Fig. 6.68 Floodplain wetland processes. (a) shows storm on a freshly emergent levee (R. Ouse). (b) shows a crevasse breakout through levees into a floodplain wetland (R. South Saskatchewan).

Prolonged occupation of an area by a river leads to the production of a channel belt occupied by active and abandoned reaches. The relatively sudden movement of a whole channel belt (not just a single reach or bend cutoff) to another position on the floodplain is termed *avulsion*. The process is recorded by abandoned channel belts preserved on floodplains or buried partly or wholly beneath them. Avulsion leaves a very characteristic imprint on the fluvial landscape. This is best illustrated by the Saskatchewan wetlands, where an avulsion in the 1870s led to the production of a vast complex of splays, wetlands, and channels in the Cumberland Marshes (Fig. 6.68a).

A river may adjust the following variables in response to independently imposed climatic or tectonic changes to runoff/discharge and slope over which the river itself has absolutely no control: cross-sectional size (*wh*), cross-sectional ratio (*w/h*), bed configuration, bed material grain size, plan-form shape (sinuosity) and size (meander wavelength), and channel bed slope. The equilibrium *graded stream* is defined as "... *one in which, over a period of years, slope, velocity, depth, width, roughness and channel morphology mutually adjust to provide the power and efficiency necessary to transport the load supplied from the drainage basin without aggradation or degradation of the channel.*" Channels are extremely sensitive to perturbations in slope, sediment load, and water discharge. These perturbations may be imposed by climate change, base-level change, and tectonics. For example, the hydraulic geometry equations imply that the magnitude of water discharge and the nature of sediment load should radically affect channel sinuosity. Many river systems around the world record major changes in channel magnitude and geometry since the last glacial maximum, commonly exhibiting a trend from large, braided, aggrading channels to large and then smaller, meandering, incised channels. These changes have occurred due to large decreases in sediment supply in response to a general decrease in runoff and increase in vegetation in the past 15,000 years. Increased temperature and humidity after the last Ice Age caused vegetation growth and substantially reduced the amount of coarse sediment liberated from drainage basins.

6.7.4 Sediment transport in the atmospheric boundary layer (ABL) over land

For the most part the land surface itself is "solid" and therefore immune to alteration by wind shear. Momentum transfer is manifest most obviously in the transport of particles from soils and/or loose sediment; we have already discussed dust storms generated in the ABL (Section 6.3).

The transport of sand modifies the surface morphology in marked ways due to the formation and migration of *bedforms*. These range in size over more than four orders of magnitude, from the centimetric-scale ripples familiar to many from the action of breezes on dry beach sand, to gigantic dunehills of sand hundreds of meters high captured by aerial and satellite images of arid-zone deserts.

Ripples and ridges form a continuous series with wavelengths 0.02–2.0 m and heights from a few millimeters to 1 m (Figs 6.69 and 6.70). Wavelength increases linearly with increasing grain size and with flow strength. Ripple migration and growth from random irregularities occurs by segregation of coarser grains that are bumped along by collisions from saltating grains in bedload transport. A stable ripple regime is reached when the crests of adjacent patches of coarser grains align and when sand transport between ripples relates to approximately the equilibrium jump length of the transported grains, itself adjusted to the magnitude of momentum flux at the bed.

Flow-transverse dunes (Figs 6.71–6.73) occur where the predominant seasonal winds of importance for sand transport are unidirectional. There is a continuum of flow-transverse forms related to the availability of sand cover. It is possible that an analogy with subaqueous dunes is apposite. This would require the wavelength of large aeolian dunes (up to several hundreds of meters) to be of the order of boundary-layer thickness, a correlation in line with known values of these parameters (Section 6.2). Dunes are frequently organized into a hierarchy of forms. *Draas* are composite duneforms of two types. In one (Fig. 6.74) the relationship is rather like the *dispersive* behavior of progressive deep-water gravity wave groups (Section 4.9) so that dunes pass through the larger form and emerge

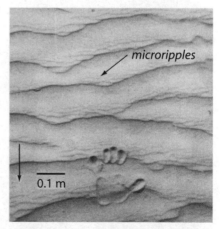

Fig. 6.69 Wind or ballistic ripples reflect the role that impacting saltating sand grains have upon their development. The coarser asymmetric crests are subject to most energetic bombardment.

Fig. 6.70 A consequence of the idea that aeolian ripples have wavelength approximately equal to the saltation jump length is that coarser bedstock should give rise to longer wavelength forms. This is borne out by observations such as here, where granule ridges ($\lambda = c.1$ m) in the foreground pass uplsope into "normal" decimetric-scale ripples under the persons feet.

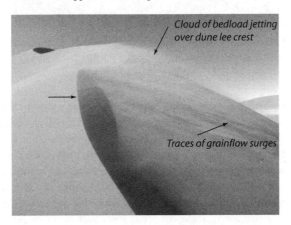

Cloud of bedload jetting over dune lee crest

Traces of grainflow surges

Fig. 6.71 Clouds of saltating grains fall out in the dune lee to produce coarse accumulations that periodically grain flow downslope. At the same time there is a continuous "rain" of finer grains.

Fig. 6.72 The sinuous crests of aklé dunes. Little is known about the physical controls upon dune wavelength.

Fig. 6.73 Dunes are frequently organised into a heirachy of forms. Here a composite draa has sinuous aklé crests delivering sand to the brinkpoint of a huge 30 m high slipface; the smaller dunes do not travel through the larger.

Fig. 6.74 Dispersive dunes; the imposing *c*.40 m high draa in centre right of this view is made up of smaller aklé dunes that rise up, overtop, and pass downwind through the larger form.

downwind. Such draas are forms of *kinematic waves*, that is, waves of mass concentration whose velocity is slower than that of the contributing dunes. A close analogy is with individual cars arriving, passing through, and leaving a slower moving stream of traffic on a single carriageway where overtaking is periodically possible. In the second type (Fig. 6.73), the smaller forms deliver sand to the brinkpoint of a high slipface so that the smaller dunes do not travel through the larger.

Flow-parallel aeolian dunes are exemplified by sharp-crested longitudinal dunes, individual examples of which may sometimes be traced for many tens of kilometers (Fig. 6.75). It has been proposed that the presence of streamwise secondary flow is of major importance in the generation of longitudinal aeolian dunes, occurring along the axis of the meeting point of pairs of oppositely rotating streamwise vortices. Finer saltating sands are thus always swept inward in broad lanes where deposition occurs and, given sufficient sand supply, the duneform grows into equilibrium with the flow. Once formed, the dunes will reinforce the secondary flow cells. Although an attractive theory, and despite many observations, it has not yet been proven that small-scale sand windrows may grow into large scale dunes. Opportunities for natural experiments are hindered by the very large scale of the effects searched for. A closer comparison in terms of scale might be made with

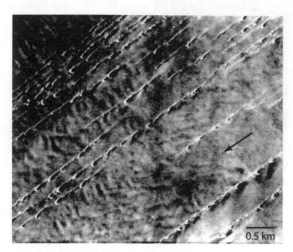

Fig. 6.75 These beaded seif dunes run parallel to the mean south-west wind blow obliquely to the longitudinal crests, nourishing first one side of the crest and then the other. They also migrate over immobile surfaces, in this case bedrock.

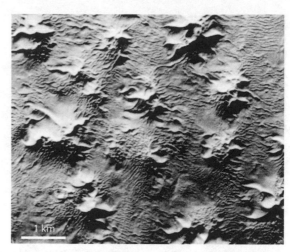

Fig. 6.76 Star dunes form when multidirectional winds cause mean rotary sand transport over immobile desert surfaces.

linear cloud formations (cloud "streets"), whose persistence and wavelength resemble linear dunes. Another explanation for longitudinal dunes is that they arise when transverse dunes are subjected to winds from two directions at acute angles to each other. One dune becomes elongated, later to become the nucleus of a new dune as the wind reestablishes itself in its former mode. The resultant dune has its long axis orientated parallel to the resultant of the two wind azimuths. This theory is broadly supported by flow visualization studies on longitudinal dunes of the Sinai desert where the oblique incidence of seasonal winds to crestlines causes leeside helical flow spirals to be set up.

Complex flow aeolian dunes include the spectacular star-shaped dunes (Fig. 6.76) known as *rhourds*, which commonly range from 500 to 1000 m wavelength and from 50 to 150 m height. The forms have central peaks about which curved crests radiate like vortex lines. They may be spaced randomly, separated by immobile rock or gravel substrates, or in rows, and seem to arise from the interaction of multidirectional regional winds with, less certainly, local winds due to convected air masses. The flow over these forms is, not surprisingly, particularly complicated.

6.7.5 Glacial ice and the cryospheric boundary layer

The frigid cryosphere makes up 30 percent of the Earth's land surface. Ten percent of this is ice cover, representing about 80 percent of surface fresh water: should all this ice and snow melt then global sea level would rise by some 80 m. There are signs from global satellite surveys that the world's ice volume is indeed contracting. Of the ice-lands, the Antarctic ice cap has about 86 percent by area, Greenland has about 11 percent and the many valley and piedmont glaciers make up the remainder. A further 20 percent of land area is affected by permafrost. In the Ice Ages of Quaternary times a staggering 30 percent of the Earth's surface was ice-covered, with vast areas of North America and Europe subjected to glacial erosion and deposition and even larger areas occupying the permafrost zone. The major environments of glacier ice are:

- *Ice-sheets* and their associated fast-moving *outlet ice streams* and coastal *ice shelves* (Fig. 6.77)
- *Valley glaciers* and their marine outlets called *tidewater glaciers* (Fig. 6.78)
- *Piedmont glaciers* – divergent, fan-like ice masses formed after a valley glacier becomes unconfined.

Moving ice is an eroding and transporting system. Motion is caused by deformation of the crystalline solid phase due to its own body force under the influence of gravity. The direction of movement is controlled by regional pressure gradients caused by the 3D distribution of ice mass and/or bedrock slope. Thus the radial flow of a mound-like ice sheet will pay little attention to local or even regional bedrock relief; ice sometimes moves uphill relative to the bedrock surface. The slow flow of glacier ice is usually measured in meters per year, with values between 10 and 200 m yr^{-1} for valley glaciers and 200–1400 m yr^{-1} for ice streams. Corresponding strain rates are also small. Although usually slow and steady, spectacular glacier surges occur periodically when ice velocity increases by an order of magnitude and more.

Antarctic ice sheet

Flow-lines diverge as
they run out to the
Ross ice stream
outflow

Fig. 6.77 Outflowing Antarctic ice streams (see also Fig. 6.83).

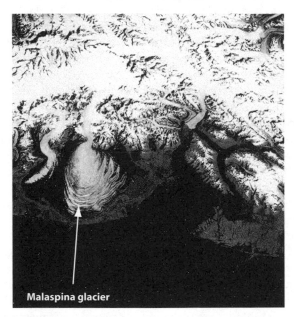

Malaspina glacier

Fig. 6.78 Radially outflowing piedmont tidewater Malaspina glacier lobe, Alaska.

Ice comprises an aggregate of roughly equigranular crystals belonging to the hexagonal crystal system; crystal size usually increasing with time and/or depth. When stressed, by burial in a glacier or insertion in a test rig in the laboratory (Section 3.15), each crystal deforms easiest internally along glide planes parallel to the basal planes of the hexagonal lattice. Since crystals in ice are not usually aligned along common axes, the polycrystalline aggregate rearranges, and recrystallization takes place during strain or flow. Natural ice crystals in the actively deforming layer of a glacier also contain a myriad of gaseous, liquid, and solid impurities, and inclusions. It is not surprising therefore that natural ice deformation is rather complex, with time-dependent behavior seen during the initial stages of application of stress. Primary, secondary, and tertiary *rheologic creep regimes* may be identified, with secondary creep representing a sort of steady state dominant in glaciers, which are usually responding to load- and slope-induced stresses in the range 50–200 kPa (0.5–2 bar). For applied stresses of this magnitude in the laboratory, the shear strain rate of creeping glacier ice is given by $du/dz = k\tau^n$ where n is an exponent ranging between 1.5 and 4.2, k is an experimental constant, and τ is the shear stress. (Compare this to the Newtonian law for fluids (Section 3.9) when $n = 1$ and $1/k$ is the molecular viscosity). n is most reliably estimated as 3, while k is partly a temperature dependent (Arrhenius) function controlled by the energy required to activate creep. It is also a function of crystal size, shape, and inclusion content. A simpler expression for ice strain derived from Antarctic field studies using borehole data yields a linear flow law for the strain rate, of the form $du/dy = k\tau$.

The basal shear stress arising from valley glacier ice sliding over a plane inclined bed may be approximately given by the familiar tractive stress equation (Sections 3.3.4) $\tau_0 = \rho g h \sin\alpha$, where ρ is the density of ice, h is ice thickness, and α is the mean valley floor (or ice surface) slope. In detail the velocity profile of an ice flow resembles that of a non-Newtonian plastic.

In Nature, the mechanical flow of ice depends on the largely unknown process of interaction between a basal ice layer, a deeply cracked (crevassed) ice column, and a solid bedrock and sediment substrate lubricated by seasonal glacial meltwater. It is the coupling of the three materials that is really the key to the whole process. Two types of kinematics for flowing glacial ice are proposed (Fig. 6.79):

1 Warm, wet ice beds lie close to the pressure melting point at the glacier sole (ignoring the conditions throughout the rest of the ice column), and the glacier slides over its bed on a slip-plane of fluid-rich and highly porous sediment. Most glaciers, including both polar ice streams and high latitude valley glaciers, seem to exist in this state. It is quite instructive to remember that the weight of over 1,000 m of Antarctic ice is held up by pore fluid pressure in a thin (5–6 m) basal deforming layer of till.

2 Cold, dry ice beds are a largely hypothetical state reached when ice lies well below its pressure melting point

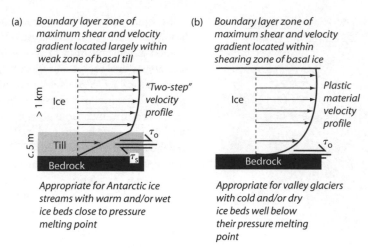

(a) Boundary layer zone of maximum shear and velocity gradient located largely within weak zone of basal till

> 1 km

Ice

"Two-step" velocity profile

τ_o

c.5 m

Till

τ_s

Bedrock

Appropriate for Antarctic ice streams with warm and/or wet ice beds close to pressure melting point

(b) Boundary layer zone of maximum shear and velocity gradient located within shearing zone of basal ice

Ice

Plastic material velocity profile

τ_o

Bedrock

Appropriate for valley glaciers with cold and/or dry ice beds well below their pressure melting point

Fig. 6.79 (a) Predominant till deformation with minor internal ice deformation; (b) Basal ice sliding and internal ice deformation.

Fig. 6.80 A glacier crevasse: product of shallow level brittle tensile deformation analogous to tension gashes in deformed rock (Section 4.14).

at depth. A condition of no-slip must exist at the ice-bed interface, with a general absence of englacial or subglacial drainage. Forward motion of such ice is therefore by internal ice creep alone, the basal ice defining a plastic flow boundary layer of differential velocity. Glacial debris is transported within the ice, with substrate erosion due to plucking and grinding effective only at the summits of protuberances on the bed.

It seems that up to 90 percent of total glacier movement may occur by basal sliding, the rest by internal deformation made manifest at shallow levels by awesome crevasse fractures (Fig. 6.80). The concept of effective stress is relevant again here, for the shear strength, τ_s, of subglacial sediment must be exceeded by that of the driving bed shear stress, τ_o, for ice flow if deformation is to occur at all. Ignoring cohesive strength, assuming that resistance is due to solid friction (ϕ) and that strength is much reduced by high porewater pressures, we may write $\tau_o \geq \tau_s$, or $\rho g h \sin \alpha \geq \Delta P \tan \phi$ where ΔP is the excess of lithostatic pressure above porewater pressure. The reduced strength allows the driving force provided by the tractive force of the glacier, actually quite small for most glaciers due to the low slopes involved, and about 20 kPa for the Antarctic ice cap, to cause deformation and steady forward motion. Direct subglacial measurements of rates of till deformation indicate values of viscosity for deforming till of between $3 \cdot 10^9$ and $3 \cdot 10^{10}$ Pa s, with yield stresses of about 50–60 kPa. Despite knowing little about the *in situ* properties of deforming subglacial sediment beds (*till*) it seems clear that both the glacier ice and the till must move along and be deformed during transit.

The process of basal sliding must also involve enhanced creep around drag-creating obstacles, pressure melting around obstacles, and direct lubrication by abundant basal meltwater. The latter comes from surface meltwaters let into the sole by crevasses and ice tunnels, ice melted by geothermal heat (e.g. spectacular Lake Vostok under the Antarctic ice cap), and ice melted by pressure at the glacier sole. The water under ice streams is often modelled as a thin (few centimeters) film but is more realistically thought to occur in a network of very shallow subice channels cut into the deforming till. It may be possible to characterize a glacier bed by some roughness coefficient or

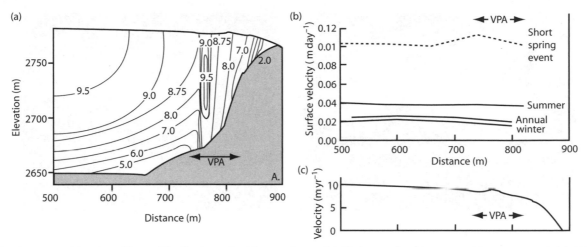

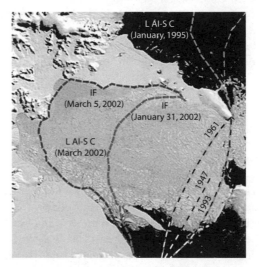

Fig. 6.81 Flow rates of the Haut Glacier d'Arolla, determined from surface and drillhole soundings (vpa – variable pressure axis). (a) Half cross-sectional mean annual flow velocities. (b) Seasonal surface velocity variations. (c) Mean annual surface velocities to show boundary layer.

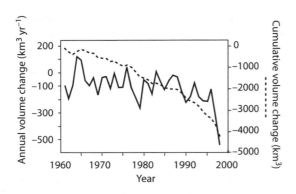

Fig. 6.82 The world's shrinking glaciers; losses from continental valley glaciers.

Fig. 6.83 Larsen Ice Shelf collapses (LA I-S C), Antarctica, and Ice front (IF) positions.

friction factor, analogous to the flow of water over sediment beds and bedforms.

Glacier flow is unsteady and nonuniform on a variety of timescales due mainly to variations in the rate of basal sliding versus internal ice deformation caused by variation of water content (Fig. 6.81). Slow winter flow occurs because meltwater is in short supply and subglacial drainage is minimal. Flow accelerates in spring and summer as more water becomes available. Recent drillhole and ground-penetrating radar results suggest that permanent through-going crevasse fractures play an important role in letting water down to the bed of temperate glaciers, and that spring meltwater throughflow may trigger seasonal renewal of bed sliding. Glaciers may also suddenly surge after years of steady slow flow and over a few months move orders of magnitude faster than in preceding and subsequent months. The process suggests that some deformation

threshold is crossed – maybe the onset of high shear strain and crevasse fracturing as the ice accumulation increases. Given the roles that basal fluid pressure (resistance to flow is strongly dependent upon water content at the glacier sole) and deforming sediment have upon glacier behavior, it seems likely that changing near-bed conditions play a crucial role, perhaps crevasse fractures trigger and grow upward (in a manner analogous to fault propagation; Section 4.15) to tap a water collecting zone. The Bering glacier of Alaska is a well-known example in which surges occur quite regularly, every 20 year or so, the most recent initiated in 1993. As in flooding rivers, surging reflects an

Fig. 6.84 Vanishing glaciers; South Cascade valley glacier, Washington State, USA.

imbalance between supply from the gathering area upstream and the geometry of the channel downstream. In addition, the icy *kinematic floodwave* that quickly travels through the system causes tectonic thickening, intense crevassing, extreme local velocity gradients, abundant discharges of meltwater, and the intense mixing of lateral and medial moraines. As the surge dissipates and the glacier resumes its sluggish phase once more, the advanced snout (but note that surges do not always reach the snout) decays back to equilibrium position. Distinctive surge moraines, ridges of ice-rafted and sheared sediment, then mark the previous maximum position of the glacier front. Major historic shrinkage and retreat of many glaciers is recorded worldwide (Figs 6.82–6.84).

Further reading

Oceanographic matters are described in H. V. Thurman's *Introductory Oceanography* (McMillan, 1994) and explained in the exceptionally clear S. Pond and G. L. Pickards' *Introductory Dynamical Oceanography* (Pergamon, 1983). For the basics of atmospheric dynamics, R. McIlveen's *Fundamentals of Weather and Climate* (Stanley Thornes, 1998) is very clearly laid out. We also like E. Linacre and B. Greert's Climates and Weather Explained (Routledge, 1997), partly because the Southern hemisphere perspective adds twists for northerners and also because of useful CD. Physical processes in terrestrial systems are well-treated by M. J. Selby in *Hillslope Materials and Processes* (Oxford, 1993), P.A. Allen in *Earth Surface Processes* (Blackwell Science, 1997), K. Hiscock in *Hydrogeology* (Blackwell Science, 2004), P. J. Williams and M. W. Smith in *The Frozen Earth* (Cambridge, 1991), W. S. B. Patterson in *Physics of Glaciers* (Butterworth-Heinemann, 1994), and J. Menzies in (Ed.) *Modern Glacial Environments, Vol. 1* (Butterworth-Heinemann, 1995).

Appendix Brief mathematical refresher or study guide

A.1 Power or exponent

A power or exponent is a number that raises another number to the value given by the product of the number with itself as many times as the value of the power indicates. Thus, x^3 is x raised to the power 3 or $(x \cdot x \cdot x)$, and x^2 is $(x \cdot x)$. For negative powers, we use the reciprocal of the positive equivalent, for example, $x^{-3} = 1/x^3$.

A.2 Logarithm

A logarithm is a number that is the power of a certain base integer to which the integer must be raised to get the number. Thus, the logarithm to base 2 of the number 8 is $\log_2 8 = 2^3 = 3$. This is also the logarithm to the base 10 of 1,000, $\log_{10} 1,000 = 10^3 = 3$. Thus we must always quote the log base used.

The exponential, e

The exponential function e is given by the infinite series

$$e = 1 + \frac{1}{1!} + \frac{1}{2!} + \frac{1}{3!} + \frac{1}{4!} + \cdots$$

where the exclamation mark means "factorial" or an instruction to multiply out the number from 1 to its value, for example, 3! means $3 \times 2 \times 1 = 6$. The series quickly converges to the value 2.718. It is used as the base for the natural or Napierian logarithm, $\log_e$ or $\ln e$.

A.3 Functions

A functional relationship is when one variable depends upon the value of another variable. Thus if we are traveling at constant speed, then our travel distance (displacement, s) depends upon time, t, passed. We write $s = f(t)$. For any case in general, $y = f(x)$, where y is the dependent variable and x is the independent variable; this means that y depends upon the value of x. For example, we saw in Section 1.10 that velocity may be a function of both position in space and time, that is, $u = f(x, y, z, t)$. Functional relationships can be linear $(y = mx + c)$, power $(y = x^n)$, logarithmic $(y = \log_a x)$, exponential $(y = e^x$, or $y = \exp x)$, trigonometric $(y = \sin^{-1} x)$, etc. A wider range of functions includes polynomial and rational functions of the second $(y = x^2 + 2x + 1)$ or third $(y = 3x^3 + 2x^2 + x + 1)$ degree, often termed quadratic or cubic, respectively.

More complicated-looking functions may also be constructed, for example, hyperbolic functions from exponential functions, like $y = \sinh x = (e^x - e^{-x})/2$ or the equation for a catenary curve, $y = \cosh x = (e^x + e^{-x})/2$, both used in wave theory (Cookie 16).

The Gaussian normal function is obtained by taking the exponential of the quadratic $y = -ax^2$ so that $y = e^{-ax^2}$.

A.4 Calculus

A.4.1 Differential calculus

The change in the y part of a function like $y = f(x)$ may be quite simple if, say, $y = 2x$. Then for every unit change in x, there is a proportional change in y of 2. In other words, the gradient of y with respect to x is a constant, 2. But for any other function of x that is not linear, the change in y is not immediately obtainable without recourse to the logic of *differential calculus*. To generalize this, we graph up a nonlinear $y = f(x)$ (Fig. A.1a). The differential calculus enables us to estimate the gradient of the function at a point, making use of an initial chord PQ on the function which we shrink to the point P so that we can say the chord PQ becomes, at the limit, as Q diminishes to P, the tangent to the curve of the function at P.

Specifically, with reference to the graphed example (Fig. A.1b) function $y = x^2 + 2$, point P is at position $x = 2$, $y = 6$, or $P_{2,6}$. Now form a right triangle PQC, placing Q as another point on the function curve with C at the normal apex. Let position C be $C_{2+h,\ 6}$, where $PC = h = \delta x$ for short. Point Q is also at x-position $(2 + h)$, with its y-coordinate as $y = x^2 + 2 = (2 + h)^2 + 2 = h^2 + 4h + 6$. The side QC of the triangle PQC is thus $h^2 + 4h$. Simple trigonometry of the right triangle gives the gradient of the chord PQ as $QC/PC = (4h + h^2)/h =$

$4 + h^2$. Now, shrinking the chord PQ to P (i.e. h or $\delta x \to 0$) the gradient at the limit at P (where $x = 2$) becomes 4 or $2x$. Thus at $P_{2,6}$, $y = x^2 + 2$ and $dy/dx = 2x$. Note that the differential gradient does not depend on the integer at the end of the function $y = x^2 + 2$.

Generally, the gradient QC/PC is expressed as $\delta y/\delta x$ for the infinitesimal gradient. The lowercase deltas signify tiny (infinitesimal) increments. At the limit, for the gradient at point, P we say dy/dx, the differential rate of change of y with respect to x, also written in various forms, including $f'(x)$ or y'. Formally, $dy/dx = \text{Limit}(\delta x \to 0)\delta y/\delta x$.

Switching from a generalized functional form to a physical expression, the function in question may be the rate of change of distance, y, with time, t, in other words, velocity, $u = dy/dt$. This is termed a first differential. Or it could be the rate of change of velocity with time, du/dt, an acceleration. Now in this case, since $u = dy/dt$, the acceleration is also $d/dt(dy/dt)$ or d^2y/dt^2. This is termed the second differential coefficient. Similarly the rate of change of acceleration with time is $d/dt\,(d^2y/dt^2)$ or d^3y/dt^3, the third differential.

There are some common standard first derivatives that are tempting to learn because we come across them quite often in physical processes. Our example above is the first

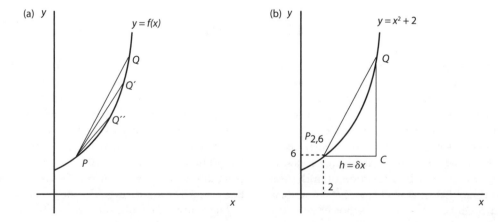

Figure 1. To illustrate the derivation of the principles of the differential calculus. 1A. For any function $y = f(x)$, as Q approaches fixed point, P, the chord PQ becomes, in the limit, the tangent to the function curve at P. This is best visualised by imagining PQ shrinking gradually via PQ′ and PQ″ to fixed point P. 1B. To illustrate the argument (see text) that the derivative of the function $y = x^2 + 2$ for an infinitesimal gradient at point P is $2x\,(=4$ for this case).

in the list and ought to be remembered if nothing else is! Try it out for $y = x^3$.

$$y =$$

$f(x)$	$dy/dx = f'(x)$	Or, equivalently
x^n	nx^{n-1}	
$\log_e x$	$1/x$	
e^x	e^x	
$\sin x$	$\cos x$	
$\cos x$	$-\sin x$	
$\tan x$	$\sec^2 x$	$1/\cos^2 x$
$\cosh x$	$\sinh x$	$(e^x - e^{-x})/2$
$\sinh x$	$\cosh x$	$(e^x + e^{-x})/2$
$\tanh x$	$\text{sech}^2 x$	$1/(\cosh x)^2$
$\sin^{-1} x$	$1/(1 - x^2)^{0.5}$	
$\cos^{-1} x$	$-1/(1 - x^2)^{0.5}$	
$\tan^{-1} x$	$1/(1 + x^2)$	
$\sinh^{-1} x$	$1/(1 + x^2)^{0.5}$	
$\cosh^{-1} x$	$1/(x^2 - 1)^{0.5}$	
$\tanh^{-1} x$	$1/(1 - x^2)$	

A.4.2 Partial differential symbol, ∂

When y depends only upon x, that is, the variation of y with x is dy/dx, a curve is defined in 2D. In 3D y also depends upon z. In such cases, $y = f(x, z)$, and a surface is defined and we say that y partially depends upon x and z, or $\partial y/\partial x + \partial y/\partial z$.

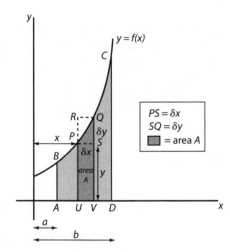

Figure 2. To illustrate the derivation of the principles of the *integral calculus*. For any function $y = f(x)$, the integration provides the area under the functional curve. This total area, for example, ABCD, can be regarded as comprising an infinite number of small strip areas like UVQP shown. So, for such a strip, as point Q approaches position P then the shaded strip of area, A, vanishes as δx goes to zero and δy also goes to zero. At this point $dA/dx = y$ and $A = ydx$. This is the *indefinite integral*. For the *definite integral*, like that total area ABCD under the curve between $x = a$ and $x = b$, see text.

A.4.3 Integral calculus

While differential calculus is concerned with the gradient of a curve (function) at a point, like P in Fig. A.1a, integral calculus calculates the area under the curve (function), like area UPQV in Fig. A.1b.

Integration (symbol $\int$), reverses the process of differentiation so that if we are given $d\alpha/dx = y$, then $\alpha = \int y dx + c = \int f(x)dx + c$. The parameter c that has mysteriously crept in is known as the constant of integration and must generally be present since the function may have an unknown integer that drops out during differentiation (like the integer 2, in our function $y = x^2 + 2$ used previously).

It is easiest to think about integration physically using velocity as an example. We know velocity is the rate of change of distance, s, over time, $u = ds/dt$. If we want to determine s, this is the integral $s = \int u dt + c$. Also we know that acceleration is the rate of change of velocity over time, $a = du/dt$. If we want to determine velocity, we need the integral $u = \int a dt + c$. We now go on to prove that all this is true and not just some mathematical gibberish.

With reference to Fig. A.1c, we first need to find any infinitesimal area under the curve of the generalized function, $y = f(x)$. We first define a tiny area under the function curve, UVQP by letting P and Q be any two points on the curve. The position of P is defined as $P_{x,y}$ and the position of Q as $Q_{x+\delta x,\, y+\delta y}$. Then ABPU is the area under the curve from $x = a$ to $x = x$; ABQV is that area from $x = a$ to $x = x + \delta x$; the increment in area, δA is UPQV. Note that UVSP $\leq \delta A \leq$ UVQR:-

$$y\delta x \leq \delta A \leq (y + \delta y)\delta x,$$

dividing by δx

$$y \leq \delta A/\delta x \leq y + \delta y.$$

Now let δx go to zero, then δy also goes to zero and the lower and upper bounds of $\delta A/\delta x$ also go to zero. Switching to infinitesimals we have the defining expression for the infinitesimal limit, as δx goes to zero,

$$\delta A/\delta x = dA/dx$$

then,

$$dA/dx = y$$

or

$$A = \int y dx = \int f(x)dx + c.$$

This is the *indefinite integral* and is the general integral solution to a given function. For a given function $y = x^n$, we have the general rule

$$\int y dx = \int x^n dx = \frac{x^{n+1}}{n+1}$$

A few common integrands are as follows:

$f(x)$	$y =$
	$\int y \, dx$
x^n	$x^{n+1}/(n+1)$
$\cos x$	$\sin x + c$
$\sin x$	$-\cos x + c$
e^x	$e^x + c$
$1/x$	$\ln x + c$

But what about a specific defined area under the functional curve, like ABCD? This is taken between the limits $x = a$ and $x = b$ (Fig. A.1c), defined by the *definite integral*

$$A = \int_a^b y \, dx = \int_a^b f(x) \, dx$$

Substituting for $x = b$ we subtract this from the value of $x = a$. Taking our example function, $y = x^2 + 2$, we thus have as its integral

$$A = \int_a^b (x^2 + 2) \, dx = \left[\frac{1}{3} x^3 + 2x \right]_a^b$$

$$= \frac{1}{3} b^3 + 2b - \frac{1}{3} a^3 - 2a$$

A.5 Scalars

A scalar quantity has only dimensions of its own magnitude, with no direction additionally specified. Easy examples are temperature (as on a weather chart), speed (as on a speeding ticket; Marta was travelling at 63 mph in a restricted 30 mph zone, it matters not for the particular offense which way she was heading), energy, or power (her car engine has 142 horsepower but no direction is specified). Here is a list of scalars: *length, area, volume, linear speed, angular speed, work done, electrical resistance, power, energy, mass, density, temperature.*

A.6 Vectors: Scalars with attitude

A vector quantity has both magnitude and direction, though the latter is not specified in the units involved, only in the direction taken with respect to given x, y, z coordinates. Thus Marta traveling at 63 mph the wrong way down a one-way street is a more serious matter than just the speed of travel; vectors add crucial information to any physical description, in this case increasing the severity of the ticket. With Cartesian coordinate axes fixed at right angles, three components to any vector are possible, in the x, y, and z directions; all vectors may be resolved into such 3D components. Easy examples are velocity (speed in a given direction: u_x, u_y, u_z) and force (mass times acceleration, positive in the direction of the acceleration). For the common case of velocity, the three different components are usually given different symbols, though these vary between disciplines; thus in this text we may state the overall velocity, u, with components u in the streamwise (x) direction, w in the vertical (z) direction, and v in the spanwise (y) direction. The reader should be careful with notation as some workers switch notation of the y and z axes and we do sometimes in this text. There is no problem with this since the axes are arbitrary anyway (Descartes is supposed to have invented orthogonal coordinate axes while watching a fly crawling across his rectangular bedroom ceiling!), but it is important always to check for consistency and usage. Here is a list of common vectors: *displacement, velocity, vorticity, force, acceleration, momentum, electrostatic force, magnetic force, electric current.*

A.6.1 The meaning of vector operator ∇

Consider elevation or temperature. In order to discuss possible variations in topography or thermal energy in space (a common situation in Nature), it is necessary to be able to succinctly state the variation of these scalars in 3D, for example, the variation of height or temperature in space. We did this with vectors above with the implicit assumption that 3D components are an essential part of the definition of vectors. We have seen repeatedly that 3D components are defined with respect to three coordinate axes. These define a *scalar field*, defined as a region of space where a scalar quantity can be associated with every point. So, temperature or elevation in a steady scalar field is a function of axes x, y, z only, that is, $T(x, y, z)$ or $h(x, y, z)$. Now a field defines a geographic spread of values, a space rate of change in fact, for the field variable; it may be greater or lesser depending on direction. We would like to be able to state this succinctly; like as your eye would seek out the steepest topographic slope for a toboggan run or a gentler slope for a more cautious ski run. We have seen already that rates of change are a concept used in differential calculus; we can simply denote the space rate of change of T or h as dT/ds or dh/ds, in 3D $(\partial T/\partial x + \partial T/\partial y + \partial T/\partial z)$ or $(\partial h/\partial x + \partial h/\partial y + \partial h/\partial z)$. The partial differentials are there because the scalar can vary in the three independent coordinate directions. This is rather clumsy to write out so we refer to *gradients* in a scalar quantity as Grad T or Grad h, or by the symbol, ∇, denoted DEL or NABLA. Thus we have ∇T or ∇h. Note that in vector geometry, a gradient is the *steepest* slope possible in a given scalar field. Now we can simply say that

$$\text{Grad } h = \nabla h = (\partial h/\partial x + \partial h/\partial y + \partial h/\partial z)$$

the quantity $\nabla = (\partial/\partial x + \partial/\partial y + \partial/\partial z)$ being the *vector operator*. In our examples, ∇T gives the heat flow direction and ∇h the topographic gradient at any point.

A.7 Tensors

Tensors are vector quantities whose vectorial components are each resolvable into three further components, giving a total of nine in all. Stress and strain are tensors (Sections 24 and 25; Cookie 22.4), as is vorticity. The concept is best imagined for stress, τ, with each main vectorial component τ_x, τ_y, τ_z further divisible into every possible combination, namely $(\tau_{xx}, \tau_{xy}, \tau_{xz})$, $(\tau_{yx}, \tau_{yy}, \tau_{yz})$, and $(\tau_{zx}, \tau_{zy}, \tau_{zz})$. Early workers in vectorial physics recognized the clumsiness of the equation sets released by such tensor notation and refer simply to the matrix τ_{ij}, with $i = 1$–3 for the vectorial components and $j = 1$–3 for the subordinate tensorial components, the repeated summation ij indicating the sum of the various combinations.

Further reading

There are very many introductory match texts and every instructor and student has their own favourite. The best that caters specifically for remedial needs in the Earth Sciences is D. Waltham's *Mathematics for Geologists* (Chapman and Hall, 1994): the book contains many worked examples from a diverse range of earth and environmental sciences. A useful calculus primer is P. Abbott's *Teach Yourself Calculus* (Teach Yourself Books, Hodder and Stoughton, 1997). It is worth seeking out second hand copies of A. G. Wilson and M. J. Kirkbys' *Mathematics for Geographers and Planners* (Oxford, 1975), it is very thorough at a more advanced level. We also like the basic college text, A. E. Coulson's *An Introduction to Vectors* (Longmans, 1969). For a more advanced treatment, try D. E. Bourne and P. C. Kendalls' *Vector Analysis and Cartesian Tensors* (Chapman and Hall, 1992).

Cookies

Cookies are good for your intellectual health.

Cookie 1   Going with the flow: Potential flow and Laplace's equation

With reference to standard 2D coordinates of Fig. 2.20, it can be shown that for any stream tube in a 2D flow, there is a magnitude of discharge ψ, known as the stream function, such that $\psi = \int u \, dz + \int v \, dx$ or, $|u| = \partial\psi/\partial z$ and $|v| = \partial\psi/\partial x$. These expressions simply state the obvious fact that velocity is higher when streamline spacing is closer and vice versa.

Velocity potential lines are (imaginary) lines of equal velocity drawn normal to streamlines. The velocity is the gradient of ϕ, that is, the streamline direction of greatest rate of change of ϕ normal to the equipotential lines ϕ = constant. Thus $|u| = \partial\phi/\partial x$ and $|v| = \partial\phi/\partial z$, in a similar way to the relationships deduced for stream functions.

The 2D velocity potential and the stream function are connected and interchangeable via the Cauchy–Riemann equations $\partial\psi/\partial z = \partial\phi/\partial x$ and $\partial\phi/\partial z = \partial\psi/\partial x$.

From the 3D continuity equation (Section 2.5; Cookie 2), $\operatorname{div} q = \nabla \cdot q = \partial u/\partial x + \partial v/\partial y + \partial w/\partial z = 0$, we can substitute for u, v, w from the velocity potential and stream function relations to get the condition $(\partial^2\phi/\partial x^2) + (\partial^2\phi/\partial y^2) + (\partial^2\phi/\partial z^2) = \nabla^2\phi = 0$ and $(\partial^2\psi/\partial x^2) + (\partial^2\psi/\partial y^2) + (\partial^2\psi/\partial z^2) = \nabla^2\psi = 0$. These expressions define Laplace's equation, the basis of all potential theory and of widespread use in the advanced mathematical physics of inviscid, incompressible, continuous, and irrotational fluids.

To understand what all the fuss is about, it should be realized that when the above conditions are violated, that is, div $q \neq 0$, then for ψ, vorticity is present and for ϕ, net fluid is departing from or arriving at the control volume. Thus our previous discussion of streamlines, stream functions, and potentials is applicable only if the fluid is ideal (no shear stresses act on the fluid) or if the fluid is irrotational. When shear stresses act or when rotation occurs then the streamlines and potential lines are no longer normal; we cannot apply ideal flow considerations very easily to boundary layers (Section 4.3).

Cookie 2   What goes in . . . : General continuity expression in 3D

Continuity can be stated for a fixed point where the local velocity and/or density change is stipulated. This is relevant to the stationary observer recording the flow as it moves past. Such a fixed observation site with respect to any moving fluid is termed the Eulerian reference frame. However, we also need to consider the changes as the flow moves from point to point, appropriate to a moving observer (you are surfing with the flow, measuring as you go). This is called the Lagrangian coordinate frame. Note that this is important, for even if density does not vary at a point, it may well vary laterally or vertically, due to features like salinity stratification or temperature gradients. The convention adopted is that the Lagrangian operator D/Dt, termed the *substantive derivative*, sometimes known as the *total derivative*, expresses both the spatial and time gradients of any variable (Section 3.2).

The control cube has sides dx, dy, dz and is fixed in the fluid volume and therefore we are considering continuity

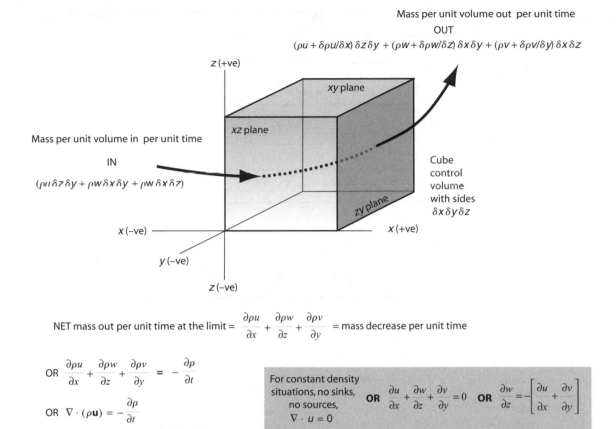

Mass per unit volume out per unit time
OUT
$(\rho u + \delta\rho u/\delta x)\, \delta z\, \delta y + (\rho w + \delta\rho w/\delta z)\, \delta x\, \delta y + (\rho v + \delta\rho v/\delta y)\, \delta x\, \delta z$

Mass per unit volume in per unit time

IN

$(\rho u\, \delta z\, \delta y + \rho w\, \delta x\, \delta y + \rho w\, \delta x\, \delta z)$

xy plane

xz plane

zy plane

Cube control volume with sides $\delta x\, \delta y\, \delta z$

z (+ve)

x (−ve)

x (+ve)

y (−ve)

z (−ve)

NET mass out per unit time at the limit $= \dfrac{\partial\rho u}{\partial x} + \dfrac{\partial\rho w}{\partial z} + \dfrac{\partial\rho v}{\partial y}$ = mass decrease per unit time

OR $\dfrac{\partial\rho u}{\partial x} + \dfrac{\partial\rho w}{\partial z} + \dfrac{\partial\rho v}{\partial y} = -\dfrac{\partial\rho}{\partial t}$

OR $\nabla \cdot (\rho\mathbf{u}) = -\dfrac{\partial\rho}{\partial t}$

OR $\nabla \cdot \mathbf{u} = -\dfrac{1}{\rho}\dfrac{D\rho}{Dt}$ for changes in time **and** space for unit volume

For constant density situations, no sinks, no sources, $\nabla \cdot u = 0$ OR $\dfrac{\partial u}{\partial x} + \dfrac{\partial w}{\partial z} + \dfrac{\partial v}{\partial y} = 0$ OR $\dfrac{\partial w}{\partial z} = -\left[\dfrac{\partial u}{\partial x} + \dfrac{\partial v}{\partial y}\right]$

This approach is used in Fig. C2.3 working (see text)

Fig. C1 General mass continuity in 3D for flow through a fixed infinitesimal volume dxdydz in space. The accompanying working thereafter uses partial differential notation, for example, $\partial u/\partial x$, etc.

at a point in space, in the Eulerian view. Let fluid pass through the cube and let us resolve the flow vector into its usual three components. Take the x-direction as an example. Mass flow IN over unit time is the product $\rho u\,\mathrm{d}y\,\mathrm{d}z$. You can check that this gives mass by dimensions. Mass flow OUT is mass flow in plus any change in ρu along dx. The full algebraic expression for this is given in Fig. C1.

Cookie 3   Ocean commotion: An application of continuity

Three-dimensional continuity for constant density gives $(\partial u/\partial x) + (\partial v/\partial y) + (\partial w/\partial z) = 0$. Let's apply this to an oceanographic example (Fig. C2) in which 2D surface velocity data for u and v are available. In this case, we rearrange the continuity equation to $(\partial u/\partial x) + (\partial v/\partial y) = -(\partial w/\partial z)$. We can now determine the left-hand side of this from the data and find the right-hand side to indicate likely vertical velocity gradient. With reference to Fig. C2 (pay close attention to the coordinate velocity component signs) we have four lots of surface velocity data for u (short for u_x) and v (short for v_y)

at points A–D located evenly and symmetrically around the center of attention at point x. We will assume that any gradients of velocity are smoothly varying (i.e. linearly) from point to point. We want to know the average vertical velocity at point x. Some clue as to this comes from sketching in the resultants of u and v, whereupon it will be seen that we seem to have surface converging flow in the region of x. Since this converging flow must go somewhere, and it cannot go upward (there are no waterspouts in the region), the mean vertical flow at x must be down. We can compute the actual

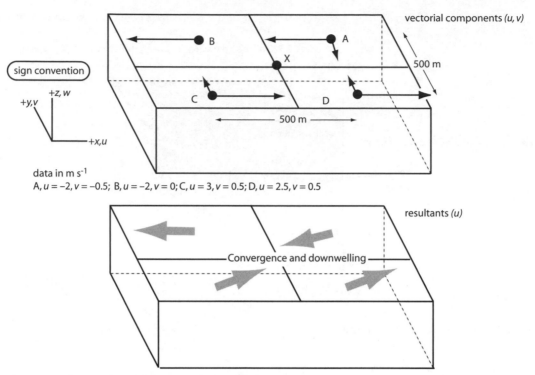

Fig. C2 Ocean surface currents: use of div $u = 0$ (see working box of Fig. C1).

value of this flow as follows:

1 Determine average surface x-velocity gradients (these are spatial accelerations (see Section 3.2), appropriate to position x halfway between AB and CD. Midway along AB, du/dx is $[(-2) - (-2)]$ m s^{-1}/500 m = 0 s^{-1}, while at the midpoint along CD, du/dx is $[(+2.5) - (+3)]$ m s^{-1}/500 m = $-1 \cdot 10^{-3}$ s^{-1}. The mean value of these two estimates appropriate to point x is $-5 \cdot 10^{-4}$ s^{-1}.

2 For the y-velocity gradients we determine halfway between BC and DA. For BC, dv/dy is $[(0) - (+0.5)]$ m s^{-1}/500 m = $-1 \cdot 10^{-3}$ s^{-1}. For DA, $dv/dy = -2.10^{-3}$ s^{-1}. The mean value of these two estimates appropriate to point x is $-1.5 \cdot 10^{-3}$ s^{-1}.

3 Solving for dw/dz, we get $-1.5 \cdot 10^{-3}$ s^{-1} as the vertical component of mean velocity gradient appropriate to point x. We infer downward motion from the convergence argument above, but also from the necessity since $w = 0$ at the surface. For a depth of 50 m this corresponds to a mean downward vertical velocity of some 0.075 m s^{-1}.

The example is somewhat an artificial one but it goes to show that vertical velocities in the ocean are appreciably less than horizontal ones. This is because the oceans are incredibly wide compared to their length, say in the ratio $h/\lambda \sim 10^{-3}$ to 10^{-4}. If we say that a vertical velocity, w, changes as $\sim uh/\lambda$ then, typically, $w \sim 10^{-3} u$.

Cookie 4 Free fall: Constant acceleration due to gravity

Many natural systems exist where acceleration, a, may be constant, that is, not varying with time, or at least approximately. Examples are systems acting under gravity, like the rise or fall of masses in air, where friction is neglected as small. The velocity as a function of time elapsed, that is, at any future time, t, under these conditions is:

$$u_t = at + u_0$$

where u_0 is the initial velocity. Because acceleration is constant the *mean* velocity, $\bar{u}$, increases linearly over the time interval and is given by half the sum of the initial velocity plus the final velocity, that is, $\bar{u} = (u_0 + u_t)/2$. The corresponding displacement is $x = \bar{u}t + x_0$, where x_0 is the initial position. Substituting for $\bar{u}$ we have $x = (u_0 + u_t)t/2 + x_0$. Substituting again, for u_t this time, we get:

$$x = at^2/2 + u_0 t + x_0 \quad \text{or} \quad \delta x = at^2/2 + u_0 t$$

where δx is the net displacement. This useful expression means that the position or displacement of any constantly accelerating mass varies as the square of time elapsed. If we want to determine position, displacement, velocity, or acceleration without reference to time then we can use $u_t = at + u_0$ to substitute for t as $t = (u_t - u_0)/a$. The expression, $x = (u_0 + u_t)t/2 + x_0$, then becomes $x = [(u_0 + u_t)/2][(u_t - u_0)/a] + x_0$. This expression can be written for velocity at any place in terms of acceleration, initial velocity, and initial position:

$$u_t^2 = u_0^2 + 2a(x - x_0).$$

The three simple expressions given above are of great use in calculating the motion of freely falling bodies under situations where we can neglect the drag force (resistance) due to fluid viscosity. Simply substitute gravitational acceleration, g, for a acting in the direction $-y$ to the center of the Earth, that is, the y-axis is vertical and positive outward from the center.

For *example*; a boulder falls from a vertical 200 m rockface during an earthquake. How long does the boulder take to reach the ground? What is the velocity at impact? What is the force exerted at impact if the mass is 20 kg?

Initial conditions at $t_0 = 0$, are $z_0 = 200$ m and $u_0 = 0$ m s^{-1}. Use $z = -gt^2/2 + u_0t + z_0$ and solve for t to get $t = 6.4$ s. Then use $u_t = -gt + u_0$, to get $u_t = -62.7$ m s^{-1}. [For such masses falling freely from rest at a given height, h, a useful simplification of $u_t^2 = u_0^2 + 2a(x - x_0)$ for the initial conditions yields $u_t = \sqrt{[2gh]}$. From Newton's second law, the force exerted at impact is $mg = -196$ N.

Cookie 5   Energizing: to show that kinetic energy $= 0.5mu^2$

Energy, the power of "doing work" by a moving body, was first used in the modern sense by Young in 1807 and extended to include potenial energy by Rankine in 1853. From Cookie 4 we have expressions for the displacement and velocity of any mass, m, acted upon by a constant acceleration like that due to gravity. From Newton's second law we know therefore that the constant acceleration, a, must be the result of a constant net force, F, acting and the expressions, $\delta x = at^2/2 + u_0t$ and $u_t = at + u_0$,

can also be written as $\delta x = (F/2m)t^2 + u_0t$ and $u_t = (F/m)t + u_0$ respectively. Solving for t in the latter expression and substituting in the former we get, after expansion and rearrangement:

$$F\delta x = 0.5mu_t^2 - 0.5mu_0^2$$

$F\delta x$ is the work done by the net force, F, along the displacement δx, a definition of the Joule unit of net work. This is thus equal to any change in kinetic energy, $\Delta 0.5mu^2$.

Cookie 6   Feeling the pressure: Pascal's Law for fluids where density is independent of pressure, that is, liquids

The basic hydrostatic equation derived in Section 3.5 and for the notation defined in Fig. 3.22 is $dp/dz = -\rho g$, so $dp = -\rho g\, dz$ and $p = -\rho g \int dz$. Integrating between the free surface at $z = 0$ and depth $-z$, where the pressures are atmospheric, p_a, and p_y respectively, $p = -\rho g\int_0^{-z} dz$ and we have $p_y = \rho gz + c$. To find the integration constant let $z = 0$, then $c = p_0 = p_a$ and for any depth, $p = \rho gz + p_a$. Thus the *absolute pressure* at any depth is equal to constant density times g times depth plus the atmospheric pressure. In effect the increasing weight force of the water column with depth causes a proportionate increase in the frequency of molecular collisions and hence a linearly increasing pressure. For many problems in oceanography and hydraulics the relatively small atmospheric pressure contribution may be neglected and the pressure is then referred to as the *gauge pressure* or expressed as a *head*, the static height of an equivalent column of fluid above any level.

Cookie 7   Gas attack: Solutions of the hydrostatic pressure equation (Section 3.5) for compressible fluids where density is dependent only on pressure, that is, ideal isothermal gases

Here we must take into account the variation of density with pressure. We make use of the perfect Gas Law

(Section 3.4), which states that $pV = RT$ for unit mass of gas. Since ρ is V^{-1}, we have $p/\rho = RT$ or $\rho = p/RT$

and the hydrostatic differential equation becomes $dp/dz = -pg/RT$ or $\int 1/p = -(g/RT) \int dz$. Assuming the effects of continuous vertical variations in temperature are small (compared say to the effects of variably unknown amounts of water vapor for example), we can integrate between pressures at certain heights, that is, pressure p_1 at height z_1 and p_2 at height z_2 so that $\int_{p_1}^{p_2} 1/p = -(g/RT)\int_{z_1}^{z_2} dz$. This becomes $\log_e(p_2/p_1) = -(g/RT)(z_2 - z_1)$ or $\log_{10}(p_2/p_1) = -0.434(g/RT)h$, where h is $(z_2 - z_1)$, the vertical height between two points where the pressures are p_2 and p_1 respectively. We see that there is a logarithmic decrease of pressure with height in an isothermal gaseous atmosphere.

We can now obtain the *atmospheric hypsometric equation* that tells us the thickness of any layer, Δz, between surfaces of equal pressure, such that $\Delta z = [(R/g) \ln (p_1/p_2)] T(y)$. This tells us that for a given pressure drop the thickness of atmosphere between two constant pressure surfaces is a function of temperature alone. This informs the derivation of the thermal wind concept in Section 6.1.

Cookie 8   Turbulent times: Reynold's accelerations, turbulent stresses, and turbulent kinetic energy

Turbulence assumes 3D eddy motions in a steady, uniform flow. If we measure velocity over time at a point in such a flow then the velocity signal will fluctuate about a steady time-mean. Although the mean fluctuation is always zero (because it is a change plus or minus about a mean value and these cancel out), the mean square of the fluctuations is not, because, as Maxwell developed, the square of the negative fluctuations becomes positive. Working in 1D for simplicity, the instantaneous flux of momentum in the x-direction is $u(mu)$, written per unit fluid volume as $u(\rho u)$, with $u = \bar{u} + u'$, where $\bar{u}$ is time mean velocity (remember the flow is steady) and u' is instantaneous fluctuation about the mean. So, we have $u(\rho u) = \rho u^2 = \rho(\bar{u} + u')^2 = \rho(\bar{u}^2 + 2\bar{u}u' + u'^2)$. Now, the mean flux of momentum over a longer time interval is $\rho \overline{u^2} = \rho(\overline{\bar{u}^2} + \overline{2\bar{u}u'} + \overline{u'^2})$ or simplified to $\rho(\overline{\bar{u}^2} + \overline{u'^2})$, since the central term in the triple term bracket involving the time mean of the fluctuation is zero. Turbulence thus involves time mean turbulent accelerations. These are caused by forces due to pressure fluctuations in the boundary layer. It has become customary to regard the mean turbulent accelerations as virtual stresses by taking the terms over to the sum of forces side of the Navier–Stokes equations of motion, $ma = \Sigma F$, where they are signed negative and termed *Reynold's stresses*. We can write similar expressions for the other two turbulent velocities (y and z components) and there are a total of six independent Reynolds stress terms in the tensor τ_{ij}. Writing the total tensor term *for unit mass* as $(\overline{u_i' u_j'})$, u_{ij} is thus

$$
\begin{matrix}
u'_{11} & u'_{12} & u'_{13} \\
u'_{21} & u'_{22} & u'_{23} \\
u'_{31} & u'_{32} & u'_{33}
\end{matrix}
$$

in subscript notation, or

$$
\begin{matrix}
u'u' & u'v' & u'w' \\
v'u' & v'v' & v'w' \\
w'u' & w'v' & w'w'
\end{matrix}
$$

in *xyz* coordinate notation.

The total acceleration term for steady, uniform turbulent flow is $D(\bar{u} + u')/Dt \equiv \partial/\partial x_j (\overline{u_i' u_j'})$. The right-hand side represents the action of velocity fluctuations on the mean flow because of the existence of eddies transporting fluid of varying velocity and thus creating space-varying velocity. The expression states that a net mean acceleration is produced in steady, uniform turbulent flows due to gradients in space of the turbulent fluctuations. So, going back to our basic point concerning accelerations and forces, net forces arise on unit fluid cells in steady uniform turbulent flows and produce rate of change of momentum. Or, more correctly, since we are viewing accelerations, we may say the turbulent acceleration requires a net force to produce it.

In the calculation of turbulent kinetic energy (TKE) production or destruction, generally for TKE transfer from the mean flow to the turbulence we have the expression $-((\partial \bar{u}_i/\partial x_j)\rho(-\overline{u_i' u_j'}))$. This is the major energy *transfer* term in the Navier Stokes equations for turbulent flows, with the negative indicating TKE *taken* from the mean flow. When $i \neq j$ so that $(-\overline{u_i' u_j'})$ is a shear stress, this usually has the same sign as the velocity gradient. No gradient in the mean gives zero product. When $i = j$ so that $(-\overline{u_i' u_j'})$ is a normal stress, the overall contribution may be either positive or negative, depending on the sign of $\partial \mathbf{u}'_i/\partial x_j$ for $i = j$.

Now, expand the terms to explore this a little further, since flows may be nonuniform in streamwise velocity and also have nonzero mean vertical velocities:

$$i = 1, \ j = 1,2,3. \quad -\left[\frac{\partial \overline{u}}{\partial x}\rho(-\overline{u'^2})\right];$$

$$-\left[\frac{\partial \overline{u}}{\partial y}\rho(-\overline{v'u'})\right]; \quad -\left[\frac{\partial \overline{u}}{\partial z}\rho(-\overline{w'u'})\right];$$

$$i = 2, \ j = 1,2,3. \quad -\left[\frac{\partial \overline{v}}{\partial x}\rho(-\overline{u'v'})\right];$$

$$-\left[\frac{\partial \overline{v}}{\partial y}\rho(-\overline{v'^2})\right]; \quad -\left[\frac{\partial \overline{v}}{\partial z}\rho(-\overline{w'v'})\right];$$

$$i = 3, \ j = 1,2,3. \quad -\left[\frac{\partial \overline{w}}{\partial x}\rho(-\overline{u'w'})\right];$$

$$-\left[\frac{\partial \overline{w}}{\partial y}\rho(-\overline{v'w'})\right]; \quad -\left[\frac{\partial \overline{w}}{\partial z}\rho(-\overline{w'^2})\right]$$

We see that TKE may have many sources in a turbulent flow, depending on the exact nature of the flow uniformity or nonuniformity.

Cookie 9 Conserve it: Forms of the Bernoulli–Euler fluid energy conservation equation

From Newton's second law, the physical arguments outlined in Section 3.12 and the notation of Fig. C3, mass times acceleration (rate of change of momentum) equals the total force, ΣF, acting in the direction of motion down a streamtube of infinitessimal length, δs.

Mass in the streamtube is $\rho(a + \delta a/2)\delta s$ and the acceleration due to nonuniformity (Section 3.2) in steady flow along the conic streamtube is $u\delta u/\delta s$. Neglecting $\delta a/\delta s$ as very small, rate of change of momentum is thus $\rho a(u\delta u)$.

ΣF has three components:

1 F_1 is the result of varying pressure along the sides of the streamtube, given by the product of the mean pressure, $p + \delta p/2$ and the change in tube area, δa, over the small distance δs, that is, $F_1 = p\delta a$, ignoring the second order term since $\delta p/\delta s$ is small.

2 F_2 results from the longitudinal change of pressure along δs acting on opposing ends of the streamtube. In the direction of motion, $F_2 = -(p + \delta p)(a + \delta a) + pa = -p\delta a - a\delta p$.

3 F_3 is the weight force component acting due to gravity, g, acting upon the mass of fluid, $\rho(a + \delta a/2)\delta s$ within the streamtube tilted at angle, θ, so that $\cos\theta = \delta y/\delta s$ and $F_3 = \rho(a + \delta a/2)\delta s g \delta y/\delta s = -\rho a g \delta y$, ignoring the second order term as $\delta a/\delta s$ is small.

Since rate of change of momentum equals the sum of forces acting, we have $\rho a(u\delta u) = p\delta a + (-p\delta a - a\delta p) +$

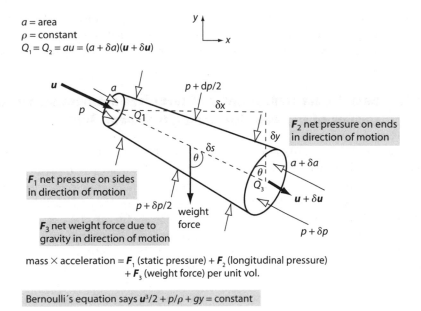

a = area
ρ = constant
$Q_1 = Q_2 = au = (a + \delta a)(u + \delta u)$

F_1 net pressure on sides in direction of motion

F_2 net pressure on ends in direction of motion

F_3 net weight force due to gravity in direction of motion

mass $\times$ acceleration = **F_1** (static pressure) + **F_2** (longitudinal pressure) + **F_3** (weight force) per unit vol.

Bernoulli's equation says **u**³/2 + p/ρ + gy = constant

Fig. C3 Euler–Bernoulli energy approach for variable velocity.

$(-\rho a g \delta y)$. Cancelling terms and changing to differential notation we have $\rho a u \, \mathrm{d}u = -a \, \mathrm{d}p - \rho a g \, \mathrm{d}y$. Dividing through by $\rho a g$ we get $(u/g)\mathrm{d}u + (\mathrm{d}p/\rho g) + \mathrm{d}y = 0$ or $(1/g)\int u \mathrm{d}u + \int \mathrm{d}p/\rho g + \int \mathrm{d}z = 0$ and for the condition of density not varying with pressure (i.e. incompressible liquid flows only) the integrals are solved to give the final Bernoulli–Euler equation for the conserva-

tion of energy $(u^2/2g) + (p/\rho g) + z = \text{constant}$. Each term in the equation has units of length and should be regarded as the quantity of energy in a fluid volume of unit weight, that is, $\mathrm{J\,N}^{-1}$. The terms are easily recognisable as (1) kinetic energy, (2) flow work energy due to pressure in the fluid, and (3) potential energy of position.

Cookie 10 Its a parabola! Equation of viscous motion for a steady Newtonian liquid flow through a conduit

For a steady Newtonian liquid flow between parallel walls there are no accelerations and no body force changes. Under these conditions the net downstream-acting viscous force $\mu \mathrm{d}^2 u/\mathrm{d}z^2$ is balanced only by the mean upstream acting pressure gradient force, $-\mathrm{d}p/\mathrm{d}x$, written for simplicity as $-p$. Using the coordinate system of Fig. 4.16 with p not varying *across* the flow (in the z-direction) and applying the boundary condition that $u = 0$ at $z = \pm a$, we have: $\mu(\mathrm{d}^2 u/\mathrm{d}z^2) = -p$ or $(\mathrm{d}^2 u/\mathrm{d}z^2) = -p/\mu = -p(1/\mu)$ and so $(\mathrm{d}u/\mathrm{d}z) = \int -p(1/\mu)\mathrm{d}z = -(pz/\mu) + c_1$. Since $\mathrm{d}u/\mathrm{d}z = 0$ at $z = 0$, then the integration constant $c_1 = 0$. We now have: $u =$

$\int -(pz/\mu)\mathrm{d}z + c_2 = -(pz^2/2\mu) + c_2$. To determine the second integration constant, c_2, since $u = 0$ at $z = \pm a$, $0 = c_2 - (pa^2/2\mu)$ or $c_2 = (pa^2/2\mu)$. Therefore finally we have $u = -(p/2\mu)(z^2 - a^2)$ or $u = (p/2\mu)(a^2 - z^2)$ and one can see that the trajectory of u across the boundary layer from $+a$ to $-a$ is a parabola, with u_{max} at $z = 0$ and $u \rightarrow 0$ as $z \rightarrow \pm a$. Reynolds first derived this expression in 1895, though the type of parabolic flow between parallel plates is often called Couette flow. Further derivations are possible for non-Newtonian liquids, different shapes of conduits (pipes, channels, etc.) and to determine volumetric discharge through these.

Cookie 11 A close thing: Solution of the viscous stress equation for flow velocity in the viscous sublayer

Experiments show that there is a linear increase of velocity with height in the viscous sublayer of turbulent flow and that turbulent stresses are negligible. Assuming therefore

that viscous stresses predominate we make use of Newton's lubricity equation, $\tau_0 = \mu \mathrm{d}u/\mathrm{d}z$ (Section 3.10). Integrating as $u = \int \tau_0(1/\mu)\,\mathrm{d}z$ we have $u = \tau_0 z/\mu$.

Cookie 12 Back to the wall: Transfer of turbulent momentum by and the "law of the wall" for steady, turbulent fluids

Further out in the flow from the viscous sublayer the rate of change of velocity with height, $\mathrm{d}u/\mathrm{d}z$, decreases. Neglecting the influence of viscosity we make the simplest possible guess as to the rate of change of this gradient, that it varies as $1/z$. Then $\mathrm{d}u/\mathrm{d}z \propto 1/z$ or $u = k\int(1/z)\mathrm{d}z$, where k is some constant, and integration gives $u = \log_e z + c$.

For a fuller physical derivation consider the spinning, rotary nature of fluid turbulence as an eddy passes through an observation volume (Fig. C4). The instantaneous velocity at the mid-point of the eddy is u. The instantaneous velocity gradient across the length, l, of the eddy is $\mathrm{d}u/\mathrm{d}z$ and the velocity of the upper and lower boundaries to the eddy motions with respect to the midpoint velocity

is $\pm 0.5l \, \mathrm{d}u/\mathrm{d}z$. This velocity transfers a mass of fluid, m, in unit time from fast to slow layers and vice versa. For cylindrical volumes of cross sectional area, a, this mass is $m = a\rho 0.5l \, \mathrm{d}u/\mathrm{d}z$ and each unit mass of fluid changes momentum by an amount $l \, \mathrm{d}u/\mathrm{d}z$. The total rate of change of momentum by both cylinders is $2(a\rho \, 0.5l\mathrm{d}u/\mathrm{d}z)(l\mathrm{d}u/\mathrm{d}z)$ and this is balanced by a shear stress acting over the ends of both cylinders of magnitude $\tau 2a$. Thus $\tau 2a = 2(a\rho 0.5l\mathrm{d}u/\mathrm{d}z)(l\mathrm{d}u/\mathrm{d}z)$, or $\tau = 0.5\rho \, 0.5(l\mathrm{d}u/\mathrm{d}z)^2$. Since eddies vary in size and strength over time and distance as they advect past an observer we must represent this expression with respect to mean velocity, mean velocity gradient, and mean eddy length

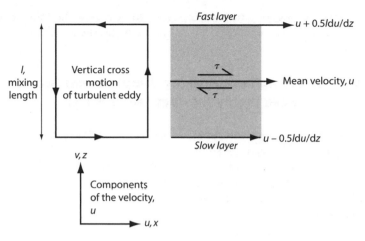

Fig. C4 Diagram for the derivation of Prandtl's "law of the wall."

by means of a constant, k. Thus $\tau = k0.5\rho\ 0.5l(\mathrm{d}u/\mathrm{d}z)^2$, or $\mathrm{d}u/\mathrm{d}z = (\tau/\rho)^{0.5}(1/(0.5k)^{0.5}l)$. Experience and experiments show that the part $(0.5k)^{0.5}\ l$ varies directly with distance, y, from the solid boundary to the flow (i.e., eddies grow in size as they move away from the bed of a river channel) where τ is designated a constant mean value at the boundary of τ_0. In fact, $(0.5k)^{0.5}l = 0.4z$ exactly, with the integer termed the *Von Karman constant* and l a characteristic *eddy mixing length*. The differential equation

then becomes $\mathrm{d}u/\mathrm{d}z = (\tau_0/\rho)^{0.5}(1/0.4z)$ or $\mathrm{d}u = (\tau_0/\rho)^{0.5}$ $\int(1/0.4z)\mathrm{d}z$. Integration gives $u = (\tau_0/\rho)^{0.5}\ 2.5[\log_e z]$ $+ c_1$. Setting $u = 0$ at positive height intercept c we have $c_1 = -(\tau_0/\rho)^{0.5}\ 2.5[\log_e 1/c]$ and $u = (\tau_0/\rho)^{0.5}$ $2.5(\log_e z - \log_e 1/c)$, or $u = 2.5(\tau_0/\rho)^{0.5}(\log_e z/c)$. Switching to $\log_{10}$ units we have the final result for the Von Karman–Prandtl "law of the wall" as $u = 5.75$ $(\tau_0/\rho)^{0.5}\ [\log_{10}(z/c)]$.

Cookie 13   Suspension of disbelief: Bagnold and Irmay momentum transfer arguments for bedload and suspension sediment transport

Concerning bedload in air, the transport rate is a cube function of the shear velocity. The flowing air accelerates each particle in transport from rest until splashdown. Solid momentum is thus gained from the air and subsequently released to the bed. Each particle of mass, m, gains velocity, U, over unit length, l and so the air loses momentum mU/l. For a total mass of particles, i_b, moving over unit area in unit time the total loss of momentum equals the drag (resisting) stress, τ, exerted by the bedload on the air, so that $i_b U/l = \tau$. Since $\tau = \rho u^\star$, where $u^\star$ is shear velocity, and both U and l are proportional to $u^\star$ we have $i_b = k\rho u^{\star 3}$ where k is an experimental constant.

Concerning the turbulent stress needed for sediment suspension, Irmay reduced the Navier–Stokes–Reynolds equations of motion for turbulent flow of constant density, ρ, and viscosity, μ, to a form for 2D channelized, steady, uniform mean turbulent flow where mean spanwise and vertical velocities are zero, that is, $\overline{v} = \overline{w} = 0$. Using standard tensor suffix notation (*See* Cookie 8): $\partial(\overline{u_i'u_j'})/\partial x_j = \partial\overline{p}/\partial x_i$ which

for $i = j = 2$ is, $\partial(\overline{w'^2})/\partial z = -\partial\overline{p}/\partial z$, where p is fluid pressure. Therefore we have the important result that in a steady uniform 2D turbulent flow of constant density and viscosity, the approximate situation in natural river channels, there may exist in every such flow vertical gradients of turbulent stress. For unit fluid volume, any net vertical turbulent stress resulting due to the rate of change of the stress with height is balanced by a corresponding gradient in the fluid pressure. If the gradient in the mean vertical turbulent stress is positive, that is, w'^2 increasing upward, then a net upward force must result since the force per unit volume in a particular direction consists of gradients of the stress components which act in that direction. Any such force has to be balanced by a corresponding vertical pressure gradient. The stress gradient equation reduces to, $\int\partial\overline{w'^2} = -(\partial\overline{p}/\partial z)\partial z$, or $\int\partial\overline{w'^2} = -\int\partial\overline{p}$ whose solution is $\overline{w'^2} = -\overline{p} + c$. To obtain c we note that at the bed, z_0, $-\overline{p} = -p_0$, the wall pressure which is a function of x alone and because $w' = 0$ here then $\overline{w'^2} = 0$ and $c = p_0$. So finally we obtain $\overline{w'^2} = -\overline{p} + p_0$.

Vertical gradients in w'^2 occur throughout turbulent boundary layers, with marked positive gradients in $\partial w'^2/\partial z$ close to the bed. It is this upward acceleration that provides the motive force for sediment suspension.

Cookie 14   Not waving but drowning: Simplest possible derivation of wave theory for deep water waves

Here we regard the wave surface as a streamline and apply Bernoulli's theory to the difference of water pressure across the trough and crest of a traveling wave. Take the harmonic water wave of amplitide, H, traveling at velocity, c, shown in Fig. C5. The speed of water flow, u, in the wave orbital motion is the distance round the wave orbit, the circumference πH, traveled over the wave period T, that is, $u = \pi H/T$. In a thought experiment in which you are the stationary observer bring the traveling waveform to a halt by imposing an opposite water flow, $-c$, then $u_1 = u - c = \pi H/T - c$ at the wave crest and $u_2 = -\pi H/T - c$ in the trough. Now, we use the notion that the wave surface is a streamline and that although the wave form no longer travels due to the counter current, there is still orbital velocity along the streamline. We use Bernoulli's equation (Section 3.12) to conserve energy between crestal position 1 and trough position 2, with the trough level $H/2$ taken arbitrarily, but conveniently, as the reference height with which to compare potential energy, that is,

$$\frac{\rho u_1^2}{2} + p_1 + \rho g H = \frac{\rho u_2^2}{2} + p_2$$

Neglecting the tiny variation of atmospheric pressure between trough and crest, $p_1 = p_2$ and dividing by ρ we get $0.5u_1^2 + gH = 0.5\,u_2^2$. Substituting $u_1 = \pi H/T - c$ and $u_2 = -\pi H/T - c$ we can expand and cancel terms to yield $c = gT/2\pi$. We see that the velocity of deep water waves depends only upon wave period or wavelength, since wave length, λ, is the product of wave velocity, c, and period, T, that is, $\lambda = cT$, then $\lambda = gT^2/2\pi$. Alternatively in terms of celerity, since $T = \lambda/c$, $c = \sqrt{(g\lambda/2\pi)}$.

Cookie 15   Groupies: Wave group velocity and energy flux in deep water waves

Because deep water waves travel at speeds determined by wavelength it is common that waves interact as they pass into each other from some source. This is a distinctive interaction rather different from that discussed for interacting shallow water waves and solitary waves (Section 4.9). For the two wave trains illustrated in Fig. C6 the small difference in wavelength causes remarkable patterns of constructive and destructive interference, with peaks and troughs in a larger waveform, the group wave, where the individual crests either reinforce each others wave height or cancel out to zero. This group wave is also called the *beat*. The simple case considered is an example of *linear wave interaction*, where the effect is achieved by wave combination (more complicated schemes of *nonlinear wave interaction* involving wave transformations into smaller and larger frequencies occurs for high frequency (less than 0.3 Hz) waves). A wave group has lower frequency and travels more slowly than the individual waves which simply appear to pass into the group at the rear and out again at the front. A nice analogy is with the *kinematic wave* phenomena of traffic concentrations along single carriageway roads; the slow-moving queue or tiresomely stationary jam receives your vehicle traveling in from behind and after some time interval you (hopefully) pass through to the front where you resume your journey at your choosen speed.

The slower speed of a group wave compared to the waves that pass through it is derived as follows. For

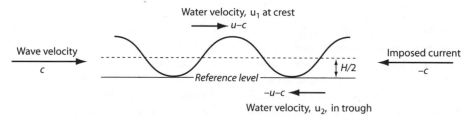

Fig. C5 Definition diagram for the artifice of bringing a deep traveling water wave of velocity, c, to a halt by applying an equal and opposite current.

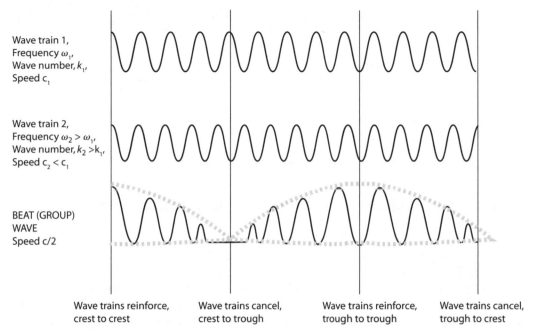

Wave train 1,
Frequency ω_1,
Wave number, k_1,
Speed c_1

Wave train 2,
Frequency $\omega_2 > \omega_1$,
Wave number, $k_2 > k_1$,
Speed $c_2 < c_1$

BEAT (GROUP)
WAVE
Speed c/2

| Wave trains reinforce, crest to crest | Wave trains cancel, crest to trough | Wave trains reinforce, trough to trough | Wave trains cancel, trough to crest |

Fig. C6 To show how linear interference between wave trains produces slower-traveling beat (group) waves.

realistic situations of group wave generation the angular frequency, ω, and wave number, k, differ little between two wave trains so that the $\delta\omega = (\omega_1 - \omega_2)$ and $\delta k = (k_1 - k_2)$. The wave group speed, c_{group} is then the ratio $\delta\omega / \delta k$ since generally for any wave, $c = \omega/k$, we can derive c_{group} as $c_1c_2/(c_1 + c_2)$. Since c_1 and c_2 differ little

this reduces to $c^2/2c$, or $c/2$, and we have the simple result that a group or beat wave travels half as fast as the waves that pass through it. In such situations the energy of the waves is all carried in the group wave, because there can be no energy at the troughs of such waves where the constituent waves cancel as they arrive or leave.

Cookie 16   Surfer: Simplest possible derivation of Airy wave theory for shallow water waves

We utilize a similar approach to that in Cookie 14 to determine the speed of a shallow water wave or a solitary wave like a bore. Here, again with the water surface representing a streamline we place u_1 at the wave crest and u_2 in front of the wave at stillwater level so that wave height is H (Fig. C7). Assuming dynamic pressures are the same along the streamline and ignoring atmospheric and capillary pressure, Bernoulli's theorem gives:

$$\frac{\rho u_1^2}{2} + \rho gH = \frac{\rho u_2^2}{2}$$

Now, the water at the crest has velocity $u_1 = (u - c)$ and in front of the wave has velocity, $-c$. Substituting gives $c^2 = u^2 - 2uc + c^2 + 2gH$. Since u is always small relative to c we ignore it and have simply $c = gh/u$.

To find u we use the continuity principle, which states that the discharge of fluid through sections 1 and 2 must

be equal. Discharge is the product of flow speed and cross-sectional area, that is, $-ch = (u - c) \ (h + H)$ or $u = c - (ch/(h + H))$, where h is water depth. Substituting for u in $c = gH/u$ we get $c = \sqrt{(g(h + H)}$ or more simply still, when wave height is much less than water depth

$$c = \sqrt{gh}$$

Full development of linear wave theory enables us to obtain a continuous function for the influence of water depth on wave speed through the expression

$$c = \sqrt{g\lambda/2\pi}\sqrt{\tan h \ (2\pi h/\lambda)}$$

in which the tan$h$ term is the hyperbolic tangent which approaches value 1 for deep water waves and tends to $\sqrt{2\pi h/\lambda}$ for shallow water, the latter giving the expression developed previously, that is, $c = \sqrt{gh}$.

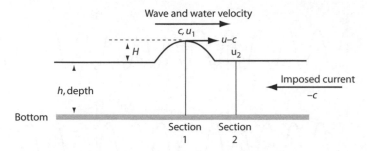

Fig. C7 Definition diagram for the artifice of bringing a shallow traveling water wave of velocity, *c*, to a halt by applying an equal and opposite current.

Cookie 17   In the rough: Static friction and dilatancy in granular masses under shear

In order to quantify the various controls upon the factors controlling static friction, first consider the effect of surface grain roughness alone. Take a mass of grains sliding over the approximately smooth surface of a solid block of the same mineral composition, for example a mass of sand resting on a sandpapered surface of about the same roughness diameter. As the solid sandpapered block is slowly tilted a simple force balance (see Fig. 4.59) for any given slope, ϕ, gives the normal force, P as $P = mg \cos \phi$, the tangential or shear force, T, as $T = mg \sin \phi$, and the ratio $T/P = \tan \phi$.

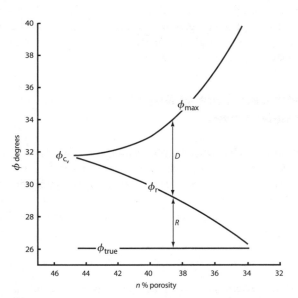

Fig. C8 Shear cell experimental data concerning the relation between various measures of the angle of internal static friction. $\phi(\phi_{max}, \phi_{CV}, \phi_{true}, \phi_r)$, and bulk mean porosity , $n\%$ for a medium-fine sand. For closely packed sands there is a larger difference between ϕ_{max} and ϕ_r, shown as the difference D, than for loosely packed sands because the applied shear stress has to do more work in dilating the particles. The two curves for ϕ_{max} and ϕ_r converge to ϕ_{CV} at high values of porosity. The difference between ϕ_{true} and ϕ_r decreases with increased packing due to less energy spent on grain fabric remoulding, shown as the difference R.

At some critical value of ϕ the grains slide down the tilted surface. In order to shear the grains over the tilted surface at this critical value of ϕ, T must exceed a certain limit. The ratio T/P is thus a friction coefficient, indicating the amount of energy that must be expended to make the grains move. For quartz grains, ϕ varies from about 31° for silt to 22° for pebbles. This is because for any given normal force the load, P_i, per individual particle increases from small to large particles (about 10^4 variation). This discussion is of a ϕ that is only appropriate for the situation described, that is loose grains resting on a solid surface. We may commonly be dealing with the friction within a mass of deposited sand that is tilted by tectonics or otherwise disturbed by slope failure. Shear planes (Section 4.14) may now form within the mass of particles. In this case, consider a volume of dry grains in a container that is tilted at ever increasing angle. At some critical angle, ϕ_{max}, some of the grains will flow off the tilted grain surface as an avalanche or grain flow. The remaining grains are now bounded by a surface resting at residual angle ϕ_r, some 5°–15° less than ϕ_{max}. These results are unaffected by the experiment being conducted under water; water surface tension effects are only relevant in damp, partially saturated sands. ϕ_{max} is the angle of initial or maximum yield and includes the effects of dilatant expansion, as well as the angularity of the grains involved; no such effect is included in ϕ_r. The actual value of ϕ_{max} is strongly dependant upon particle packing, varying between about 40° for tightly packed particles to 32° for loosely packed particles (Fig. C8). This is because maximum work has to be done by expansion in close rhombohedral packing. This expansion is generally termed *dilatancy*, from the Latin *dilatare*, to spread out, amplify, extend, widen; defined in the dictionary as "*The property of dilating or expanding, specifically that of expanding in bulk with change of shape, exhibited by granular masses, and due to the increase of space between their rigid particles when their position is changed*." The stress relation at the point of shear is $T = P \tan \phi_{max}$ and in order for any natural grain aggregate to shear the applied shearing stress must exceed 62–84% of the normal stress due to the static body force. As porosity increases the difference between ϕ_{max} and ϕ_r steadily decreases, till at maximum theoretical (cubic)

porosity of about 46 percent the two values converge to a limiting value, ϕ_{cv} in which no dilitant behavior is involved. ϕ_{cv} is thus the final or constant-volume friction coefficient appropriate when initial variations in shape and porosity are reduced by remoulding and repacking during shear.

Cookie 18   Collisions: Bagnold's derivation of intraparticle stresses in shearing Newtonian fluid-particle suspensions

In 1906, Einstein extended Newton's analysis of viscous shear resistance to include the effects of very dilute suspensions of solid particles; very dilute meaning that the particles had no effect one upon the other. For natural cases where particles are sufficiently close so that interaction or physical collision occurs, making use of Reynold's concept of dilitancy (Section 4.11) and the general ideas of kinetic theory of molecular gases (Section 4.18), Bagnold imagined (Fig. C9) a dispersion of a mass of rigid spheres of uniform density, D, and density, σ, previously arranged in canonball-style (rhombohedral) packing (Section 4.11). Here, during shear, the mean distance between particle centers is increased from D to bD and the resulting free distance between individual particle rims is denoted, s, we have $b = s/D + 1 = 1/\lambda + 1$, where $\lambda = D/s$, is *linear concentration*. L is ∞ for $s = 0$, 22.5 for rhombohedral packing, and 8.3 for the loosest cubic packing. The lower the value of λ below 8.3 the less likely that intraparticle interactions will take place. For dispersions where there is no slip between fluid and solid (the fluid and solid densities are equal, that is, neutral buoyancy occurs) and which are maintained in a state of uniform, constant shear strain, dU/dz, where U is the particle velocity, the kinetic energy per unit volume of the system is maintained by frictional

losses involving either intraparticle collisions, or a general 3D oscillation or both, due to pressure effects at particle near approaches and recessions. For the case where particle collisions dominate (grain inertia conditions) the grains of an upper layer, B, are sheared over a lower, A, at mean relative velocity $dU = kb\,dU/dz$ where k has some value between $\sqrt{0.66}$ (rhombic packing) and $(0.5)\sqrt{2}$ (cubic packing). Each grain in layer B makes $f(\lambda)\,dU/s$ collisions with an A grain in unit time, where $f(\lambda)$ is unknown. Since the number of grains in unit area in the xz plane (spanwise plane) in each layer is $1/b^2D^2$, and at each collision each B grain experiences a total momentum change of $2m\delta U\cos\alpha$ (for elastic grains, see kinetic theory, Section 4.18), where α is an unknown collision angle, the repulsive pressure in the z-direction is

$$P_z = \frac{1}{b^2 D^2}\frac{f(\lambda)\,dU}{s}2m\delta U\cos\alpha$$

$$\text{or}\quad P_z = k\sigma\lambda f(y)D^2\left(\frac{dU}{dz}\right)^2\cos\alpha,$$

where k is some proportionality constant. There is also an intraparticle particle shear stress (in addition to any fluid shear stress) of magnitude $T_{xz} = P_z\tan\alpha$.

Cookie 19   Stereographic projections

Stereographic projection is one of the most useful tools to solve geometric problems relating to the orientation of planes and lines, and it is in widespread use in structural geology, crystallography, and seismology. It is a spherical projection since all geometrical elements involved, both the coordinate system and the plotted geometrical elements, are confined in a sphere (Fig. C10a). Important elements in this sphere are the equatorial plane, which is the stereographic projection plane, the center of the sphere, and the zenith or projection point, which is the highest point of the sphere. One of the hemispheres is selected to work within; in structural geology the southern hemisphere is the

favorite (Fig. C10b). To produce stereographic projections a stereonet, which is the projection or equatorial plane in the 3D world, is used. There are different kinds of stereographic nets to solve different problems , for example, the Wulff net is an equal-angle net (Fig. C10c) and the Schmidt net is an equal-area net. Stereonets show a network of intercrossing curved lines which are projections of a series of planes (great-circles) having a common N–S strike line, dipping at different angles from 0° to 90° (Fig. C10d1), plus a series of concentric semi-cones (small-circles) having a common N–S axis (Fig. 10d2). Major great and small circles (thick lines) are drawn 10° apart, with additional

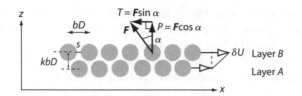

Fig. C9 Definition diagram for Bagnold's theory of particle shear collisions and the production of normal and shear components of an intragranular stress.

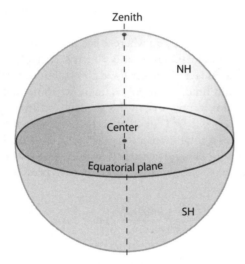

Fig. C10a

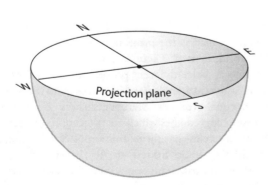

Fig. C10b

thinner circles drawn every 2° (thin lines). Note that in the Wulff net used in this Cookie both sets of projection curves cross at 90° (Fig. C10c) with the N–S and E–W axes as straight lines corresponding to vertical planes at 90°.

Projecting a plane in the stereonet

In Fig. 10e1 a set of parallel planes (i.e., strata, lava flows, or intrusions) are represented along with their orientation as strike and dip. Strike is defined as the horizontal angle between the geographic north and a horizontal line contained in the plane (line of strike). Dip is the vertical angle between the line of maximum inclination (line of dip) and its horizontal projection, which is at 90° with respect to the line of strike. Figure C10e2 shows a 3D sketch of how such a plane intersects the lower hemisphere of the projection sphere. The intersection is a semicircle, which is the geometric element projected in the stereonet. To understand how this method works it is essential to always visualize the relations of the geometrical elements that we want to represent in 3D and in the projection sphere. The semicircle is projected to the equatorial plane of the stereonet by tracing straight lines from different points of it toward the zenith (Fig. C10e3). The lines intercept the equatorial plane at points; when joined these give a circular arc (Fig. C10e4) called the cyclographic trace (CT). Finally, Fig. C10e5 shows a 2D plan view of the projection plane with the resulting stereographic projection of all the equally orientated planes in Fig. C10e1 which plot as the cyclographic trace. Vertical planes show up on the stereonet as straight lines, whereas horizontal planes always plot in the outer primitive circle (PC). Figures. C10f1–C10f4 show various steps in representing an inclined plane on the stereonet using the example in Fig. C10e. The plane has an orientation of 78°/54° S (an azimuth of 78° – from the north toward the east – and dip to the southern sector). To use the stereonet, we first put tracing paper over it, fixed by a drawing pin right at the center of the stereonet, so the tracing paper can be rotated freely. North has to be clearly marked on the tracing paper for reference. Second, we proceed to make a mark on the primitive circle counting 78° from the north toward the east (Fig. C10f1). The tracing paper is then rotated to put the mark over the N–S axis allowing the strike line to be orientated in the N–S position. Note that the N mark in the paper rotates too (Fig. C10f2). Now, another mark is needed for the dip. Dip angles are measured over the E–W axis from 0° at the primitive circle (from both E and W ends) to 90° at the center of the stereonet (F3). At this point it is important to visualize the problem in 3D

Equal-angle Stereonet

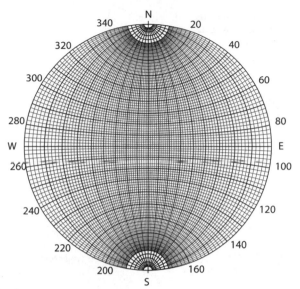

Fig. C10c

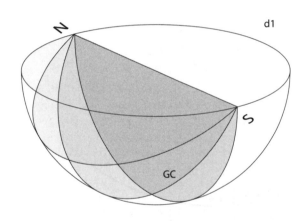

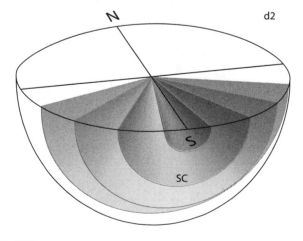

Fig. C10d

as shown in Fig. C10e2-C10e4. Note that the cyclographic trace (Fig. C10e4) is located in the same part of the circle as the dip line (see Fig. C10e2). Look at the position of North on the tracing paper (Fig. C10f3); in order to plot the cyclographic trace correctly we have to start at the E end in this case and measure the dip angle toward the center as the plane is dipping south. Finally, plot the cyclographic trace following the great circles in the stereonet (Fig. C10f4) and rotate the tracing paper to see the final result (Fig. C10f5).

Projecting a line in the stereonet

Imagine lineations on a plane, such as slickenlines on a fault plane or elongated current structures. There are two main ways to orientate the lineations. First, we can measure their trend and plunge (Fig. C10g1). The trend is the horizontal angle between geographic north and the vertical projection of the line onto a horizontal plane. The plunge is the vertical angle between the line and the same projected line on the horizontal plane. The line can be also orientated in space by giving the attitude of the plane in which the line lies and the rake (R), which is the inclined angle, measured over the plane, between the line and the horizontal line (Fig. C10g2). A sketch for the definition of trend and plunge of a line is shown in 3D in the projection sphere (lower hemisphere) in Fig. C10h1 and the stereographic projection in 2D over the stereonet in Fig. C10h2. Similarly, Fig. C10i1 depicts the rake of a line over a plane in 3D and Fig. C10i2 its projection in the stereonet. All lines pass through the center of the projection sphere and cut at two points, one in the upper hemisphere and the other in the lower hemisphere. The projection of a line is thus a dot. As we are using the lower hemisphere as the working area, the lower point (the intersection between the line and the lower hemisphere) is plotted. Figures C10j1–C10j4 show the procedure to plot a line in the stereonet by using trend and plunge. We will use the example in Fig. C10g1, where a line with a trend of 48° and a plunge of 35° S occurs on a plane. First, the line's trend is marked in the primitive circle (Fig. C10j1) at 48°, counting the angle from the north toward the east. Then the tracing paper is rotated to position the mark in one of the main axes, the N–S in this case (Fig. C10j2); the lines trend is now aligned along the N–S axis. The plunge is measured along the axis in which the trend lies from the primitive circle (0°) to the center (90°) from both ends. To choose the right end to count the dip and place the mark we have to again visualize the 3D sketch in Fig. C10h1. Note that the dot representing the projected

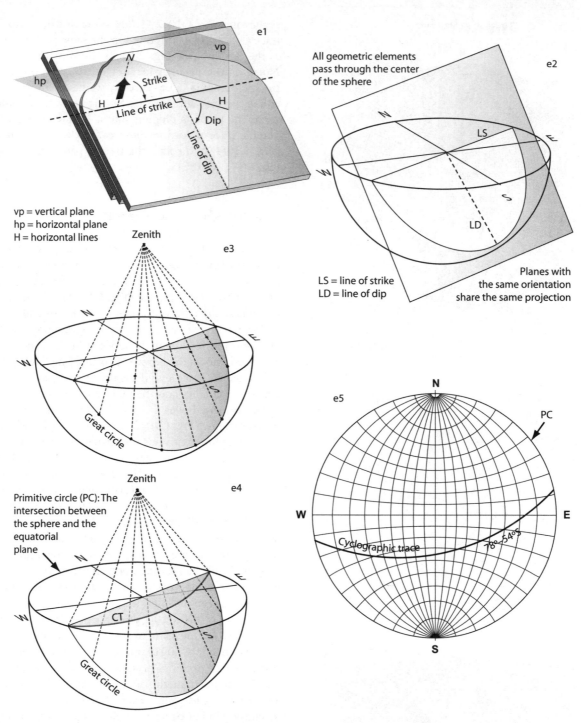

e1

vp

N

Strike

Line of strike

H

H

Dip

Line of dip

hp

vp = vertical plane
hp = horizontal plane
H = horizontal lines

All geometric elements
pass through the center
of the sphere

e2

N

LS

E

W

S

LD

LS = line of strike
LD = line of dip

Planes with
the same orientation
share the same projection

Zenith

e3

N

E

W

Great circle

Primitive circle (PC): The
intersection between
the sphere and the
equatorial
plane

Zenith

e4

N

E

W

CT

Great circle

e5

N

PC

W

E

Cyclographic trace

78°-54°S

S

Fig. C10e

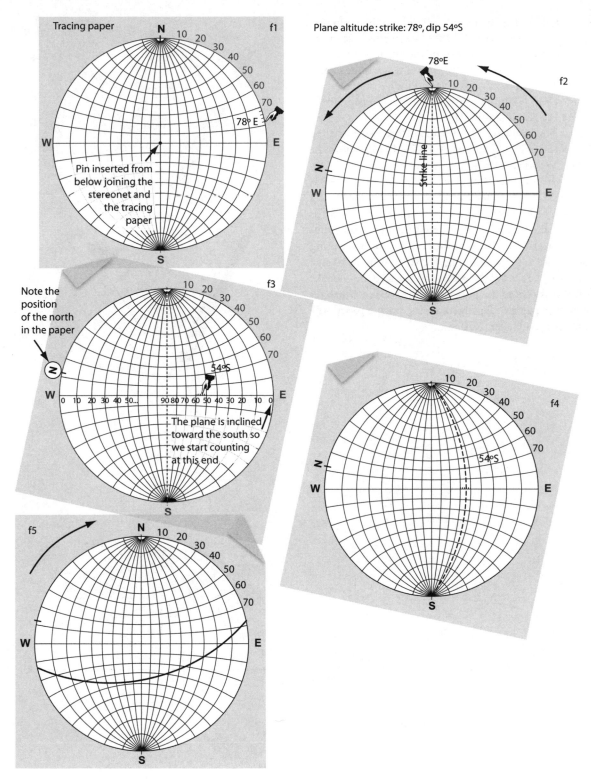

Plane altitude : strike: 78º, dip 54ºS

Tracing paper f1

Pin inserted from below joining the stereonet and the tracing paper

78º E

78ºE f2

Strike line

Note the position of the north in the paper f3

54ºS

The plane is inclined toward the south so we start counting at this end

f4

54ºS

f5

Fig. C.10f

line is located in the quadrant where the line plunges (compare Figs. C10h1 and C10h2); so in this case (note the position of the rotated north in the tracing paper) we have to measure 35° from the south end toward the center (the tick marks are the minor circles). Plot a dot representing the line (Fig. C10j3) and rotate the tracing paper back to visualize the result (Fig. C10j4).

To plot a line knowing the attitude of the plane and the rake of the line (see Figs. C10i1 and C10i2), the plane has to be plotted first. We will use the same plane as before (78°, 54°S). The rake of the line over the plane is 45°S. A sketch in 3D is shown in Fig. C10k1. Once the plane is plotted, the tracing paper is rotated till the strike line is positioned in the N–S axis (Fig. C10k2); then the rake angle is measured from the primitive line toward the center along the cyclographic trace using the minor circles as tick marks (Fig. C10k3). As in the plunge and trend case, we start at the south end because the line rake is toward the SW (note the position of the rotated North on the tracing paper again). Finally rotate the tracing paper back to see the final result (Fig. C10k4).

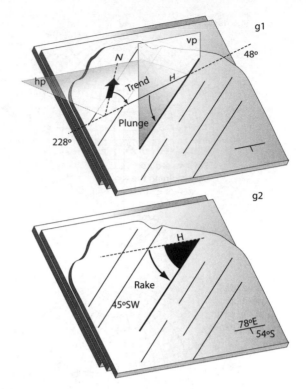

Fig. C10g

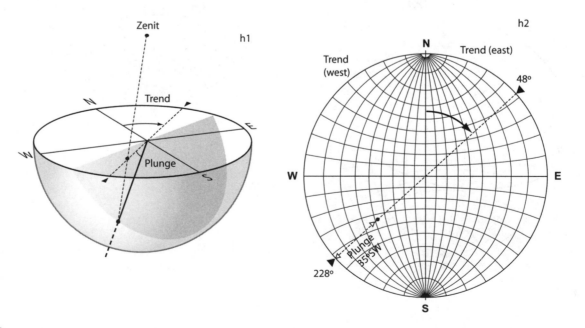

Fig. C10h

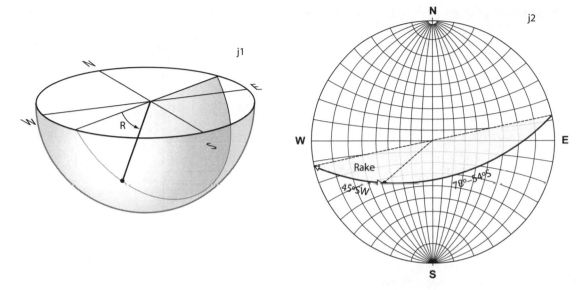

Fig. C10i

Cookie 20   Getting the Hots: Heat conduction

There are many examples in nature when a more-or-less constant heat source must propagate heat energy through matter by conduction. In such cases Biot and Fourier determined that where the matter is of uniform composition and properties, the heat flow rate, Q (in Watts of thermal energy), is invariant with time and simply proportional to the temperature difference, $-\Delta T$ between the heat source and the measuring position, that is, $Q \propto -\Delta T$. The minus sign is there because temperature difference is conventionally taken as the difference or gradient between the heat source and distant matter, that is, it is negative: since conductive heat flow is always positive, the sign is needed to give that result. Also the heat flow rate depends directly upon the area of the matter in question, and indirectly upon the distance, y, the heat flows, so that $Q \propto A/y$. Overall, therefore, for 1D flow we have $Q = -k(A/y)\Delta T$, where k is the constant *thermal conductivity* property of the matter in question. In differential form, heat flow or flux *per unit area* is thus $q = -k \mathrm{d}T/\mathrm{d}y$. This is the Fourier equation and immediately yields a useful result for average heat flux through the Earth crust of 75 mW m^{-2} where the observed temperature : depth function $\mathrm{d}T/\mathrm{d}y$ is of order 25°C km^{-1} and k is typically 3 W m^{-1} K^{-1}.

Other natural thermal systems exist when the flow of thermal energy is neither constant with time (a source body of thermal matter, like a lava flow or isolated magma chamber e.g., is cooling) and in which there may be a thermal source generating heat. To be completely general now we work in a 3D infinitesimal element and consider both time variations and source fluxes. With reference to the infinitesimal cubic volume for a system of internal energy, U_i, we have the simple continuity relation, heat in − heat out = internal energy change. Heat in is $Q_x + Q_y + Q_z + Q_g$. Heat out through the infinitesimal element is $Q_{x+\mathrm{d}x} + Q_{y+\mathrm{d}y} + Q_{z+\mathrm{d}z}$. Internal energy change over the time increment is $\mathrm{d}U_i/\mathrm{d}t$. Here are the various terms:

1 Q_x is the heat flux in through the x-direction over the infinitesimal area normal to the x-direction $\mathrm{d}y\,\mathrm{d}z$. We get this from the Fourier equation above as $Q_x = -k\mathrm{d}y\mathrm{d}z(\partial T/\partial x)$. The term has a partial differential sign because in 3D the temperature may vary in x, y, or z directions, and indeed over time.

2 $Q_{x+\mathrm{d}x}$ includes the heat in from (1) plus any change in the heat flux in the x-direction along the x-direction, or $Q_{x+\mathrm{d}x} = -[k(\partial T/\partial x) + (\partial/\partial x)(k(\partial T/\partial x))\mathrm{d}x]\mathrm{d}y\mathrm{d}z$.

3 Similar arguments yield the heat flow in the y and z directions.

4 Q_g is the heat energy generated within the cubic volume by q_g, the heat energy per unit volume, so $Q_g = q_g\mathrm{d}x\,\mathrm{d}y\,\mathrm{d}z$.

5 The rate of change of internal energy over time in the volume $\mathrm{d}x\mathrm{d}y\mathrm{d}z$ is the action of the time change in temperature, $\partial T/\partial t$, on the product of the density, ρ, and specific heat, c, so that $\mathrm{d}U_i/\mathrm{d}t = \rho c\,\mathrm{d}x\,\mathrm{d}y\,\mathrm{d}z(\partial T/\partial t)$.

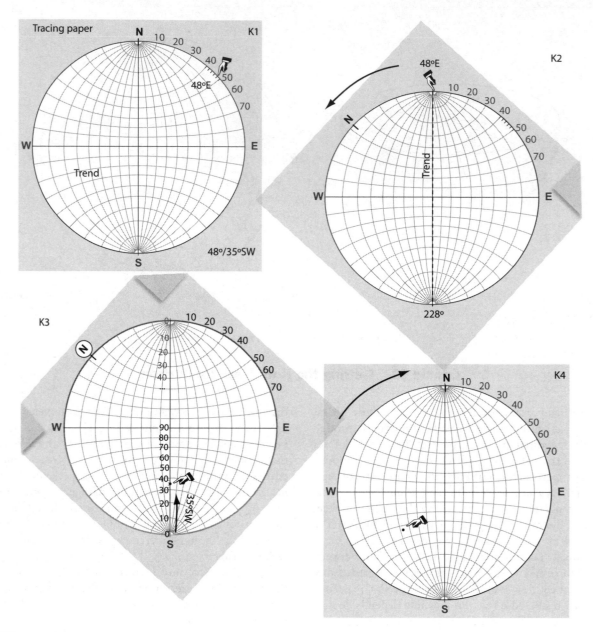

Fig. C10j

The continuity equation then becomes, after substitution:

$$\frac{\partial}{\partial x}\left[\left(k\frac{\partial T}{\partial x}\right) + \frac{\partial}{\partial y}\left(k\frac{\partial T}{\partial y}\right) + \frac{\partial}{\partial z}\left(k\frac{\partial T}{\partial z}\right)\right] + q_{\mathcal{B}} = \rho c\frac{\partial T}{\partial t}$$

And for matter of constant k

$$\left[\frac{\partial^2 T}{\partial x^2} + \frac{\partial^2 T}{\partial y^2} + \frac{\partial^2 T}{\partial z^2}\right] + \frac{q_{\mathcal{B}}}{k} = \frac{1}{\kappa}\frac{\partial T}{\partial t}$$

where κ is the thermal diffusivity $= k/\rho c$. In mathematical shorthand (useful for these 3D applications) the left-hand side in brackets is $\nabla^2 T = \nabla \cdot \nabla T$.

For situations where there is no change in the temperature with time we get *Poisson's equation*

$$\left[\frac{\partial^2 T}{\partial x^2} + \frac{\partial^2 T}{\partial y^2} + \frac{\partial^2 T}{\partial z^2}\right] + \frac{q_{\mathcal{B}}}{k} = 0$$

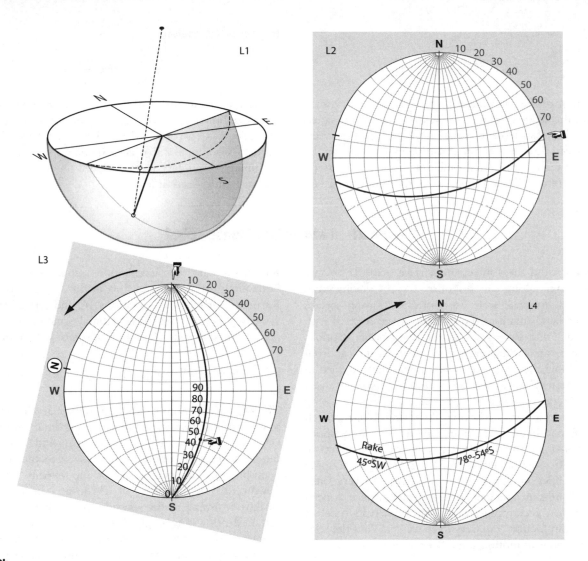

Fig. C10k

and when there is no internal heat generation, *Laplace's equation*

$$\left[\frac{\partial^2 T}{\partial x^2} + \frac{\partial^2 T}{\partial y^2} + \frac{\partial^2 T}{\partial z^2}\right] = 0$$

We now seek solutions to the simpler of these differential equations.

The 1D steady state conduction of heat with some internal heat generation (maybe due to radioactive heating or an electrical current) is from Poisson's equation:

$$\frac{\partial^2 T}{\partial x^2} + \frac{q_g}{k} = 0$$

For the situation where the surface is at $x = 0$, the temperature is T_0 and the heat flux, $-q_0$. A first integration

gives:

$$\frac{\partial T}{\partial x} = -\frac{q_g}{k}x + c_1$$

where c_1 is the first constant of integration. At $x = 0$, $(\partial T/\partial x)k = q_0$ and hence $c_1 = q_0$ and we have:

$$\frac{\partial T}{\partial x} = -\frac{q_g}{k}x + q_0$$

A second integration gives

$$T = -\frac{q_g}{k}\frac{x^2}{2} + q_0 x + c_2$$

where the second integration coefficient is found by realising that $T = T_0$ at $x = 0$. Substituting we get $c_2 = T_0$ and, finally

$$T = -\frac{q_g}{k}\frac{x^2}{2} + q_0 x + T_0$$

Derivation of the 1D time dependant conduction case is beyond the scope of this introductory text but the equation can be recast to express the rate of change of temperature with time as, $\partial T/\partial t = \kappa(\partial^2 T/\partial y^2)$. This can be solved for temperature after some time, t, ambient or starting temperature, T_1, and source or initial temperature, T_0,

to give the expression

$$\frac{T - T_1}{T_0 - T_1} = \text{erfc}\frac{y}{2\sqrt{\kappa t}}$$

where erfc stands for a special function, the complementary error function whose value may be got from advanced texts (such as Turcotte and Schuberts tome advertised in Further Reading, p. 235). The expression is used to derive the thermal boundary layer thickness in an oceanic lithosphere plate in Section 5.2.

Cookie 21 Twister: The Ekman spiral

Turbulent wind shear mixes the surface ocean layers and drives water along in response to the momentum exchanged. In small scale experiments (blowing surface floats) or field situations the water is driven leeward with the wind. On a large scale, in the open ocean, Coriolis accelerations make the water flow direction diverge from the windflow, right in the northern hemisphere and left in the southern. This was first systematically noted by Nansen in the Arctic seas where he measured icebergs moving 20–40°C to the right of the prevailing wind. Ekman solved this problem theoretically by making use of the equations of motion for a rotating fluid system with the effects of constant Coriolis acceleration, $f = 2\Omega \sin\phi$, and upper boundary friction, F, taken into account. The very small effect of pressure gradients due to water surface slope is ignored (i.e., geostrophic component of velocity is zero), as is bottom friction (i.e., very deep water). Thus the relevant equations of motion simplify to:

Coriolis force + Friction force = mass × acceleration

For steady-state (no accelerations) conditions involving horizontal turbulent water flow in the xy plane, constant Reynolds' turbulent accelerations, a_z, and vertical (z-axis) velocity gradients only, the horizontal equations of motion per unit mass of water (the *Ekman equations*) are then:

$$fv + a_z\frac{\partial^2 u}{\partial z^2} = 0 \quad \text{and} \quad -fu + a_z\frac{\partial^2 v}{\partial z^2} = 0$$

The Ekman equations can be solved (not developed here) to demonstrate that water velocity is maximum at the surface in a direction 45° clockwise to wind velocity in the northern hemisphere (anticlockwise in the southern hemisphere). The magnitude of the velocity decreases downward but increases in clockwise turning (anticlockwise turning in the southern hemisphere) until at a certain depth a vanishingly small velocity points upstream relative to the surface Ekman velocity. This depth defines the thickness of the wind-driven surface current known as the *Ekman layer*.

Index